Watersheds in Marxist Ecofeminism

Watersheds in Marxist Ecofeminism

By

Pamela Odih

CAMBRIDGE SCHOLARS
PUBLISHING

Watersheds in Marxist Ecofeminism, by Pamela Odih

This book first published 2014

Cambridge Scholars Publishing

12 Back Chapman Street, Newcastle upon Tyne, NE6 2XX, UK

British Library Cataloguing in Publication Data
A catalogue record for this book is available from the British Library

ISBN (10): 1-4438-6602-4, ISBN (13): 978-1-4438-6602-6

Dedicado a Río Virilla, porque el espíritu del río ha inspirado orgullo reflexiva en mi propia imaginación sociológica

A share of royalties will be donated to the charity United Nations Children's Fund (UNICEF). Established in 1946, UNICEF is dedicated to furthering the rights, wellbeing, prosperity and creative potential of children.

CONTENTS

LIST OF ILLUSTRATIONS

Figure P.i., "Ye who sing of atoms, rest, and motion, say, why Time, Sets in this toy a larum to my heart" (Ann Yearsley *Soliloquy*, 1796). Photographic Image, Isis Sculpture, The Serpentine, Hyde Park London, April 2014

Figure P.ii., "The moon now from old Avon's stream…And as she leaves the weary flood, Dark grow the vales, dark grows yon wood, E'en fancy takes her flight" (Ann Yearsley *Extempore*, 1796). Photographic Image, South London, River Thames, September 2013

Figure P.iii., "Oh ! Sensibility! Thou busy nurse…" (Ann Yearsley *Poems Addressed to Sensibility*, 1787). Photographic Image, South London, River Thames, September 2013

Figure P.iv., "Thy valued image in my aching mind…The blow that sever'd two congenial souls" (Ann Yearsley *Poems Addressed to Sensibility*, 1787). Photographic Image, South London, River Thames, September 2013

Figure P.v., "Come thou unconquer'd pow'r! that aid'st the line, And boldly bidd'st the wild idea rise, Rush on my sense! Swift o'er my tranquil soul Breathe thy strong influence, till her deepest springs Are all in motion set" (Ann Yearsley *The Bristol Marine Society*, 1787). Photographic Image, West London, River Brent Tributary of the River Thames, February 2014

Figure P.vi., "Lo! The calm Sea, Like me, inactive, waits the breath of Heav'n" (Ann Yearsley *The Bristol Marine Society*, 1787). Photographic Image, West London, River Brent Tributary of the River Thames, February 2014

Figure P.vii., 'Tremendous scene! When the unwieldy hulk Sleeps on the breast of Ocean, nor obeys The eager efforts of despairing man. Bereft of her tall mast, and friendly sail … She disobeys, or runs to wild misrule' (Ann Yearsley *The Bristol Marine Society*, 1787). Photographic Image, South London, River Pool, which flows through Lewisham and Greenwich into the River Thames, February 2014

Figure P.viia., 'Tremendous scene! When the unwieldy hulk Sleeps on the breast of Ocean, nor obeys The eager efforts of despairing man. Bereft of her tall mast, and friendly sail … She disobeys, or runs to wild misrule' (Ann Yearsley *The Bristol Marine Society*, 1787). Photographic Image, South London, River Pool, which flows through Lewisham and Greenwich into the River Thames, February 2014

Figure P.viii., “While Glory, nurs’d within her merchants arms, Shall blaze refulgent on a wond’ring world” (Ann Yearsley *The Bristol Marine Society*, 1787). Photographic Image, South London, River Pool, which flows through Lewisham and Greenwich into the River Thames, February 2014

PREFACE

Interviewer: Can you tell me about it? [The river restoration project]
Interviewee: *The project is to improve the river after the flood defence work they done about forty years ago now, where they straightened the channel. In the late sixties, it all flooded and there was this knee jerk reaction to put this sort of flood defence in, which doesn't really work.*

Interviewer: What was the flood defence that they put in, in the late sixties?
Interviewee: *You're looking at it now, this straightened channel which all it does is push the water down quicker and creates bottle necks everywhere else and floods other places. So, erhm, but we can't change this, so all we are trying to do is improve the channel by putting structures in the river, so the river meanders within the channel; and by doing that you create better habitat, for the fish, for the bugs, for the birdlife.*

Interviewer: Why did they put in straight channels before, in the sixties?
Interviewee: *This was the thinking of the time, was if you had a big straight channel then the water would run through very quickly. What you have to understand was that Lewisham was underwater by a meter, back then, it rained hideous amounts over the period of a week and this entire place was flooded up to street level; above street level. So they thought, just straighten the channel; stick it all in concrete and shove it out as quickly as they can down to Deptford.*

Interviewer: Where in Deptford?
Interviewee: *Deptford Creek into the Thames, that way… But the thinking now is to actually hold the water back and let it come out slowly, more so than throw it all out.*

Interviewer: And what's inspired that thinking; the change in thinking towards the meandering?
Interviewee: *Because it doesn't work. The system they thought was the solution is not the right solution. In terms of its, kind of, not conservation as such the bio-diversity of the river is completely ruined, by the aspect of doing this kind of structure, these concrete works. It doesn't work, it just creates more problems in terms of flooding; and they've slowly worked it out; this costs a lot of money; this concrete only has a limited life-span on the river. So in twenty, thirty years time this will all have to be broken out anyway, so the idea now is to have a more sustainable system. And use the*

parks where the rivers can naturally flood and hold the water up and then, you've got a, you know, you've got a more natural flood defence system that was always there in the first place, which was flooding fields, which is quite natural.

Interviewer: And how is the project sponsored? And organised?
Interviewee: *I work for [River Conservation Charity Three], I'm the project manager. We get, we get all sorts of funding from loads of different people really. Erhm, some from the borough, some from landfill taxes, lottery money - years ago...So there's grants and applications, you put in applications in for grants, funding.*

Interviewer: What about the Environment Agency?
Interviewee: *They give us some money. They give us a bit of money every year to play with, and they're quite happy.*

Interviewer: What are you supposed to do to actually enable sponsorship from the Environment Agency? What performance indicators do they require?
Interviewee: *Erhm well the structures are paid for by the [Environment] Agency that we've been putting in, so all the berm structures; all the low level shelves and all that sort of stuff. That ticks boxes in terms of wildlife. Erhm we have something called the WFD, which is the Water Framework Directive coming from Europe; and that requires every water body in this country to be of a 'good' ecological standard. And one of the key indicators of measuring that is fish; and at the moment we don't have a lot of fish. By putting these structures in we are creating havens where fish can hide and you know keep out of the flow a little bit; you know when the flow's too high. By creating a more meandering river in that sense, not only do you slow the waters down a little bit. But you also create a habitat, for, a refuge if you like, for fish.*

Interviewer: You said the Water Framework Directive: How does that directly impact on this project again? So the European Water Framework Directive: is that through the Environment Agency that it impacts, or is it directly impacting on the project?
Interviewee: *No, the Agency [is] the delivery agent for the Water Framework Directive, which is... it's quite a task. In urban rivers it, you know you've got a lot of problems in urban rivers. We're not ever sure we're ever gonna reach that standard. Although this one is reasonably ok, the Ravensbourne and the River Pool is 'good'. Erhm, so that one ticks the box, whereas this one doesn't [the latter is the Ravensbourne and the former is the River Pool]. This is the Ravensbourne and that's the River Pool.*

Interviewer: Over there, oh the two of them meet?
Interviewee: *Yeah, so they confluence here [pointing to a meeting of the rivers]. So from here onwards it's the Ravensbourne, so the River Pool is from here up to there. It's all the same system it is the Ravensbourne catchment. They are slowly trying to improve the river in a sense. You can see what it is like, it is all concrete and rubbish really. But further down there it's all gravel bed and further up here, it is gravel so this is concrete for a mile up to Bellingham and there's no fish in there. There's fish in that one [pointing to the River Pool].*

Interviewer: So to actually be able to achieve a 'good' in terms of the Water Framework Directive, so are the rivers actually meeting that? Is the objective to achieve a 'good' or is there a bigger objective?
Interviewee: *I think 'good' is where they're gonna stop at. You know I think there's 'bad', 'good' and 'excellent', I'm not quite sure what, I can't remember what all of them are. But 'good' is the minimum standard they wanna get to. Like I said, in London, in any urban river, it's going to be very difficult because a lot of the storm [water] run[s] off the roads; that just brings pollutants into the river and erhm you know that's always going to hold you back. So you can never ever get it, you know like an 'excellent' river. I defy anyone who can have an excellent river, you know, in modern times; because even if you was in the country you're getting diffuse pollution off the land, you know you're getting all the, you know all the fertilizer phosphates nitrates, that drains in. So you got to talk to the farmers about changing their practices because it's affecting the quality of the river. You know, you've got things like this sort of stuff: you can stop the cattle... damaging the river by putting a fence up, that makes a difference quite quickly. You know, so there [are] things you can do in the country, you know, you can do that. But you know in a city, you'd have to change every single drain and divert that water somewhere else. And in some places they are doing that, they're called SUDs: Sustainable Urban Drainage Systems. Where they chuck the water into a lagoon; and it all settles out; and it trickles out, you know; and you can put reeds in that and it'll remediate the water, and the soil and all that and you could do that. But then again there's not a lot of places where you can put the SUDs really, so there's one or two we can put along here but we don't know who owns the land and we would have to divert the water into it; not the river just the drains. If we could do that, that would be great, that would take out a lot of the silt and the filth, a lot of it is from rotten tyres and stuff like that. It's not just the oil of the engines off cars; you know break-pads, I've been told on cars all that dust that come off that, you know, that all ends up in the river.*

Interviewer: So if the objective is not to achieve the 'excellent', what is your main motivation?
Interviewee: *What we would like to see is a more natural river system, the pollution we might have to live with I think. I think realistically if we can restore rivers to a semi-natural state by removing all the concrete, putting bends back in and having little... lanes in the parks and all that I think that's a good place to start. Then slowly, slowly you never know we might run out of oil and might never use our cars again and we'll never pollute our roads; and that won't wash into the rivers so. I think, I think that will be a good start. Restore the rivers to what they used to look like; put bends back in; hold the water up; catch the water and just let it out gently and then you know, you've got a nice green corridor. One of the good things about this river is it's got a very good cycling track along; because it's a flat, it's a flat landscape. All the rivers should have cycling tracks along, keep the cyclists off the road put them along the rivers...*

Interviewer: And the Environment Agency's helping you to achieve this?
Interviewee: *The Agency well, we help the agency in a sense. My job, yea, they're not helping us, we're helping them.*

Interviewer: In what way?
Interviewee: *By doing the structures; by keeping the rivers clean; we pull out a lot of rubbish you know things like that, so by building these structures and creating refuge for fish and stuff like that; so that should help them meet their targets. You know, from my perspective as a [River Conservation Charity III] operative is to get people involved in rivers. That's where we come from as a charity. You know, for much of the time people are just disengaged from the environment especially rivers; and we provide an opportunity for people to go in, play and erhm and go and do something good in it really. First thing pull out the rubbish; then maybe build one or two structures; do some water quality testing. Basically educate people about rivers. You know for years they have been behind fences and that's all being ripped out. You go down Ladywell now, it's fantastic down there. All the kids can just run straight into the water...*

Interviewer: So they actually built concrete banks? And it was because it was meant to just straighten the river? And that's what they wanted to do? But it's like over forty years now and so it hasn't broken down?
Interviewee: *Not yet, but some of it is looking a bit shabby. They'll probably keep the bits by the bridges, but the rest of it they'll probably take out eventually. They've widened it as well you see, so what you could do is leave that in there and just build a new bank inside it, you could leave that structure in there and then soften the edges. So one of the things you don't get in there: you don't get marginal vegetation along the river's edge, so the fish don't hide. So that plant there was along the river's edge, all the way round, like that one over there say, so that's where all the fish are they*

hide in there. You come down here sometimes you see a white egret, it looks like a heron… anyway they pop along and they go along there and you can see them using their foot to actually scare out the fish. If [you] sit here long enough you see a King Fisher.

Interviewer: And so you are building these natural systems to actually try and slow down the river?
Interviewee: *Yes to slow down the river and actually to speed up the river in the summer; because they are low flow structures, so as I was just saying: since we built these structures we've had nothing but rain. But in the summer there'll be half this much water. And the idea of the structures is to narrow the river down so the water during the summer goes through there much quicker because it's much narrower.*

Interviewer: And that's what motivates you to want to work with the charity; because you want to actually to address these issues?
Interviewee: *Yea and one of the other things that [hasn't] been done down here; there's been no maintenance. So forty years ago they came in [and] straightened everything. And just let things grow. And what happens is now, there's so much tree down here it's actually what makes it too dark in the channel and you're not getting the growth of plants around the edges where you would like; because it is too dark. So we've been taking out a lot of the trees, cutting them back a bit to get some more light. Not all of them, just some. But you know, you need to manage trees, you can't just leave them. You can't just think, I'll plant a tree, you know, they need a lot of management. For years, it's only in the last fifty years we've not managed trees in this country. For 2,000 years before that we did. We didn't kill them we just cut them down for fire wood and let them grow again. They call it pollarding, you know they chop the top half off and then use that for your firewood, but they won't kill the tree and then it grows again and they come back and chop another bit off. You know and that's what they used to do; and this is what pollarding is and that's why you never used to get trees falling down, but now people don't do that any more; and they just let trees grow thinking it's the natural thing they should be doing, which it probably is; but in the landscape we are living now it's not good, they just get blown down…We lost one of the big poplars over there recently, we lost a lot, a few trees come down. Surprised we ain't lost a tree; no one said we got a tree down at the moment. Couple came down over Christmas they chopped [them] up. You got to manage these things.*

Interviewer: And what sorts of things hinder the management of these things? What sort of things facilitate and what sort of things make it more work?
Interviewee: *Money [he laughs]. Money really; and with cut backs, you know what do you do?*

Interviewer: What types of cut backs are affecting you?
Interviewee: *Loads; lot of the councils, and they put everything to contract, so they want to get as much as they can out of their contract time and things like that. But there's a limit what you can take out; ultimately, in the end, if you ain't got people to do the work then it ain't ever going to get done. One of the, one of the first things that, know when you have a recession and things like that, one of the first things that happens is they cut back on the environmental services. I mean we're very fortunate in Lewisham I think they are being very good so far. Not sure what's going to happen this year. But the team ain't very big, in the council, which is a bit of a shame. You've got to look at; they've got their bio-diversity department and things like this sort of stuff; that's really important. What ever you do you want to try and maximise all you have, without doing too much for it or doing enough for it. It's very difficult; cutting down tress and getting a tree surgeon down for a day is a lot of money. All I have to do is give them a saw and they'll just chop it down anyway [referring to the volunteers]...There's a lot of interest in rivers, we do community events, advertising you can get anywhere between half a dozen people to sixty, depends on the weather; depends on the time of the year. You know just by going through it every so often, just keeping it clean. That's a bonus. A lot of people think they're just rubbish; I mean we used to pull out shopping trolleys, motorbikes, everything out of these rivers; but because we've been doing it for so long a lot of that big stuff doesn't come back anymore. Rubbish begets rubbish. So if people don't clear it out then that's what happens. The funny thing about rubbish in rivers is, no one has to take it. There's no law in this country that requires any authority to have to take rubbish out of a river. That's why we set up as a charity... [Working alongside other civic initiatives] cleaning rivers in London, because no one had to clean them. Years ago we'd do an event for two hours, you'd be astonished the amount of rubbish, you know van loads, van loads. You'd be lucky now to get a quarter of a van load in a day. Over then in two hours, with twenty people or so, you'll easily take out two or three van loads. Just you know, that's how much less there is now.*

Interviewer: So it's really important to work with communities then I guess?
Interviewee: *That's what all these are; all these are volunteers all local...*

Interviewer: I was really interested in what's happening; because the Environment Agency's been [in the media] rivers over the last couple of weeks have been a big thing?
Interviewee: *People have woken up. You know; it's like, I think, I saw, I missed the Panorama programme; I'll get it on my Ipad later. I only caught the end of it; and they said you've got to be frank. You can't just tell these people everything's going to be ok, because it's never going to be ok. You know, they've got a new type of flooding now: saturation flooding.*

Which is fantastic you know, never even heard of it before. With the water's coming up underneath their houses; you know that's not rivers bursting its bank. You can't mitigate against that; there's nothing you can do about that; apart from move your house! [laughter]...I think because we've not had flooding in this country for such a blemming long time people don't know. All the[se] thing[s] about the Somerset Levels, what's that about? I mean that's been flooding ever since King Arthur was down there, got buried down there; that's Avalon isn't it? You know; you know, it's nuts. It's just a marsh landscape that used to be the sea, two thousand years ago, that was under water. You know, what do they expect? I mean for goodness sake, you know it's below sea level guys, work it out. You know, you can't blame the [Environment] Agency for that amount of water. It's just bonkers; you know all these people in Staines who live along the river banks, up the river: well yes you live on a floodplain. It might be a very nice place to live and you want to spend a million pounds on your house; but it's going to flood. You know; and that Panorama programme just said to them: if you can not cope with the fact that you house is going to flood then you need to be thinking about where you live. Right? And I think this is a wake up call for many people frankly; because they are consistently, consistently building on floodplains. You know the Agency put a flood protection scheme right; all of a sudden the developer says we can build on that land now. Right; and the Agency will always say to them 'no you can't'. Or say 'you can't', 'we don't recommend it'. They can't say no 'you can' or 'you can't' that's not their job right. The recommendation from the Agency is do not build a house on a floodplain. They'll say we've got a flood defence system engaged here; but that's only for a one in five year flood or a one in twenty year flood. It is the magnitude of flood; the floods we are looking at now [referring to the winter floods 2013-2014 in England] is a one in a hundred year flood. Right that's how rare they are. But it might not be, you might get another one of those ...or two of those one hundred year floods back to back. So then the magnitude of the flood has changed...So it's only based on an assessment based on the current weather conditions right? So they don't know what the weather's going to do next year right? We know for a fact we are going to get more rain, that's what global warming really means. Global warming really means more heat, more water in the air and that's just going to transfer into rain we know about that right? ... It's interesting people don't get it. Actually working in places like this, people have that debate, will talk about it and people understand. By working in the river I think you understand what a river system is; and people just don't get it. I was talking about Lambeth, the other day, about breaking out the [river] Effra, Brixton High Street you know. There's a river underneath there you know; get it out! You know it's going to happen one day, eventually, you know. You have to understand that you know, I remember seeing a map of London; if that was a map of London [he places one glove on the raised concrete bank] right and that was a map of the river system [he places the other glove on top of the first] in London it's exactly the same. So that tells you doesn't it, people live by the river. You look at it, it's identical almost identical. Where the housing overlays the river system right, it just goes to show: it's not the river's fault that it's flooding. That's what it does. It's our fault for living next to a river. You've got to cope with it.

Interviewer: Do you think there is a system that would be particularly able to manage the urban river? Is there a new way of thinking about it; a different way of thinking about it?
Interviewee: *It's what you do with flood water isn't it. Yeah they are doing that. A lot of the parks are being converted into floodplains, yeah. So the floods will be allowed to flood and hold water and that's about it really; that's all you can do.*
...
Interviewer: What did you think of all the hysteria around the recent floods?
Interviewee: *Idiotic really you know. Like I said, I think it's people so disconnected from the environment, in which they live. I wouldn't blame the media because they had the right to report what they saw. I don't think the government helped. The press in particular were quite rubbish I thought; you know picking up on these stories about the Agency being rubbish and this and that; I mean them guys have been killing themselves for two months. You know, they want someone to blame don't they. It's deflection isn't it; government deflection? Blame the Quango, and I'm glad Chris Smith actually came out and actually stood up and said: well actually what we do is what you tell us to do; the guide lines that we work to are set by government not us, you know: protect life; protect property; farmland comes last; so don't tell us about the Somerset Levels why we ain't done nothing down there when we've got the whole rest of it. Only 5,000 houses got flooded, you think about the amount of rain that we had right; which was, is a fantastic amount...it was phenomenal. For that two months since before Christmas, and it's just a fantastic amount of rain. You know, and I just think, you can't mitigate. You can't do anything about that. But considering the amount; I mean it was country wide it wasn't just one area. I can understand it, if it was just one area; but it wasn't. It was country wide. And then, how much they could protect; if they protected one and a half million properties from flooding, five thousand got done...five thousand homes compared to one and a half million. To me that's successful. You've got to put it in perspective, you can't protect them all.*

Interviewer: Should the rest be protected by insurance?
Interviewee: *I don't know they can't get it now can they? Some guy on the radio yesterday, he had a small flood; it got paid out and they wouldn't insure him again. So now he's got twelve to fifteen thousand pounds to his property. There you go. (Project Manager, London River Thames Conservation Charity III, February 2014)*

Figure P.viiia., "While Glory, nurs'd within her merchants arms, Shall blaze refulgent on a wond'ring world" (Ann Yearsley *The Bristol Marine Society*, 1787). Photographic Image, South London, River Pool, which flows through Lewisham and Greenwich into the River Thames, February 2014

Figure P.ix., "The present is not valu'd; restless man Lives for the past, and future, fix'd his eye On op'ning prospects that shall never end, Till, in the vast pursuit, the rover falls" (Ann Yearsley *Familiar Epistle to a Friend*, 1787). Photographic Image, South London, River Thames, September 2013

Figure P.x., "With sorrow mute, I take my pen – thy plaintive flute, So charms my thinking soul" (Ann Yearsley *Extempore*, 1796). Photographic Image, South London, River Thames, September 2013

Figure P.xi., "Unfold the charts of never-ending space, How would my spirit strike the eager wing, To claim her new creation!" (Ann Yearsley *Elegy,* 1787). Photographic Image, West London, River Brent Tributary of the River Thames, February 2014

Figure P.xii., "Long, dear Idea, gentle Love's soft nurse, Lay silent, inexpressive in the mind; Long did the Spirit wrestle with its force, Till, dress'd by Art, it rises unconfin'd. Lo, the tints of Clara flow; Thoughts embodied, ardent glow; Gently breathes the pleasing form, And passions truly painted warm" (Ann Yearsley *Ode to Miss Shells, on Her Art of Painting*, 1787). Photographic Image, South London, River Pool, which flows through Lewisham and Greenwich into the River Thames, February 2014

ACKNOWLEDGEMENTS

I would like to express a heartfelt thank you to Professor David Knights who continues to inspire my academic development. Professor Barbara Adam's prolific contribution to the study of time and society has been inspirational, and I would like to pay tribute to her work.

Heartfelt appreciation and thank you to Susan Dunsmore; her editing assistance, kindness and support have been a tremendous encouragement. Genuine appreciation to Camilla Harding (Commissioning Editor), and Amanda Millar (Editorial) at Cambridge Scholars Publishing. My colleagues at Department of Sociology, Goldsmiths University of London have been supportive: Dr. Brian Alleyne, Prof. Vikki Bell, Prof. Vic Seidler and Prof. Bev Skeggs.

Big thank you to my friends Andrea Reay and Ivalee Harris. Special heartfelt thank you to my friend Andy and the enchanting cultural delights of the Edinburgh Festival 2014. Love and best wishes to my family. Ultimately, I am grateful for the time of my interviewees; thank you for speaking so candidly during the ethnographic interviews.

ACKNOWLEDGEMENTS (ILLUSTRATIONS)

Figure 2.2., "Volunteers Remove Plastic Bottles from the Banks of the Polluted Waters of Virilla River in Ciudad Colon, About 55 Kilometers South of San Jose on April 10, 2010" (Getty Images). Photographic Image by: Yuri Cortez/AFP/Getty Images, Created 10 April 2010, Editorial Image: 98394301. Permission of Getty Images.

Figure 2.3., "According to Environmentalists the Virilla River of Costa Rica is One of the Most Polluted in Central America" (Getty Images). Photographic Image by: Yuri Cortez/AFP/Getty Images, Created 10 April 2010, Editorial Image: 98394314. Permission of Getty Images.

Photographs by Author Pamela Odih:

Figure P.i., "Ye who sing of atoms, rest, and motion, say, why Time, Sets in this toy a larum to my heart" (Ann Yearsley *Soliloquy*, 1796). Photographic Image, Isis Sculpture, The Serpentine, Hyde Park London, April 2014

Figure P.ii., "The moon now from old Avon's stream…And as she leaves the weary flood, Dark grow the vales, dark grows yon wood, E'en fancy takes her flight" (Ann Yearsley *Extempore*, 1796). Photographic Image, South London, River Thames, September 2013

Figure P.iii., "Oh ! Sensibility! Thou busy nurse…" (Ann Yearsley *Poems Addressed to Sensibility*, 1787). Photographic Image, South London, River Thames, September 2013

Figure P.iv., "Thy valued image in my aching mind…The blow that sever'd two congenial souls" (Ann Yearsley *Poems Addressed to Sensibility*, 1787). Photographic Image, South London, River Thames, September 2013

Figure P.v., "Come thou unconquer'd pow'r! that aid'st the line, And boldly bidd'st the wild idea rise, Rush on my sense! Swift o'er my tranquil soul Breathe thy strong influence, till her deepest springs Are all in motion set" (Ann Yearsley *The Bristol Marine Society*, 1787). Photographic Image, West London, River Brent Tributary of the River Thames, February 2014

Figure P.vi., "Lo! The calm Sea, Like me, inactive, waits the breath of Heav'n" (Ann Yearsley *The Bristol Marine Society*, 1787). Photographic Image, West London, River Brent Tributary of the River Thames, February 2014

Figure P.vii., "Tremendous scene! When the unwieldy hulk Sleeps on the breast of Ocean, nor obeys The eager efforts of despairing man. Bereft of her tall mast, and friendly sail … She disobeys, or runs to wild misrule" (Ann Yearsley *The Bristol Marine Society*, 1787). Photographic Image, South London, River Pool, which flows through Lewisham and Greenwich into the River Thames, February 2014

Figure P.viii., "While Glory, nurs'd within her merchants arms, Shall blaze refulgent on a wond'ring world" (Ann Yearsley *The Bristol Marine Society*, 1787). Photographic Image, South London, River Pool, which flows through Lewisham and Greenwich into the River Thames, February 2014

Figure P.ix., "The present is not valu'd; restless man Lives for the past, and future, fix'd his eye On op'ning prospects that shall never end, Till, in the vast pursuit, the rover falls" (Ann Yearsley *Familiar Epistle to a Friend*, 1787). Photographic Image, South London, River Thames, September 2013

Figure P.x., "With sorrow mute, I take my pen – thy plaintive flute, So charms my thinking soul" (Ann Yearsley *Extempore*, 1796). Photographic Image, South London, River Thames, September 2013

Figure P.xi., "Unfold the charts of never-ending space, How would my spirit strike the eager wing, To claim her new creation!" (Ann Yearsley *Elegy,* 1787). Photographic Image, West London, River Brent Tributary of the River Thames, February 2014

Figure P.xii., "Long, dear Idea, gentle Love's soft nurse, Lay silent, inexpressive in the mind; Long did the Spirit wrestle with its force, Till, dress'd by Art, it rises unconfin'd. Lo, the tints of Clara flow; Thoughts embodied, ardent glow; Gently breathes the pleasing form, And passions truly painted warm" (Ann Yearsley *Ode to Miss Shells, on Her Art of Painting*, 1787). Photographic Image, South London, River Pool, which flows through Lewisham and Greenwich into the River Thames, February 2014

Figure P.xiii, "Unfold the charts of never-ending space, How would my spirit strike the eager wing, To claim her new creation!" (Ann Yearsley *Elegy,* 1787). Photographic Image, Thames Head, Inception of the River Thames, Gloucestershire, April 2014

Figure P.xiv., "Long, dear Idea, gentle Love's soft nurse, Lay silent, inexpressive in the mind; Long did the Spirit wrestle with its force,

Figure I.9a., "Upon that great solemnity repair'd The river gods…Unconquerable souls" (Gentleman of Cambridge, *The Praises of Isis, 1755*). Photographic Image, River Isis, April 2014

Figure I.9b., "Upon that great solemnity repair'd The river gods…Unconquerable souls" (Gentleman of Cambridge, *The Praises of Isis, 1755*). Photographic Image, Thames Head Gloucestershire, April 2014

Figure I.9c., "I wander thro' each charter'd street, Near where the charter'd Thames does flow" (William Blake *London*). Photographic Image, King George III and Father Thames, forecourt of Somerset House, May 2014

Figure I.9d., "I wander thro' each charter'd street, Near where the charter'd Thames does flow" (William Blake *London*). Photographic Image, Father Thames, forecourt of Somerset House, May 2014

Figure 1.9e., "I wander thro' each charter'd street, Near where the charter'd Thames does flow" (William Blake *London*). Photographic Image, Father Thames, forecourt of Somerset House, May 2014

Figure I.9f., "To rise from nature up to nature's God" (Gentleman of Cambridge, *The Praises of Isis, 1755*). Photographic Image, Thames Head Gloucestershire, April 2014

Figure I.10., "Time present and time past, Are both perhaps present in time future" (T.S. Eliot *Burnt Norton*). Photographic Image, South London, River Thames, September 2013

Figure I.11., "And the pool was filled with water out of sunlight" (T.S. Eliot *Burnt Norton*). Photographic Image, South London, River Thames, September 2013

Figure I.12., "And every moment is a new and shocking valuation of all we have been". (T.S. Eliot *East Coker*). Photographic Image, South London, River Thames, September 2013

Figure I.13., "What might have been and what has been, Point to one end, which is always present" (T.S. Eliot *Burnt Norton*). Photographic Image, South London, River Thames, September 2013

Figure P.1., "Time past and time future" (T.S.Eliot *Burnt Norton*). Photographic Image, South London, River Thames 2013

Figure 1.1., "Investing form with lucid stillness, Turning shadow into transient beauty", (T.S. Eliot *Burnt Norton*). Photographic Image, South London, River Thames, September 2013

Figure 1.2., "Ere young Ideas spring; to local Thought" (Ann Yearsley, *On Genius Unimproved*, 1787). Photographic Image, South London, River Thames, September 2013

Figure P.xiii, "Unfold the charts of never-ending space, How would my spirit strike the eager wing, To claim her new creation!" (Ann Yearsley *Elegy,* 1787). Photographic Image, Thames Head, Inception of the River Thames, Gloucestershire, April 2014

Figure P.xiv., "Long, dear Idea, gentle Love's soft nurse, Lay silent, inexpressive in the mind; Long did the Spirit wrestle with its force, Till, dress'd by Art, it rises unconfin'd. Lo, the tints of Clara flow; Thoughts embodied, ardent glow; Gently breathes the pleasing form, And passions truly painted warm" (Ann Yearsley *Ode to Miss Shells, on Her Art of Painting*, 1787). Photographic Image, River Isis, Inception of the River Thames, Gloucestershire, April 2014

PROLOGUE

MARX AND ECOFEMINISM: CONCEPTUAL FLOWS AND METABOLIC RIFTS

Introduction

In *Difference between the Democritean and Epicurean Philosophy of Nature* Karl Marx states, of the nature of time: "Since in the atom matter, as pure relationship to itself, is exempted from all relativity and changeability, it follows immediately that time has to be excluded from the concept of the atom, the world of essence. For matter is eternal and independent only insofar as in it abstraction is made of the time moment" (Marx 1841/2006:131). Within Marx's materialist dialectic, nature is intrinsic to metabolic exchange. Nevertheless, Marx defined nature as the "inorganic" body of human beings; "the universality of man is in practice manifested precisely in the universality which makes all nature his inorganic body" (Marx 1844/2007:74). When one encounters this statement, embedded within the text of Marx's *Economic and Philosophic Manuscripts of 1844*, it initially appears that nature is "man's inorganic body", inasmuch as humans are unable to deny bodily fragility; humans depend on natural processes for survival; i.e., nature is our "direct means of life" (ibid: 74). On further initial inspection it is evident that Marx believed the "inorganic" basis of nature also to arise from the imperative need for man to labour in a way that situates nature as "the material, the object, and the instrument of his life-activity" (ibid). For these reasons Marx argued that:

> Nature is man's inorganic body – nature, that is, in so far as it is not itself the human body. Man lives on nature – means that nature is his body, with which he must remain in continuous intercourse if he is not to die. That man's physical and spiritual life is linked to nature means simply that nature is linked to itself, for man is a part of nature. (Marx 1844/2007:74)

Figure Prol.i. "Instructive Spirit, hail! For thee I call mild contemplation" (Ann Yearsley *Soliloquy*, 1796). Photographic Image, River Isis, Gloucestershire, April 2014

Continuous with this conception of a metabolic interaction with nature was Marx's belief that the growth of industrial capitalism precipitated a degenerating estrangement of human beings from their self and their life activity. For the essential species-being of humans involves "creating an objective world" through "practical activity" applied to "working-up inorganic nature" so that "nature appears as his work and his reality" (ibid: 74-75). Consequently, "the object of labour is...the objectification of man's species life; for he duplicates himself not only, as in consciousness, intellectually, but also actively, in reality, and therefore he contemplates himself in a world that he has created" (ibid: 75). It is evident here and elsewhere, that the archetype of nature as "man's inorganic body" situates the active human subject as the principal agent in the humankind-nature dialectic (Salleh 2001: 443). In this fashioning of nature, the environment is a passive object for instrumental practical applications. This should not be comprehended "as a denigration of nature's value"; for Marx formulated nature as integral to the cultivation of the human self (ibid: 445). Nevertheless, for many generations of ecofeminist writers Marx formulated a (hu)man-centric concept of nature, epistemologically continuous with the way that the "masculine/feminine, productive/reproductive dichotomies, have functioned as an instrument of domination" (ibid). For those feminists committed to a dialectical materialist approach to ecofeminism, one defined as embodied materialist, Marx's construction of nature as "man's inorganic body" is immensely problematic. Of particular concern is Marx's tenet that humans actualise their species-being through productive and creative pursuits. In productive labour, human beings are differentiated from nature and all other animate beings. For embodied materialist ecofeminists "Marx's prioritisation of productive over reproductive labour, construct[s] women as belonging not to 'history' but to 'nature'" (ibid: 447). This is because productive labour privileges economic transactions over the embodied relations of reproductive labour (the latter is associated with unpaid familial and social relations). Reproductive labour is unremunerated and yet the absence of such care work would irrevocably hinder vital processes of social reproduction. To this end, ecofeminists castigate "the very foundations of historical materialism, with its purportedly transhistorical concepts of history, nature and productive labour" (ibid). In operation of this pursuit, embodied materialist ecofeminists have challenged the material incongruities underlying Marx's conception of the nature of time.

This prologue provides a dialogue between the eighteenth century rural poet Ann Yearsley and Karl Marx's nineteenth century foray into the nature of time in the ancient Greek atomist philosophy of Epicurus (ca. 341-270 BC). Part one of this prologue provides a Marxist ecofeminist reading of time, atomist philosophy and nature as formulated in Ann Yearsley's poem *Soliloquy*. Partly informed by Ann Yearsley's poetics, part two of this prologue critically engages with Karl Marx's atomist-inspired nature of time evident in his key texts published in 1844, 1857 and 1867. The central objective of this exercise is to cultivate a creative space for a conceptual dialogue between classical Marxism and contemporary ecofeminism.

Significance of Epicurean Physics to Ann Yearsley's Poetics on the Nature of Time

This prologue provides an ecofeminist critique of Marx's account of the nature of time. The eighteenth century poetry of Ann Yearsley provides for an extended trajectory of ecofeminist engagement. Ann Yearsley's poem *Soliloquy* is illustrative of the early antecedents of ecofeminist critique of the nature of time in ancient Greek atomist philosophy. Part one of this prologue situates Ann Yearsley's *Soliloquy* as an early challenge to the Epicurean *clinamen*, and in so doing provides an alternative perspective through which to challenge the marginalisation of gender in Marx's nature of time. Published in 1796, *Soliloquy* is featured within Ann Yearsley's volume of poems entitled, *The Rural Lyre*. The poem commences with Ann Yearsley pondering the nature of time and existence:

> How patiently toils on this little watch!
> My veins beat to its motion. Ye who sing
> Of atoms, rest, and motion[1], say, why Time
> Sets in this toy a larum to my heart.
> O sacred Time! thy moment goes not down
> But I go with it! Sixty coming hours
> Are with us poor expectants of more price
>
> …

[1] 'Mechanical philosophy is that which undertakes to account for the phenomena of nature from the principles of mechanics, taking in the consideration of motion, rest, figure, size, &c. This is also called the corpuscular philosophy' (Yearsley 1796:43)

Shall beat around me, when I may repose,
Or be as thou art! I have read the code
Of statutes form'd by man for future worlds;
And found his plan, so pompously display'd,
One lot of heterogeneous fragment. Man
Adores in fancy, violates in fact,
Laws serving his frail being. Yon pale moon
Forsakes the mountain top, to bring us round
Her renovated splendour; nature works
Obedient and unseen forever: we
May meet in spheres remote - If not, farewel!
I feel and know, those wishes can arise
But from affections growing with my life,
Mingling with hope, oppress'd by fear. The change
Fulfill'd in thee may chill me; ev'ry thought
Oblit'rate; vision, fancy forms, be doom'd
To sink, like beaming glory in the west;
Whilst space contracts on my weak eye, and heav'ns,
By human artists coloured, fade away,
As life goes gently from my beating heart.

...

Grant this could be - the import were no more
Than as an atom 'mid the vast profound
Impell'd, not swerving from the whole. Suppose,
This frame dissolving, to the busy winds
My ashes fled dividing: shall I know
To mourn? - How like my brethren I display
Conjecture without end! - Impatient pow'r
Of thought! where wouldst thou fly? Return, return!
Nor lose thy strength in phrensy, nor resign
The form I love - This watch is down! Ye points,
Attun'd to motion by the art of man,
As tell-tales of his doings, can ye mark
Eternity by measur'd remnants? No.
Fallacious in your working, ye would say,
With us, the life of man is but a day.
(Ann Yearsley (1753-1806) *Soliloquy* from *The Rural Lyre* 1796:43-44)

It is notable that Ann Yearsley commences her *Soliloquy* with meditations on time and materiality. Indeed, the actual inception of *Soliloquy* recounts a brief dialogue, February 27th 1795, involving Yearsley and her son:

Author to her son.	Go you to bed, my boy.
Son.	Do you write to-night?
Author.	I do.
Son (laying his watch on the table).	See, how late!
Author.	No matter – You can sleep.

(Ann Yearsley (1753-1806) *Soliloquy* from *The Rural Lyre* 1796:43)

We are informed in J.M.S. Tompkins' (1938) analysis of *Soliloquy* that the watch, as a timepiece, in the 18th century was considered to be a privileged luxury among the agrarian poor. Consequently Yearsley's son "laying the watch on the table" symbolically frames the watch. But this object is not only "a pleasant sign of comparative prosperity" Tompkins (1938:88). For the abstraction of the watch causes Ann Yearsley to contemplate beyond her immediate maternal activities, as she is drawn into the impulse of "the watch ticking in the silence" (ibid). The result is poem that begins with a discourse on the nature of time:

How patiently toils on this little watch!
My veins beat to its motion. Ye who sing
Of atoms, rest, and motion, say, why Time
Sets in this toy a larum to my heart.
(Ann Yearsley (1753-1806) *Soliloquy* from *The Rural Lyre* 1796:43)

Of significance is Ann Yearsley's use of neoclassical atomist philosophy to express her uncertainties of the nature of things. In her earlier poems Yearsley had dwelled significantly on the hardship and the personal tragedies of rural life in the 18th century. Born in humble circumstances in Clifton Gloucestershire, where her mother was a dairywoman, Ann Yearsley is often referred to as the "*Lactilla, Milkwoman*" of Clifton. West-country milk sellers were often self-employed "with a respectable traditional position in their home villages" (Waldron 1996:14); and thus, when Ann Yearsley followed her mother in the trade, her adolescence was relatively distant from the disempowering position of the wage-dependent rural labourer. Indeed, in contrast with the impoverished literacy of the rural labourers, Ann Yearsley's mother – Mrs. Cromartie, a moderately learned woman - had taught her daughter Ann to read and encouraged "her daughter to think about the great issues of life and death" (ibid). In her poetic reflections and indivertible portrayal of the intersectionalities of class and gender, one may consider Ann Yearsley as belonging to the historically significant tradition of "writing maids". According to Carolyn Steedman's (2005:9) *Poetical Maids and Cooks Who Wrote*, writing maids "were particularly useful … for the purposes of reckoning and accounting" within the household. In similar respects the writing maids enabled the transfer of knowledge of the household, through their ability to diary (diary maids) the domestic and dietary particularities of a household (ibid).

Ann Yearsley's poetry reflects the pragmatics and sensibilities derived from having to negotiate the complex rhythms and negotiations of domestic labour in the private sphere. Embodied and embedded in care duties the time of reproductive labour exceeds "the empty mechanical measure of clock time" (Felski 2002: 21). Elsewhere, feminists have described women's time as cyclical: distinctively rotating with the ebb and flow of daily chores and women's embodied affinity with biological, reproductive rhythms (Davies 1990). Rita Felski's (2002:21) *Telling Time in Feminist Theory* describes the time of feminist literature as: "time as redemption, time as regression, time as repetition, and time as rupture". Felski describes feminist literary traditions as characterised by a time "suffused with meaning" and "saturated with affect" (ibid). Feminist literary theorists who portray women's time as cyclical, according to Felski (ibid: 26), presuppose that "women … are less invested in breaking with the past". Conversely, the linear temporality of economic labour is disembodied in its calculative and means-end future orientation. For many feminists, the cyclical time of reproductive labour is valorised as "a sign of women's association with immanence rather than transcendence" (ibid: 25). The feminist focus on cyclical feminine time is easily comprehended and is just as easily challenged, for its dualistic epistemology produces arbitrary contrasts between linear and cyclical time. While I have elsewhere emphasized the necessity to deconstruct the hegemony of linear time (Odih 2007), there is some relevance, here, to the methodology of analysing an intertwining of cyclical and linear time. Thus, for example, Felski (2002:26) describes how "the task for feminist theory is surely to connect repetition and change rather than to sever them".

Returning to Ann Yearsley's rural verse: the poem *Soliloquy* vividly depicts the intertwining of cyclical and linear time as is evident in the line "How patiently toils on this little watch! My veins beat to its motion" (1796:43). Yearsley's desire to intertwine the cyclical and linear reflects the emphasis in *Soliloquy* on immanence as opposed to transcendence.

Ann Yearsley's poetry often returns to the perturbations of "time and death and the power of memory" (Waldron 1996: 246). It is evident that the Yearsley family were unable to keep at bay the repeated losses and misfortunes involved in rural agriculture. Ever since 1699, the entire manor of Ann Yearsley's birthplace, Clifton, had been owned by the

British Society of Merchant Venturers "a body cheerfully and unashamedly dedicated to the making of money and the promotion of trade and commerce" (Waldron 1996:19). Keen to capitalise on an expansion in Clifton (linked to the appreciating commercial viability of the Hotwells spa on the banks of the river Avon), these merchant venturers, in the eighteenth century, began to change the conditions of small leaseholding (ibid). By the late-eighteenth century, the rapacious effects of these changes in landownership were encroaching into the meagre livelihood of small leaseholders. Ann had since married John Yearsley (rural labourer) and when pregnant with her sixth child the family was impoverished. Indeed it is an often recounted historiette that Ann Yearsley's genesis as a professional poet can be traced to the "shocking conditions on the outskirts of Bristol in the harsh winter of 1783-1784" (Ferguson 1995:46). Legend has it that the austere winter of 1783-1784 crippled Clifton's agriculture and perilously impoverished the Yearsley family. Mary Waldron's (1996:20) *Lactilla, Milkwoman of Clifton* describes how the austere winter conditions perilously affected the Yearsley family: "their only sources of income were labouring, … milk selling, which may have become unprofitable as a result of the changes in land ownership and land use … and the efforts of Mrs Cromartie in taking in washing". Unable to meet the avaricious conditions of small leaseholding, and deprived of their income by the exceptionally severe winter conditions of 1783-1784, the family risked almost certain demise from starvation (ibid). Indeed, "Old Mrs. Cromartie had not been able to tolerate the privations she had suffered and died; she was buried on March 23, 1784" (ibid: 21).

These cataclysmic conditions are evocatively asseverated by Ann Yearsley and "infused her poetry with powerful feelings…and an evolving ideological perspective" (Ferguson 1995:47). Elsewhere, it has been suggested that her later poetry dwells less on the frustrations and anguish of Yearsley's earlier years and cultivates a detached meditative maturity (Waldron 1996:246). Indeed, the poem *Soliloquy* is particularly intriguing in its invocation of eighteenth-century scientific philosophy, and "the poet speaks not as a struggling victim of circumstances but as an experienced messenger, confident even amidst her uncertainties" (ibid). It is Yearsley's maturer aptitude and ability to articulate immanence and the embodied materiality of rural life, "amidst her uncertainties" (ibid), that engenders my belief in the importance of Yearsley's prose for marking the genesis of an embodied, materialist, ecofeminist meditation on time and the nature of things.

In tune with contemporary embodied materialist ecofeminist praxis Yearsley articulates amatorial devotion to exploring the nature of time: "O sacred Time! thy moment goes not down. But I go with it! Sixty coming hours. Are with us poor expectants of more price" (Yearsley 1796:43). While this extract of verse appeals to the idea of time as linear succession, it is evident that Yearsley also sought to voice a daring speculation of mechanical atomistic philosophy and the properties of the atom that lie at the foundation of classical Greek atomistic philosophy, as is evident in the following extract:

> Grant this could be - the import were no more
> Than as an atom 'mid the vast profound
> Impell'd, not swerving from the whole. Suppose,
> This frame dissolving, to the busy winds
> My ashes fled dividing: shall I know
> To mourn? - How like my brethren I display
> Conjecture without end! - Impatient pow'r
> Of thought! where wouldst thou fly? Return, return!
> (Ann Yearsley (1753-1806) *Soliloquy* from *The Rural Lyre* 1796:46)

Ann Yearsley's emphasis towards an embodied engagement with the realm of necessity, rather than seeking to transcend the necessities of nature, the body and reproductive labour, is particularly evident in *Soliloquy*. Just as atoms are "Impell'd, not swerving from the whole" (ibid), so the destiny of nature can exceed the calculative judgement of clock time:

> Could thou declare how long the storms of fate
> Shall beat around me, when I may repose,
> Or be as thou art! I have read the code
> Of statutes form'd by man for future worlds;
> And sound his plan, so pompously display'd
> One lot of heterogeneous fragment. Man
> Adores in fancy, violates in fact,
> Laws serving his frail being. Yon pale moon
> Forsakes the mountain top, to bring us round
> Her renovated splendour; nature works
> Obedient and unseen forever: we
> May meet in spheres remote – if not, farewel!
> (Ann Yearsley (1753-1806) *Soliloquy* from *The Rural Lyre* 1796:45)

Immanence is a persistent motif in *Soliloquy,* especially in its "loving appreciation of the beauty of the observable world" (Waldron 1996:247). According to Kerri Andrews (2013:104): "Yearsley finds desolation and despair as she contemplates the eternal and unchanging nature of the universe". While it is evident that Yearsley's *Soliloquy* has a sepulchral, contemplative tone, it is also clear that the poem is suffused with agency and inspirational ecofeminist challenges. According to Andrews the atomistic philosophical ideas evident in Soliloquy "were shaped by contemporary scientific theories about the earth and the wider cosmos" (ibid: 107). Conversely, it is suggested here that Ann Yearsley's *Soliloquy* challenges the transcendent nature of time in atomistic philosophy. Moreover, Ann Yearsley is formulating questions about time and materiality that would, in the 19th century, preoccupy Karl Marx's magnum opus entitled, *Difference between the Democritean and Epicurean Philosophy of Nature*.

Significance of Epicurean Physics to Karl Marx's Nature of Time

> Since in the atom matter, as pure relationship to itself, is exempted from all relativity and changeability, it follows immediately that time has to be excluded from the concept of the atom, the world of essence. For matter is eternal and independent only insofar as in it abstraction is made of the time moment. (Marx 1841/2006:131)

Central to Marx's dialectic of time and nature is the concept of indeterminacy. According to Epicurus, as atoms swerve and repel each other "there is a new oblique motion created" which exceeds the "natural necessity" of parallel descent in a straight line through infinite void (McCarthy 2003:19). Democritus' deterministic parallelism presupposed that each atom descending in a straight line relinquishes its autonomy to the formulae of gravity and nature, i.e. "the motion of falling is the motion of non-self-sufficiency" (Marx 1841/2006:112). Significantly, Marx perceived Democritus' account of motion to have over-emphasised the material properties of the atom and in so doing inaccurately subjugated motion to the determinant laws of blind necessity. For Marx was "opposed to seeking an external and blind cause of declination outside the principle of the atom itself", and this was continuous with his contraposition to "the necessary and deterministic world of Aristotelian physics" (McCarthy 2003:19). Conversely, Epicurus' account of declination, in the descent of the atom from a straight line, is heralded by Marx to have disrupted the necessary and deterministic world of Democritus' blind necessity. Epicurus' formulation of the swerve opened up the possibility of explaining the self-evident existence of generation through different and new associations of atoms. Just as atoms release themselves from the binding relativity of the straight line, so Epicurus' physics swerved away from the deterministic bonds of fate's infinite succession of causation. Indeed, indication of Marx's enthusiasm for the methodological implications of *clinamen*, for understanding the multifaceted materiality of social existence, is evident in the following extract:

> Lucretius therefore is correct when he maintains that the declination breaks the fati foedera [bonds of fate] and, since he applies this immediately to consciousness, it can be said of the atom that the declination is that something in its breast that can fight back and resist. (Marx 1841/2006:113)

Evidently *clinamen* represented for Marx not only the possibility of the occurrence of chance, but also the possibility of the existence of indeterminate chance. As Marx briefly articulates, "the declination of the atom occurs *without chance*" (ibid). This, of course, implies that the atom can exist independently from other atoms and that in such a unitary state it is separate from extraneous forms of determination. But this presents a problem for Marx, a problem elsewhere described in the following terms: "If Lucretius can ask how nature 'created' things, Marx has asked how a motion, whether the fall or the swerve, can lack any materiality or relativity" (Stanley 2002:100). For once it is conceived that the swerve constitutes, in spatial terms, the negation of the straight line "then we must ask how and in what way self-sufficiency can be related to motion as materiality" (ibid). To this end Marx stated that:

> To inquire after the cause of this determination means therefore to inquire after the cause that makes the atom a principle – a clearly meaningless inquiry to anyone for whom the atom is the cause of everything, hence without cause itself. (Marx 1841/2006:114)

Marx uses these insights to identify a further level of, conceptually useful, conditions of *clinamen*. Democritus' focus on the natural motion of atoms was also an emphasis on the "distinct material entity" of the atom (McCarthy 2003:19). Consequently Epicurus' rejection of the principle of blind necessity was also a negation of an empiricist materialist account of the universe as governed by material laws and distinct determinable entities (ibid). Atoms swerve virtually imperceptibly from a straight-line motion of parallel descent; thus *clinamen* challenged the mechanical principles of Democritus' atomist philosophy. Marx was dissuaded by the pursuit of a material causality basis for declination "outside the principle of the atom itself" (ibid). Or as Marx specifically expressed it, we need to examine the "soul of the atom" (Marx 1841/2006:114). In correspondence with Epicurus, Marx abstracts the movement of the atom from dependence on external forces outside the atom itself. Swerve is not the product of the determination of other atoms, for its declination is an act of self-sufficiency. To this end "the atom defines itself from within its own principle or concept as a distinct material entity – a particular object in nature" (McCarthy 2003:19). While for Democritus the atom's form "begins and ends with its materiality", for Epicurus the essence of the atom exists in spiritual form (ibid). To this end, Epicurus discounts and refuses to preference material conditions over subjective representations and appearances. Neither the atom nor its anarchic declinations are the determined outcome of empirical conditions indexed to an objective material world. Rather, Epicurus postulates that we enquire into the subjective properties of the atom to appreciate its inherent indeterminacy. Having abrogated empirical reality as a determinant of the atom, Epicurus swerves away from materialism and steers towards absolute idealism (Fenves 1986:438). And this introduces a substantial paradox into Marx's dialectical materialist reading of Epicurus. For Marx's reluctance to invoke idealism as the embodiment of the atom's inner properties encouraged an emphasis on self-determination as the material basis of *clinamen* as effectively summarised in the following extract:

> The concept of the atom is therefore realised in repulsion, inasmuch as it is abstract form, but no less also the opposite, inasmuch as it is abstract matter; for that to which it relates itself consists, to be true, of atoms, but other atoms. But when I relate myself to myself as to something which is directly another, then my relationship is a material one. (Marx 1841/2006:116-117)

What is of particular importance here is precisely Marx's account of the nature of time. In recent times it has been observed that, in the concept of *clinamen,* "Epicurus speaks of the movement of atoms in the smallest continuous time and again of their lack of movement in a uniform direction in the times which are distinguishable in thought" (Rist 1972:46). By implication, "there are indivisible minima of time – times 'distinguishable in thought' – and that we cannot say that an atom is moving in any one direction in them" (ibid). Epicurus therefore postulated the existence of multiple temporalities; but not all of the universe's properties were ascribed with this characteristic. Indeed, both Marx and Epicurus presupposed that in the atom matter is eternal; insofar as it is "pure relationship to itself" it is excused "from all relativity and changeability". To this end "time has to be excluded from the concept of the atom, the world of essence" (Marx 1841/2006:131). This of course implies that "matter is eternal and independent only insofar as in it abstraction is made of the time moment" (ibid). One need only reflect momentarily on the characteristics of the atom, as defined in atomist philosophy, to appreciate the complexity of these entry-level suppositions. The atom is presupposed to be indivisible, the smallest unit in the universe, and indestructible. Consequently it is disregarded from possessing qualities that are relative and changeable i.e. the qualities of time. According to Marx, Democritus and Epicurus concurred on this issue; for both assumed that time is "removed from the world of atoms" (ibid: 132). Nevertheless, significant differences are evident in the manner in which Democritus and Epicurus philosophised the nature of time. Marx observes that Democritus discounts time; it "has neither significance nor necessity for the system" (ibid). Marx states that Democritus introduces the category of time into his philosophic discourse "in order to negate it [*aufzuheben*]" (ibid). To this end, the coming into being of things and their ebbing away is not a property endowed by the atom. Democritus' negation of the temporal incites philosophical meditations on the necessity of absolutes in time: "time itself offers proof that not everything need have an origin, a moment of beginning" (ibid). And this realisation precipitates in Marx's analysis a more incisive recognition of profound contradictions in the pursuit of absolute beginnings. As he expresses it:

> The imagining intellect that does not grasp the independence of substance inquires into its becoming in time. It fails to grasp that by making substance temporal it also makes time substantial and thus negates its concept, because time made absolute is no longer temporal. (Marx 1841/2006:132)

According to Marx, Democritus deals with the exclusion of time from the universe of the atom by projecting time into the realm of introjection: "the self-consciousness of the philosophising subject" (ibid). Marx conjectures that this is an inadequate resolution to the question of time; for it does not enable time to be apprehended as in "contact with the world itself" (ibid). Conversely, Epicurus' advance to the exclusion of time from "the world of essence" projects philosophical conjecture into "the absolute form of *appearance*" (ibid). In Greek philosophy the world of appearance is obverse to the objective quantitative world of essence. Appearance is considered in terms of our experience of the world as mediated through the senses of sight, touch etc. For Marx "when appearance is considered ...in terms of its concept, essence is totally separated from existence, so that it descends to the level of subjective semblance" (ibid: 133). Within this realm of sensorially experienced qualitative manifestations, "time is determined as accidens of the accidens" (ibid: 132). This latter phenomenon is defined in terms of "the change as reflecting in itself, the change as change" (ibid). Marx observes that, in the writing of Epicurus, "this pure form of the world of appearance is time" (ibid). To this end the expression of things as composition is "merely the passive form of concrete nature, time its active form" (ibid). This is because composition is the aggregation of atoms, which although elided remain independent from each other: "the composition behaves indifferently and materially towards its essential foundations" (ibid: 133). Conversely time interpolates between essence and appearance:

> Time, in contrast, is in the world of appearance what the concept of the atom is in the world of essence, namely, the abstraction, destruction and reduction of all determined being into being-for-itself. (Marx 1841/2006:133)

Having conceived of time as manifest in the phenomenon of change, which is apprehended in the world of appearances, Marx returns to Epicurean atomism so as to formulate our apprehension of time at the level of human consciousness. According to Marx "conscious sensuousness" is the form in which we apprehend change at the level of appearances. From this perspective Marx argues that "human sensuousness is therefore embodied time, the existing reflection of the sensuous world in itself" (ibid). Time as so defined "emerges when the accidentals of bodies, perceived by the senses, are thought of as accidentals" (ibid). Marx's affirmation of the Epicurean formulation of time as sense perception is further evident in the following statements:

> Indeed, time being the abstract form of sensation, according to the atomism of Epicurean consciousness the necessity arises for it to be fixed as a nature having a separate existence within nature. The changeability of the sensuous world, its change as change, this reflection of appearance in itself which constitutes the concept of time, has its separate existence in conscious sensuousness. Human sensuousness is therefore embodied time, the existing reflection of the sensuous world in itself. (Marx 1841/2006:134)

Marx describes how Epicurus, in his letters to Herodotus, illustrates the sensuousness of time as emerging through occurrences that can be seen to be present and pass away (ibid). Time exists, therefore, through our sensuous experience of the world "for sensuous perception reflected in itself is time itself, and there is no going beyond it" (ibid). Marx summarises "the interconnections between sensuousness and time" as thus:

> [T]ime is revealed in such a way that the temporal character of things and their appearance to the senses are posited as intrinsically One. For it is precisely because bodies appear to the senses that they pass away…Therefore: just as the atom is nothing but the natural form of abstract, individual self-consciousness, so sensuous nature is only the objectified, empirical, individual self-consciousness, and this is the sensuous. Hence the senses are the only criteria in concrete nature, just as abstract reason is the only criterion in the world of the atom. (Marx 1841/2006:135)

Marx's account of time as experienced through the self-conscious apprehension of sensuous nature is consistent with an Enlightenment mind-body dualism within which nature is subjugated to the reasoning cognition of the mind. When we compare Marx's account of time as "conscious sensuousness" with Ann Yearsley's account of the pragmatics of time embodied in the reproductive labour of women, it is evident that Marx's nature of time emphasises transcendence at the expense of immanence. Marx's (1841) observations on the nature of time had direct implications for his (1844 1857, 1867) later conception of nature as "man's inorganic body", much to the annoyance of present-day generations of Marxist ecofeminists. Before exploring the latter I shall firstly detail how Marx's foray into atomist physics provided the infrastructure for examining the social and economic determinants of the nature of dialectical materialism.

Marx's Ecology - Paradox of Two Natures

It is commonly asserted that Marx's dialectical materialism designates relations between nature and humans in Promethean terms, whereby the historical trajectory of advances in modes of productive technology ultimately entails the subjugation of nature to human rationality. There is evidence to suggest that alongside this concept of nature as dystopian, Promethean, Marx's position on the question of nature also articulated echoes of romantic eco-centrism. Explanations for this oscillation are enigmatically sparse. Nevertheless several writers bring to the fore Marx's dual view of nature as being consistent with his critique of Hegel's dialectic. Of significance here is Alfred Schmidt's (1971) *The Concept of Nature in Marx*. According to Schmidt, the Germany of the 1850s provides some indication of why Marx's consideration of human relationship with nature is surprisingly manifold and dimensionally complex. During the 1850s, obscurantist opposition to technical progress in Germany was based around "the deification of the immediacy of elemental nature" and the wish "to maintain pre-capitalist forms of production" (ibid: 132).

Continuous with this sentiment existed an ideology of nature which deplored the destructive advancement of capitalism and in which, "nature was glorified as a refuge in face of capitalism's ever more ruthless pillaging" (ibid). According to Schmidt (1971) some insight into Marx's theoretical consciousness is gained when one considers these ecological debates to have been in the foreground of Marx's material investigation. Measured against this view, Marx's position on the question of nature was also indebted to prominent philosophical debates concerning the dialectic materiality of freedom from necessity. Of particular significance were the synergies and disjunctures between Marx's and Hegel's concepts of freedom. Hegel conceived of human freedom as actualised through the unfettering of humans from the capricious control of necessity. Freedom "is a conception completely freed from the necessity of Nature, a conception which leaves that necessity behind and operates spontaneously above it" (Hegel 1998:156).

Marx also conceived that freedom, at its most basic level, resided with rationally mastering the realm of necessity. In the writings of Marx the life of the species consists materially in the fact that the human being (as with animals) "lives on inorganic nature" (1844/2007:74). Humans are least like animals when their creativity of action loosens them from the shackles of immediate necessity, providence, and happenchance, in the material conditions of existence. Axiomatic to the concept of immediate necessity is that of "natural needs". Marx comprehended "natural needs" as including nutrition, clothing, shelter etc. Marx treated natural needs as a "universal concept and general abstraction respectively" (Fraser 2000:148). Natural needs exist across human societies and the necessity of their satisfaction constitutes the "first premise of all human existence", that human beings "must be in a position to live in order to be able to 'make history'" (Marx and Engels 1846/2007:48). Basic to the activity of life existence, "the first historical act" involves "the production of the means to satisfy these [natural] needs, the production of material life itself" (ibid). While it might seem that the satisfaction of natural needs relates to drives of bygone years, Marx is keen to ascertain that this is both "an historical act, a fundamental condition of all history" and also "must daily and hourly be fulfilled merely in order to sustain human life" (ibid). Marx continuously elucidates the changing composition of natural needs, and dialectical shifts in the form of their satisfaction, as part of his broader analysis of "the contradictory movement of human beings in the shaping and making of their world" (Fraser 2000:146). Given that the form by which natural needs are gratified has divergent structural properties, it is evident that a socio-historical analysis of the satisfaction of need can render transparent the dialectical infrastructure of a given society (ibid: 147). It is, therefore, unsurprising that Marx's analysis progresses from the "general to the determinant abstraction" so as to interrogate the composition and economic infrastructure of need satisfaction in a given society and at a given historical interval. Continuous with this presupposition is Marx's treatise that, in capitalist modes of production, the particular manifestation of the satisfaction of natural needs "leads to a complex and overbearing system of need satisfaction" (ibid: 148). Central to Marx's conceptual and procedural strategy is the discovery of "inner connections" of need-forms (ibid).

The Nature of Human Drives

According to Marx, while "an animal's product belongs immediately to its physical body", humans can produce independent of immediate material necessity (1844/2007:74). This is not to assume that humans are dispossessed of animate drives; rather, Marx argued that "humans have '*vital powers*', which exist within them as '*drives*'" (Fraser 2000:149). Some indication of Marx's thinking is evident in the following extract from *The German ideology*:

> Consciousness is at first, of course, merely consciousness concerning the immediate sensuous environment and consciousness of the limited connection with other persons and things outside of the individual who is growing self-conscious. (Marx and Engels 1846/2007:48)

Consequently, where human beings confront nature, by self-reliantly "working-up of the objective world", they prove their selves as "a species-being" (Marx 1844/2007:76). Clearly part of Marx's conception of needs presumes the possibility of "a kind of creative superabundance over what is materially essential, that which overflows the measure and becomes its own yardstick" (Eagleton 1999:6). The satisfaction of needs does not entail the cessation of activity, but rather "production is his active species life" (Marx 1844/2007:76). Through the action of production humans overcome obstacles and extrinsic limitations and, in so doing, transform material nature. Rooted in the essential material conditions of this process is the unfolding of human nature conceived as a "dialectical movement of self-enriching alienation" (Walicki 1988:11). Marx assumed that human beings reproduce themselves through the process of exteriorising their creative capacities in the act of physically transforming the objective world. Axiomatic to this contention is the proposition that humans are natural beings "equipped and endowed with objective (i.e., material) essential powers" (Marx 1844/2007:155). Continuous with Marx's presumption of "essential [vital] powers" is the expectation that the human "should have real natural objects of his essence" (ibid). For it is only by relinquishing part of oneself, into the objects of our creation, that the latent potential of our creative attributes becomes consciously realised.

Consequently, it is through the process of self-enriching alienation that "nature appears as his work and his reality" (Marx 1844/2007:76). It is important to appreciate that "self-enriching alienation is a dialectical movement through alienation to self-enrichment" (Walicki 1988:11). For it is evident that Marx was reluctant to bind our consciousness to the material world in mechanistic terms, but rather held that the duplication of the self in the production of objects is also a dialectical moment of self-reflection and contemplation. Suffice it to say that the universal existence of human drives is axiomatic to the manifestation and gratification of natural needs. Walicki's (1988) concept of "self-enriching alienation" captures much of Marx's dialectical framing of the materiality of human drives and their irrevocable link to the natural world. For drives are both essential to the human being and coexistent with the external existence of "the object of these drives" (Fraser 2000:149). Ontologically speaking, Marx sets up a complex dyad in which essential vital power (needs and drives) map onto material objects in the external objective world. Through these objects needs are expressed and affirmed, and procedures are adopted to satisfy them. Consequently, Marx contends that "man's physical and spiritual life is linked to nature", and this at its most basic means "that nature is linked to itself, for man is a part of nature" (Marx 1844/2007:74). It follows from Marx's supposition that human drives exist in reciprocal relation to the material world of objective nature. And this observation has particular significance for ecofeminist critiques of Marx's account of the nature of freedom and freedom from necessity. For it is evident that in Marx's account of nature it is gendered social subject as praxis which is formulated as the defining agent, i.e. the subject that actualises his or her humanity through working-up and transforming nature. Returning to the earlier discussion of Epicurus, if from Epicurus we gain the assumption that nature is an idealistic construct of subjectivity, "from Marx we arrive at the materialist conclusion that nature is a construct of social praxis" (McCarthy 2003:23).

The Nature of Freedom from Necessity

Ironically the superabundance over nature, which Marx considers the authentication of humanity, is entangled in the fundamental conditions of human existence. Both Marx and Hegel proposed a dialectic in which axiomatic to freedom is the ability of human beings to rationally control and direct the material conditions of existence. For it is only when vast proportions of society are released from the daily toil of immediate necessity that their creative attributes are free to cultivate philosophical grandeur. While in Hegel's *Phenomenology of Spirit* this pre-supposed that "greater and greater areas of nature come under human control", Marx's dialectics oscillates between control of nature and a continual embattled reciprocity (Schmidt 1971:136-137). Marx reasons that a principal condition of freedom is the practical action of human beings who are continuously embattled in the process of overcoming the constraints of necessity. It is precisely the idea of continuous metabolic exchange with nature that distinguishes Marx's dialectic "from Hegel's ultimate idealism, the material of nature is never totally incorporated in the modes of its theoretico-practical appropriation" (Schmidt 1971:136). Axiomatic with this presumption is Marx's belief that human consciousness is cultivated through interchange with the material world:

> Men are the producers of their conceptions, ideas, etc., - real, active men ...Consciousness can never be anything else than conscious existence, and the existence of men is their actual life-process. (Marx and Engels 1846/2007:47)

Metabolic Exchange with Inorganic Nature

Marx, in accentuating the material basis of consciousness, inverts Hegel's philosophical account of humanity's metaphysical idealistic spirit. Moreover, Marx consistently articulated the belief that authentic humanity is actualised through encounters with nature. Marx's initial formulation of this interdependence between humans and nature was documented in the *Economic and Philosophic Manuscripts of 1844*. Marx argued that the material conditions of bodily existence elide all beings with nature. Humans are continuous with other animals, in that "the life of the species, both in men and in animals, consists physically in the fact that man (like the animal) lives on inorganic nature" (1844/2007:74). Within Marx's materialist dialectic the "nature", intrinsic to metabolic exchange, is defined as the "inorganic" body of human beings; "the universality of man is in practice manifested precisely in the universality which makes all nature his inorganic body" (ibid). Viewed through Marx's lens, nature is "man's inorganic body", inasmuch as humans are unable to deny bodily fragility and their dependence on natural processes for survival, i.e. nature is the human being's "direct means of life" (ibid). The "inorganic" substance of nature, also, arises from the imperative need to labour in a way that situates nature as "the material, the object, and the instrument of his life-activity" (ibid). For these reasons Marx argued that:

> Nature is man's inorganic body – nature, that is, in so far as it is not itself the human body. Man lives on nature – [which] means that nature is his body, with which he must remain in continuous intercourse if he is not to die. That man's physical and spiritual life is linked to nature means simply that nature is linked to itself, for man is a part of nature. (Marx 1844/2007:74)

Continuous with this conception of a metabolic interaction with nature, was Marx's belief that the growth of industrial capitalism precipitated a degenerating estrangement of human beings from their selves and their life activity. For the essential species-being of humans involves "creating an objective world" through "practical activity" applied to "working-up inorganic nature" so that "nature appears as his work and his reality" (ibid: 74-75). Consequently, "the object of labour is...the objectification of man's species life; for he duplicates himself not only, as in consciousness, intellectually, but also actively, in reality, and therefore he contemplates himself in a world that he has created" (ibid: 75). As stated previously, alienation is an essence of human existence, for part of our species-being is to objectify ourselves in the objects we produce.

Nevertheless in capitalist society, capital's privation of production means labour's material transformation of the objective world returns to the worker in an obsequious, impervious form to which the labourer is subjected; and the labourer no longer encounters his or her reflection in the objective world of human creation. To this end, Marx argued that the destructive propensity of capitalism resides with the alienation of humans from the objects of their production. For this entailed tearing and rupturing humans from their species life and transforming the creative interchange between humans and nature into a means to an end i.e. "a means to his individual existence" (ibid: 76). In capitalist production the estrangement of labour reaches unparalleled extremes, as the expansion of commercial calculation into the labour process is further expression of how humans are broken, divided from their humanity, as estranged from their bodies "as it does external nature", alienated from their "spiritual essence" (ibid), and ultimately subjugated to the conflicting rationalities of capitalist production.

Nature of Use-Value

Significantly, Marx argued that the "the general character" of the production of use-value is not altered by the mode of production. Consequently, in the first instance Marx considers "the labour-process independently of the particular form it assumes under given social conditions" (ibid). Self-evidently, Marx's ontological approach defines productive labour as involving manual applications upon the entities of production, which induces their material transformation. Productive labour, at its simplest, involves interventions into material existence accomplished through human practice, conceived as a metabolism of nature, and mediated through nature. In this sense:

> It is the necessity of bringing a natural force under the control of society, of encompassing, of appropriating or subduing it on a large scale by the work of man's hand, that first plays the decisive part in the history of industry. (*Capital Vol. 1,* 2003:481)

Considered to be a cornerstone of Marx's thought, the concept of "use-value" is premised on the materialist conjecture that productive labour involves the material transformation of nature as object. Indeed, "Labour is, in the first place, a process in which both man and Nature participate, and in which man of his own accord starts, regulates, and controls the material re-actions between himself and Nature" (ibid). Evidently, Marx was keen not to subordinate the realm of nature entirely to the social. For humans are labouring beings, naturally motivated towards securing the means of subsistence and accrescence. As "materially embedded beings who are a part of nature", human beings' existence requires the physical transformation of the natural environment "by the subject as a condition of its own human reality" (Mawhinney 2000:56). The locus of this process involves a life-activity in which the human being "opposes himself to Nature as one of her own forces, setting in motion arms and legs, head and hands, the natural forces of his body, in order to appropriate Nature's productions in a form adapted to his own wants" (ibid). It is through this metabolic exchange with nature that human subjects cultivate their nature, i.e. "By thus acting on the external world and changing it, he at the same time changes his own nature" (ibid). Extending nature into the logic of identity enables Marx to advance the supposition that productive labour exercises physical energy and mental processes:

> He not only effects a change of form in the material on which he works, but he also realises a purpose of his own that gives law to this modus operandi, and to which he must subordinate his will. And this subordination is no mere momentary act. (*Capital Vol. 1,* 2003:174)

Nature of the Labour-process

Here, as elsewhere in Marx's oeuvre (c.f. Grundrisse), a persistent antinomy emerges in which the humanized account of nature gives way to the objectification of nature. This is very true of Marx's distinction between the "elementary factors of the labour-process" (*Capital Vol. 1,* 2003: 174). The first of these factors refers to the rudimentary objects of nature: "The soil (and this, economically speaking, includes water) in the virgin state in which it supplies man with necessaries or the means of subsistence ready to hand, exists independently of him, and is the universal subject of human labour" (ibid). According to Marx, nature in its elementary form is freely bequeathed to humans:

> All those things which labour merely separates from immediate connexion with their environment, are subjects of labour spontaneously provided by Nature. Such are fish which we catch and take from their element, water, timber which we fell in the virgin forest, and ores which we extract from their veins. (*Capital Vol. 1,* 2003:174)

Production, in its elementary form, involves human beings interacting with objects "spontaneously" yielded by nature, so as to secure the means of subsistence. Marx's conception of "raw material" perceives additional dimensions in the elementary relationship between living productivity and the natural world. Where the subject of labour has been filtered through previous stages of production, Marx defines this substance as "raw material". While "all raw material is the subject of labour", it is the case that "not every subject of labour is raw material: it can only become so, after it has undergone some alteration by means of labour" (ibid). This subtle shift in emphasis provides Marx with a platform from which to introduce the conceptual nature of "an instrument of labour". In the process of production, "the labourer interposes between himself and the subject of his labour" instruments which serve "as the conductor of his activity" (ibid). If we disregard deliberation of "such ready-made means of subsistence" as recherché crops, in which one's own limbs operate as instruments of labour, "the first thing of which the labourer possesses himself is not the subject of labour but its instrument" (ibid: 175). In the organisation of production, it is not the objects made, but the process through which they are created and the instruments used "that enables us to distinguish different economic epochs" (ibid: 175).

Nature of Capitalism's Ecological Disjuncture

In *Capital Vol.1* Marx presents nature almost as a series of degrading archetypes, dialectically in transition. Evolutions in the labour process signify a relation, not of metabolic equilibrium, but of ecological disjuncture. This is particularly evident in capitalist commodity production, a system in which labour becomes a commodity i.e. "the capitalist buys labour-power in order to use it; and labour-power in use is labour itself" (*Capital Vol. 1,* 2003:173). The intrinsic worth of labour is thus objectified into the fetish of the equivalences of commodity exchange. But it is also evident that "in order that his labour may re-appear in a commodity, he must, before all things, expend it on something useful" (ibid). Consequently the labourer is compelled to convert its natural propensities into a mere instrumental apparatus of material survival, i.e. "in degrading spontaneous activity, free activity, to a means, estranged labour makes man's species life a means to his physical existence" (1844/2007:76). In capitalist society, humans are goaded to produce both by the immediacy of physical necessity and by the alienation of estranged labour. In these circumstances:

> Estranged labour turns...Man's species being, both nature and his spiritual species property, into a being alien to him, into a means to his individual existence. It estranges man's own body from him, as it does external nature and his spiritual essence, his human being. (Marx 1844/2007:76)

This is why Marx rounded against capitalist modes of production and positioned communist society as providing the opportunity to emancipate the creative use-value of human beings from the beguiling commodity fetishism of exchange-value. For nature, and by extension the human body, is the source of social life. The full potential of human beings resides in emancipating their natural productive propensities to creatively transform the world through self-enriching alienation. Central to the realisation of this freedom is the development of material conditions in which associated prodcers govern metabolism with nature in a rational way. According to

Marx, it is only in these circumstances of contiguity with nature that humans actualise their species being:

> [S]ince for the socialist man the entire *so-called history of the world* is nothing but the begetting of man through human labour, nothing but the coming-to-be of nature for man...Since the real existence of man and nature has become practical, sensuous and perceptible – since man has become for man as the being of nature, and nature for man as the being of man – the question about an alien being, about a being above nature and man – a question which implies the admission of the inessentiality of nature and of man – has become impossible in practice. (Marx 1844/2007:113-114)

Intriguingly the preceding extract not only encapsulates a further example of oscillations in Marx's account of nature, it also provides a glimpse into what Marx perceived as the redemptive properties of technologically innovative transcendence of necessity. For it is evident that Marx believed that "the true realm of freedom, the development of human powers as an end in itself, begins beyond it, though it can only flourish with this realm of necessity as its basis. The reduction of the working day is the basic prerequisite" (*Capital Vol., 3* 1981:959). Suffice to say, Marx envisaged advances in technology as providing, in a context where the means of production are democratically owned, the conditions for time to be free from inefficient entanglements with the realm of necessity. It is apparent that Marx perceived the reclaimed free-time to form the material base for the cultivation of the self. Conversely in capitalist regimes of accumulation workers are unable to satisfy their natural needs directly, because the mode of production is privately owned. Consequently the labourer is forced to satisfy his or her natural needs through the mediation of the waged labour market. "Expenditure of the wage on goods means that they are positing their needs as necessary – 'necessary needs' –within the alienated system of capital" (Fraser 2000:152). When one considers necessary needs to be "the form that the satisfaction of 'natural needs' takes in capital" it is evident that their need-form is formulated exponentially with the value of the worker's labour power (ibid). What is significant here is that, in capitalism, the satisfaction of needs propels individuals into exploitative mediations with the object-form of nature.

Embodied Materialist Ecofeminist Nature of Time

> [M]aterialist ecofeminism takes us beyond the market and waged labor by incorporating the sexual division of labor and the ecological limits to production and reproduction into historical materialism and the critique of capitalism. (Dordoy and Mellor 2000:44)

It is conventional to identify the inception of ecofeminism with Francoise D'Eaubonne's writings in the 1970s. Following through this continental European trajectory it is evident that ecofeminism has tended to propagate from congeries of: feminist activism, ecology, peace movements and response to ecological devastation such as "the meltdown at Three Mile Island [which] prompted large numbers of women in the USA to come together in the first ecofeminist conference…in March 1980 at Amherst" (Mies and Shiva 1993:13-14).

Ecofeminists perceive the exploitation of nature to be bound-up with the prerequisite of patriarchal relations of power, to exploit the reproductive labour of women. From this conviction emerges the rallying contestation: "social domination and the domination of nature are interrelated" (Salleh and O'Connor 1991:131). Ecofeminists give prominence to environmental degradation in terms of "the connection between patriarchal violence against women, other people and nature" (Mies and Shiva 1993:14). And this endeavour raises momentous issues concerning "corporate and military warriors" and their "aggression against the environment"; for ecofeminists have perceived this "almost physically as an aggression against our female body" (ibid). Such impassioned sentiment invites Marxist ecological activists to be more reflective about gender and the conceptual terms of Marx's political ecology; i.e. about the dialectical reciprocation of nature and human existence. While it is imperative that ecological understanding interrogates the market form of environmental degradation, it is equally as important to expand the notion of economy so as to encompass "the multiple contributions of women's reproductive labour to the maintenance of capital" (Salleh 2010:203).

To this end, Marx's Enlightenment confidence in rationality and technology as vehicles to transcend time-consuming freedom from necessity has been correctly and justifiably challenged. Of significance has been the attack, within Marxist sociology, from "embodied materialist eco-feminists". Rooted in the economic categories of Marxism, materialist ecofeminism is "embodied" because it enfolds, into the analysis of production, an ethic of care-centred reproductive labour (Salleh 2003:65).

And it is "materialist" because it argues that "the subordination of women and the degradation of the natural world are historically and materially related" (Dordoy and Mellor 2000:41). In addition to Marx's structures of political economy, embodied materialist eco-feminism maintains that the nurturing and caring activities, integral to the social reproduction of economic life, are co-existent with the environment; both exist as rapaciously exploited natural resources (ibid: 44). To this end ecofeminist Marxist debate has turned to the "provisioning" of social reproduction; and the manner in which reproductive labour is associated with "human immanence, which creates the social time and space that allows for transcendence of those conditions (temporarily) by those who extract this time and space from the labour (and resources) of others" (ibid: 46).

In part two of this paper, I argue that Marx's dialectics of nature presupposes that human species-being is cultivated in competitive contiguity with the natural world. This observation led Marx, in his early writings, to describe the human-nature relation in romanticised metrical terms. Evidence of this is amply provided in the *Economic and Philosophic Manuscripts of 1844*; indeed consider the following extract:

> The domination of the land as an alien power over men is already inherent in feudal landed property...But in feudal landed property the lord at least appears as the king of the estate. Similarly, there exists the semblance of a more intimate connection between the proprietor and the land than that of mere material wealth. The estate is individualised with its lord...It appears as the inorganic body of its lord. Hence the proverb *nulle terre sans maître*[2], which expresses the fusion of nobility and landed property. (Marx 1844/2007:61)

According to Marx, transcendence of ecological relations involves the use and development of technology so as to achieve the most efficient evolution away from the realm of necessity, and this innovation precipitates man's acceleration towards "the reduction of the working day" (*Capital Vol.3* 1981:959). But the conception of free-time here reinforces fundamental contradictions in Marx's materialist analysis. In contrast to the transcendent ego's cultivation of free-time through the instrumental application of scientific technology, the time of reproductive labour is endlessly "spatially embedded because it is, of necessity, local and communal, centred on the home" (Mellor 2009:255). Reproductive labour time, unlike productive labour, is not remunerated; consequently it

---

[2] Translated as 'there is no land without its master' (Marx 1844/2007:61)

exceeds and is foudroyant of Marx's economic categories (Odih 2007). As embodied materialist ecofeminists express it, "humanity is immanent within the ecological whole not transcendent over or beyond it" (Dordoy and Mellor (2000:43).

Timely Reflections

In the formulations of gendered time evident in Mary Mellor (2009) and Ariel Salleh's (2003) writings, reproductive labour time is far removed from the transcendent uninhibited energy of the Epicurean *clinamen*. The embodied time of care-centred relations *swerve*, at their peril. To this end, Mary Mellor (2009:255) states that "women's work represents the fundamental reality of human existence, the body's life in biological time, the time it takes to rest, recover, grow up and grow old". Keeping focus on the time necessitated in the transformation of nature and natural resources leads Mellor (ibid) to consider a broader framing of biological time; one which explores the interconnections between the local eco-systems of the private sphere and the macro-economics of trans-local ecological systems. By virtue of this constellation Mellor (ibid) formulates the concept of "ecological time", which "is the time it takes to restore the effects of human activity, the time-cycle of renewal and replenishment within the ecosystem". Mellor (ibid) convincingly highlights a material basis for the economic exploitation of ecological time. While emphatically agreeing with the principle of this argument, it is necessary to avoid essentialism and dualistic categories in the formulation of the concept of reproduction time. To conceive of reproductive time in terms of women's bodies and the phenomenology of women's daily activities reproduces Enlightenment epistemologies, as well as the modernist practices which have authorised the unrelenting subjugation of nature/women to rational scientific technology. It is my contention that the political effectiveness of ecofeminist conceptions of reproductive times is disarmed by slipping into dualistic categories of male vs female time. An alternative conception is evident in Rosa Luxemburg's (1913/2003) *The Accumulation of Capital*. Here she argues that primitive accumulation is co-existent with capitalist accumulation rather than, as Marx suggests, a period prior to Industrial Capitalism. As Luxemburg (ibid: 345) expresses it:

> [C]apitalism in its full maturity also depends in all respects on non-capitalist strata and social organisations existing side by side with it.

Continuous with Luxemburg's (ibid) assertion, it is my contention that reproductive time is more effectively broadly conceptualised as part of the qualitative times of social reproduction (Odih 2007). In so doing it becomes apparent that these qualitative times form "non-capitalist strata" integral to the global spread of capitalist accumulation. It is from this conceptual basis that the nature of time in historical materialism needs to account for the manner in which "the valued economy has gained its power and ascendency through the marginalisation and exploitation of women, colonised peoples, waged labour and the natural world increasingly on a global scale" (Mellor 2009:255-256).

INTRODUCTION

ECOFEMINIST WATERWAYS

Interviewer: How did you arrive here?
Interviewee: *I heard about it ages ago and I thought it was such a great idea, because I'm always appalled by the rubbish that is just thrown into the water. And well I live in London so obviously I only attend these events if it's on the Thames, because it feels closer to home. But, erhm and yeah I think it's a great idea. Unfortunately it's quite sad that we have to do that. But I think it's a great idea.*

Interviewer: And what do you gain from being here?
Interviewee: *Erhm, just the satisfaction that actually, I'm contributing to my local community…I'm from London so and I think it's just so disgusting that people think that they can just chuck away whatever they want whenever they want.*

Interviewer: And what do you think about the fact [that] there's lots of plastic?
Interviewee: *Yeah, erhm. Well, this is a, it's all over the world now you know the plastic. And we seem to be, not to be able to do without it. But it's not so much that we use plastic. It's that we dispose of it the wrong way. And that is shocking to me.*

Interviewer: And do you think a community is developing around this type of event?
Interviewee: *I don't know, well it's the first time I attend such an event so I can't really speak from experience. I don't feel that people will actually really bond and make, you know, friendship out of it. I think it's, to me it's more a personal contribution to keeping the waterways cleaner and keeping, in general the land cleaner. You know we can do that in our own street. You know we can do that in parks; we can do that you know. So maybe it will encourage other charity groups to develop that elsewhere. And I think it's great for the kids as well, they get outdoors and you know they have that sense of responsibility. I think it's brilliant that parents bring their children. (Female Volunteer, South London, River Thames Foreshore, September 2013).*

Figure I.1., "Exquisite thy mental pleasure, Common transports are not thine; Far surpassing vulgar measure, All thy joys are near divine" (Anne Yearsley *To a Sensible but Passionate Friend* 1787). Photographic Image, Inception of the River Thames, Gloucestershire, April 2014

Figure I.2., "So the darkness shall be the light, and the stillness the dancing" (T.S. Eliot *East Coker*). Photographic Image, South London, River Thames, September 2013

(Her)Story of the River Thames

Rising in the Cotswolds in Gloucestershire, the River Thames majestically streams through a magnificent arc, from England's southwest through to its east, before debouching into the North Sea. Earliest recordings indicate the Roman Emperor Julius Caesar named the river, Tamesis: "in his *De Bello Gallico* [Julius Caesar] says that at the time of his invasion of Britain the river was called Tamesis" (Herbert 1966:35). Popular etymology suggests that: "The first part of the name is found in the descriptive title of many British rivers, as Tame, Teme, Tamar, Thame; the second part seems to mean merely 'water'" (Phillips 1871:i). An intriguing cultural etymology coadunates the formative waterways of Tamesis; for it is argued that: "its original name is Isis, and that it has no pretensions to the name Thames, till its confluence with the little river Tame, in the vicinity of Dorchester, in the county of Oxford" (Combe 1794:1). Contiguous with this description, the appellation of the Tamesis/ Thame-Isis/Thames is ascribed to the lower regions of the river, immediately below the confluence of the River Thame and the River Isis; the latter rises in the vicinity of south-west Cirencester, near a village called Kemble; streaming through Oxfordshire. The appellation of Isis as the source of the Thames traces an intriguing waterway; A.P. Herbert's (1966) *The Thames*, describes the "statutory existence" that: "Five Acts of George II and George III, from 1750 to 1842, have 'the River Thames and Isis' in their long titles" (ibid:35). Moreover, the Thames Conservancy Act 1894, which established governance of the entire River Thames, stipulated that "The expression 'the Thames' means and includes so much of the Rivers Thames and Isis respectively as are between the town of Cricklade in the County of Wilts and an imaginary straight line drawn from the entrance to Yantlet Creek in the County of Kent" (ibid).

Poetry and local historical records provide an intriguing pronouncement of Isis as the source of the River Thames. Peter Ackroyd's (2007) *Thames Sacred River* identifies the earliest recorded association between the Thames and Isis: Ranulphus Higden was a Chester monk who lived in the early fourteenth century and authored "the *Polychronicon*", part of which states "*Tamisia videtur componi a nominibus duorum fluminium, quae sunt Thama et Ysa aut Usa*", which Ackroyd translates as "Thamesis seems to be composed from the names of two rivers, that are the Thama and the Ysa or Usa" (ibid:26). According to Ackroyd:

> It is more than probable that the Ysa and Usa of Higden's account are in fact from the Celtic isa or esa. In the fourteenth century, then, the people along the banks of the river knew it from its Celtic term. Perhaps it was always known as the Ysa…It was easy enough for Ysa to become the resplendent Isis. (ibid:26-27)

Figure I.3a., "Thames [Thamesis, so called from the meeting together of the Rivers Tame and Isis or Ouse] the chief River of Great Britain" (Nathan Bailey 1731). Photographic Image, Inception of the River Thames, Gloucestershire, April 2014

Figure I.3b., "Thames [Thamesis, so called from the meeting together of the Rivers Tame and Isis or Ouse] the chief River of Great Britain" (Nathan Bailey 1731). Photographic Image, Inception of the River Thames, Gloucestershire, April 2014

Figure I.3b., "Thames [Thamesis, so called from the meeting together of the Rivers Tame and Isis or Ouse] the chief River of Great Britain" (Nathan Bailey 1731). Photographic Image, Inception of the River Thames, Gloucestershire, April 2014

Figure I.4., "Faire Isis to bring forth, the mother of great Tames" (Michael Drayton quoted in Fearnside 1834). Photographic Image, River Isis, Gloucestershire, April 2014

William Camden's sixteenth century eulogy to the River Isis describes: "The Isis, commonly called the Ouse, rises from a continual spring in the south side of this county. This is that Isis which, after receiving the Tame, and by adding the names together, is called Tamisis, the chief of British rivers" (William Camden quoted in Fearnside 1834:2). William Camden is also noted to have said of the Isis: "...which we may truly say, as ancient writers did of the Euphrates in the East, that it both plants and waters Britain" (quoted in Hall 1859:4). William Camden charmingly proposes the appellation of Isis in verse:

> Thus sang the Goddess, straight Tamisis stream,
> Proud of the late addition to its name,
> Flows briskly on, ambitious now to pay
> A larger tribute to the sovereign sea.
> (Quoted in Fearnside 1834:2)

Regarding the ancestry of the name Isis, it is evident that the River Isis in its ancient history has often times been called Ouse; an appellation ascribed to several principal English rivers (Fearnside 1834:4). In ancient Hebrew, one can discern that: "the word *Us*, pronounced Ouse, denotes speed – haste; and in the early days of the church, the monkish brethren may have thought this word not an inappropriate term to represent the quick passage of the current" (ibid). And in ancient poetry there is evidence to suggest that Ouse was called Usa; and thus the Ouse of Gloucestershire originally had the appellation Usa, which transfigured into Ysa, Ise, and latterly Isis as "the Romans translated Ouse, Isis" (ibid). The influence exerted by ancient Egyptian and Greek spiritual observances is described by George Sarton (1952) in his *Ancient Science Through the Golden Age of Greece*. Sarton emphasises intersectional ties linking maritime finance capital flows and "religious festivals" from e.g. "the center … of the Ionian [maritime] league … in Delos" (ibid:224), geodesy and earth-mother worship (ibid:125-126). According to Sarton, the cult of the goddess Isis commenced its "foreign conquests in the seventh century, if not before. Herodotus says that the women of Cyrene worshiped her. The diffusion of the Egyptian religion was much intensified when the Naucratis settlement was established in the Delta in the sixth century, and from that time on it increased steadily. Temples and inscriptions to Isis and other Egyptian gods can be found in many of the Islands, even in the sacred Delos" (ibid). It is an intriguing feature of the River Thames' popular etymology that it is intertwined with the celebration of an iconic deity drawn from Egyptian and Greek mythology. The Greek philosopher Diodorus conjectured that in Egyptian mythology, the earth was considered to be the progenitor of all things, and Egyptians honoured Isis as Mother Earth. The inundations of Egypt's River Nile were revered; for it was believed to be initiated from Isis's tears of mourning. The association of Isis with nurturing and motherhood is a consistent theme in literatures on Egyptian mythology. Garry Shaw (2014:46) describes Isis as intimately associated with "magic, motherhood and love". Reginald Eldred Witt's (1997:15) *Isis in the Ancient World*, describes the Egyptian goddess as "One of the most sublime deifications of Motherhood…the female embodiment of the Nile's annual reawakening".

According to Sarton (1952:126), Isis's cult of motherhood conjoins with the worship of the Greek goddess Demeter as part of the graduated coming together of Egyptian and Greek gods. Indeed it is necessary, when appreciating ancient Greek culture, to attribute importance "to the sacred mysteries" and recognise that: "Those mysteries which constituted the inner life of religion were to a large extent of foreign origin" (ibid). Thus the celebrations of Isis as part of the Greek Eleusinian mysteries were conceived originally in Egyptian mythology, as Sarton (ibid) indicates: "Demeter, the glorification of motherly love" (cf. Isis)". In the goddess worship of Isis, motherly love assumes universal breadth; for it is pronounced that: "Isis says of herself, 'I am everything which existed, which is now and will ever be, no mortal has ever disclosed my robe'" (ibid:152). George Sarton's account of Isis's worship in the Greek island Delos and evocation of "the timeless gods" as evident in "the inscription in the Iseum of Sais" in which "Isis says of herself" (ibid:152) correspond with Rufus Richardson's (1898) challenge to the tendency for Herodotus to conflate the earth-mother(s) of the Eleusinian mysteries with Demeter. As Richardson (1898:227-230) asserts: "in 1895, when the Greek excavators at Eleusis found, in a grave containing vases of the very oldest class, some scarabs and a statuette of Isis. … Instead of the bald, dead equation, Demeter=Isis, we have proof of a stream of influence which, beginning as far back as Psammetichus, flowed into Greece from the older civilization". The earth-mother Egyptian mythology of Isis relayed a boundless cycle of rejuvenation, entangled in a mesh of divine objectifications of the mundane and extraordinary permutations of nature. Just as the myth of Isis symbolically mediated the River Nile's time of inundation, so it is that Isis's medicinal and curative skills have come to intimate the "perpetual resurgence" of the River Thames (Ackroyd 2007:26). Ackroyd's *Thames: Sacred River* quintessentially describes "reaches of the Thames that effortlessly summon up memories of the Nile".

John Phillips' (1871:i) *Geology of Oxford and the Valley of the Thames*, eloquently describes how Oxford dignitaries often referred to "the well-known spring three miles west of Cirencester in the Parish of Kemble, [as] the 'very head of Isis'". Alexander Pope verses tribute to Greek and Egyptian mythology in his poetic vision of "The winding Isis, and the fruitful Thame" (Pope 1826:9). Compassionate mother goddess and breather of life into earth: "Isis represents the water as feminine. It is the water as the female principle, circling like amniotic fluid…mingling with clay, it creates shape and form" (Ackroyd 2007:29). Michael Drayton in the cultivated topographical poem *Poly-Olbion* anthropomorphises Isis as the maternal mother of "great Tames":

> But Cotswold, be this spoke to th' onely praise of thee,
> That thou of all the rest, the chosen soyle should bee,
> Faire Isis to bring forth, the mother of great Tames,
> With whose delicious brooks, by whose immortal streames,
> Her greatnesse is begun. (Quoted in Fearnside 1834:8)

Edmund Spenser's (1590/1758) majestic allegorical poem *The Faerie Queene* features the knight Britomart entering the temple of Isis and this *meare* signifies maternal love and nurturance as a general rather than a particularized aspect of the goddess Isis's reality:

> Britomart comes to Isis church,
> Where she strange vision sees
> …
> Isis, whom they likewise made
> A Goddess of great power and sovereignty
> And in her person cunningly did shade
> That part of Justice which is equity,
> Whereof I have to treat here presently.
> Unto whose temple whenas *Britomart*
> Arrived, she with great humility
> Did enter in, ne would that night depart;
> But *Talus* mote not be admitted to her part.
> …
> There did the warlike Maid herself repose,
> Under the wings of Isis all that night;
> And with sweet rest her heavy eyes did close,
> After that long days toil and weary plight.
> Where whilst her earthly parts with soft delight
> Of senseless sleep did deeply drowned lie,
> There did appear unto her heavenly spright
> A wondrous vision, which did close imply

> The course of all her fortune and posterity.
> (Spenser 1590/1758:233-235)

Edmund Spenser's (ibid) depiction of Britomart and Isis situates these female characters within an allegory of society in which their accomplishments demonstrate the possibilities "of the proper female role in the restraint of male power" (Levy 1987:224). In Egyptian mythology Isis's wanderings exude restorative force, devotion and exaltation; in Spenser's depiction of Britomart's odyssey her valiant crusade is analogous to "the female principle of Nature" as opposed to "a destitute and brutish manner of living" (Feinstein 1968:535). Just as in the mythology of Isis the order of things becomes "euenly weyed in the balaunce of equyte and right" (Diodorus quoted in Feinstein 1968:533), so it is that in Spenser's poetry Britomart "afresh appeared the glory of the later world to spring"; but this female principled vision has to contend with the pertinacity of London's river, which in its confluence assumes a more masculine form: "that with the waves Of wealthy Thamis washed …the Troian brute did first that citie fownd, And Hygate made the meare thereof by West, And Overt-gate by North: that is the bownd Toward the land; two rivers bownd the rest." (1590/1825:292). It is evident therefore, that intertwining these reverential acclamations of Isis as Mother Goddess is a spiritual belief that Isis connotates "that the river is a living thing …It has its own organic laws of growth and change" (Ackroyd 2007:29). Suffice to say, the River Isis's popular etymology is associated with the nurturing goddess Isis; and therefore anthropomorphises as a loving, nurturing and irrepressibly wise mother; as is indicated in the following extract from Thomas Tickell's (1707/1808) poem *Oxford a Poem. Inscribed to Lord Lonsdale 1707*:

> Thy mother Isis…,
> A subject not unworthy of thy verse;
> So Latium fields will cease to boast thy praise,
> And yield to Oxford, painted in thy lays:
> And When the age to come, from envy free,
> What thou to Virgil give'st shall give to thee,
> Isis, immortal by the poet's skill,
> 'Shall, in the smooth description, murmur still'
> New beauties shall adorn our sylvan scene,
> And in thy numbers grow for ever green.

Figure I.5., "To shew that Isis doth the Moon portend" (Spenser 1590/1758:233). Photographic Image, River Isis, Gloucestershire, April 2014

Figure I.5a., "To shew that Isis doth the Moon portend" (Spenser 1590/1758:233). Photographic Image, River Isis, Gloucestershire, April 2014

Governing Riverscapes in National and International Political Economies of the Sign

Although extraordinary in mythological stature, the River Thames is unremarkable in its length; for it flows 215 miles from source to sea, compared to the Mississippi's flow of 2,560 miles (Herbert 1966:19) and the Yangtze's 3,500 miles (Ackroyd 2007:3). The River Thames' drainage basin is a terrain of land, commensurate to its size at 5,264 square miles, into which converges the flows of water, sediment and natural energy (ibid). The average rate of river flow, calculated in gallons at Teddington (the interface of the tidal and non-tidal waters) is 1,145 million gallons per day, calculated over an entire year (ibid). It is estimated that the velocity of the river's current advances "between ½ and 2 ¾ miles per hour" (ibid). It is significant to differentiate the tidal and non-tidal Thames. The section of water in which the Thames enters into the North Sea is known as the Thames Estuary: from Teddington downstream debouching into the North Sea. As with any estuary, the Thames Estuary is bell-shaped, with fresh river waters entering at the attenuated end, in addition to any river tributaries that feed into the river course (Wheeler 1979:6). Tidal waters refer to the twice daily tidal flow into the river, with the volume of water varying according to the obstructions of tidal sluice and the volume of fresh water flowing downstream (ibid). This tidal motion creates a complicated water system "unlike a freshwater river which flows downstream all the time, in an estuary the same water flows back and forth for days and even weeks, before it reaches the sea" (ibid). In the geographical science of rivers, the structural form as created by the flowing of river water through its landscape is called fluvial morphology and integral to this is the "geomorphic cycle which is a cycle of erosion passing through several stages" (Lane 1955:2).

In a geomorphological approach to river flow, one commences with the notion that the fluvial morphology begins as a shear horizontal land surface which, over time, becomes warped; a phenomenon precipitated by motions in the earth's surface. As water continues to stream through the land's surface, the transportation and deposition of sediments into the sea, vertically etches into the streambeds. In due course, and after considerable time, the elevation of the river's surrounding land surface is considerably lowered, "the end of the cycle is reached when the whole surface has been reduced to a very gently sloping plain called a peneplain, and thus comes back to the condition in which the cycle started" (ibid). Fluvial morphology evolves into the generation of floodplains as the lateral migration of the river channel leads to the erosion and sedimentation of surface area alongside the river channel; "This occurs in any stretch of the river when the sediment supplied by the river upstream, the tributaries within the stretch, and the erosion of the banks and valley sides is equal to the transportation capacity of the river" (ibid:3). Lane (ibid) describes: "It then ceases its rapid cutting down, but the valley widens out as the material from the banks and valley sides continues to be carried away". Floodplains are, therefore, part of the river's natural geomorphic cycle; but where the floodplain's capacity is exceeded by excessive streaming from the catchment area, hazardous flooding occurs.

An engaging history of the River Thames and its catastrophic flooding events is provided in Anthony Milne's (1982) *London's Drowning*. According to Milne (ibid:29), from 1703, the science of river systems advanced towards the discovery of "monthly patterns in the weather cycle". It was increasingly evident that the recurring phenomenon of "westerlies" and "easterlies" had begun to affect the fluvial processes of the River Thames. These meteorological conditions are further exacerbated by a denaturing of the River Thames' hydrological cycle. It is best to comprehend this latter feature by framing the river in terms of "a catchment ecosystem, which consists of a river, or lake, together with its associated drainage area" (Petts and Foster 1985:8). The river catchment approach segments the river into a definable waterway whereby: "Each catchment (=watershed) is delineated by a topographical divide that isolates it from adjacent catchments, the only interaction being the migration of mobile biotic populations and some groundwater transfers" (ibid). Comprehension of river ecosystems has increasingly advanced the framework of "the catchment hydrological cycle". In this respect, rivers are first and foremost hydrological cycles, "it is the stochastic nature of the runoff process" that is the source of their dynamic properties. (ibid:9). Evidently the catchment hydrological cycle has "a single input, precipitation, and output by evapotranspiration and stream-flow" (ibid: 20). From this it can be discerned that "the runoff component of the catchment hydrological cycle represents the excess precipitation after accounting for losses due to evapotranspiration" (ibid). It is a natural feature of a river's hydrology that during its annual discharge maximum, the water inundates the river's floodplain. When observing the effects that cities have on river flooding, it is evident that there are three forms of flood risk facing London: excessive river water, adverse tidal surges and excessive surface water.

London's Flooding; Voluntary Sector River Conservation Interventions

The land surface upon which London is built, especially the low-lying Thames Valley region, largely consists of chalk escarpment, which has low absorbency with regards to its ability to soak up rainwater into the ground (Banks and Stanley 1990:19). London is particularly vulnerable to surface water flooding: its burgeoning mosaic of impermeable paved streets and built-up urban surface areas means that stormwater surface runoffs saturate the ground, exceeding the drainage capacity of London's ageing sewage network and rapidly overflowing the river system. The latest Intergovernmental Panel on Climate Change (IPCC) identifies that "Due to sea-level rise projected throughout the 21st century and beyond, coastal systems and low-lying areas will increasingly experience adverse impacts such as submergence, coastal flooding, and costal erosion" (2014:16). The IPCC's forecasts indicate that low-lying areas will be subject to an increased likelihood of tidal flooding and fluvial flooding. In 2014 London's Mayoral office published *London's Response to Climate Change*, which states:

> Fluvial flood risk is likely to increase significantly through the century, as a result of climate change. Predictions of increases in peak flows of up to 40 per cent would mean that we would have to expect increased flood risk on all of London's tributary rivers. (Mayor of London 2014a:175-176)

During the winter months, December 2013 – January 2014, the United Kingdom (UK) experienced a spate of extreme weather: "the largest storm surge in 60 years"; "stormiest December on record"; "wettest January in 250 years"; "lowest depression since 1876" (Environment Agency 2014a:1). The extreme weather system produced flooding from tidal surges in which the Thames' tidal cycle combined with the acute weather storm to force aberrant volumes of water into the Thames Estuary (London Assembly 2014d). Immense tidal surges and prolonged rain water inundation into swollen rivers and saturated land surfaces produced a catastrophic level of flooding, especially in the southern regions of England. The Environment Agency erected over 2,800 km of floodwalls, and other flood defences protecting over 1.3 million homes (ibid: Parliamentary Papers 2014a). The Environment Agency estimated that 24,000 households throughout Greater London were at risk of flooding (London Assembly 2014a:1). Flood defences to protect London include the Thames Barrier and its associated defences. Operational since 1982, the Thames Barrier stretches 520 meters, traversing the Thames near the docklands in Woolwich and defends 125 square kilometres of central London, from extreme tidal surges. In December 2013 it "was raised to protect London from the largest tide recorded at Southend since it became operational" (Environment Agency 2014b, Parliamentary Papers 2014a:3). It is estimated that during this extreme weather period "1,700 homes [across England] had been flooded and the Thames Barrier was raised nine times to protect £200bn worth of property in London" (ibid:4).

Prolonged torrential rain, tidal surges and storms continued into January 2014; "On 30 January 2014 the Met Office said that parts of England have had their wettest January since records began" (ibid:4). Indeed an Environment Agency spokesperson is reported to have said: "This is an exceptional event, it was the highest rainfall in January since 1776 and we think it is likely December, January and February will be the highest for 250 years" (ibid: 4-5). Alongside briefings from the Environment Agency, a vast array of organisations became directly involved in flood response of particular significance: "the Environment Agency, local authorities, voluntary bodies, charities, and business" (Parliamentary Papers 2014a:4). The Prime Minister accompanied by a burgeoning media scrum "visited the site of flooding in Oxfordshire near his own constituency of Witney on 11 January 2014" (ibid). Media broadcasting reached a heightened level of spectacle, progressing southwards into Hampshire, Dorset and Somerset. On 27 January 2014, the Rt. Hon Owen Paterson, the Secretary of State for Environment, Food and Rural Affairs, visited the Somerset Levels "where 20,000 acres of farmland had been underwater for a month or more" (ibid). Media spectacle engulfed this event as Owen Paterson encountered public frustration and criticism of the government's response to the floods. Local anger coalesced into a spectacle of frustration directed at the Environment Agency and its apparent failure to have maintained a defensive programme of dredging key rivers. Dredging refers to a process of mechanically removing silt and other sediment from the riverbed and adjacent sides of a river channel.

Dredging quickly galvanised the linguistic repertoires of protest; its neglect by successive administrations of the Environment Agency appeared to symbolise a disregard for riverside communities. In its defence, the agency highlighted the vast sums invested in flood risk defence measures: "Nationally, we spent £45 million in the last financial year on improving the flow in rivers, including dredging…in some circumstances, dredging is not the best long term or economic solution compared with other flood risk measures such as building walls or providing storage" (Environment Agency 2014c:1). A curious feature of the public insistence on dredging the major affected rivers was the public's refusal to be satisfied by the argument of economic rationality. The Environment Agency's duty of care was visibly in disharmony with its public; on 31 March 2014 the Environment Agency announced "Dredging of one of the major rivers in the Somerset Moors and Levels will begin today" (Environment Agency 2014d:1). In the immediate wake of the December 2013-January2014 flooding crisis, the direction of flow of resolutions increasingly place emphasis on river restoration. Thus, for example, the London Assembly on Friday 14 February 2014 announced:

> There are 24,000 homes at risk of river flooding, and we could help protect them by restoring many rivers and flood basins to their natural states, increasing their capacity to absorb and hold water. This can also benefit the local ecology and create new accessible open spaces for the local community.
>
> This Assembly notes that the Mayor planned to restore 15km of local rivers by 2015 to reduce flood risks, and that 14km have already been delivered. There is potential for hundreds of kilometres to be restored, and the River Restoration Centre has already identified 63 projects around the capital which could reduce flood risks to thousands of homes. These would be delivered by the Environment Agency, boroughs and through local community action. Past projects have varied in cost between £104,000 and £18 million. (London Assembly 2014b:1)

Defining river restoration in terms of refurbishing their "natural states" is done in the spirit of a new era in which the implementation of the European Union's Water Framework Directive has encouraged a reframing of river ecology in terms of the requirements of a neoliberal political economy (Smith, Clifford and Mant 2014). The London Assembly (2014b) in their recent pronouncement on river restoration frequently refer to a "natural state" against which the current condition of rivers can be compared. It is evident that a paradigmatic shift has occurred in which the design and maintenance of rivers have veered away from the scientific engineering solutions that pervaded river maintenance up until the 1960s and 1970s. During the era of hard-engineering solutions: "Rivers were viewed as channels to convey water as rapidly as possible to the sea, or a resource to be exploited for abstraction purposes" (London Assembly 2014b:1). Rivers were straightened into concrete confines and tunnelled into enclosed channels (culverts). It was assumed that these engineering solutions would lay down flood protection: the channels contain the river within a confined area and are intended to accelerate the flow of water downstream (London Assembly 2011: 43). These engineering interventions invariably produce the opposite effect as "the accelerated water movement can increase the downstream flood risk" (ibid); "[artificial channels] can also increase flood risk, both alongside their banks (a narrow channel, or one with weirs, can only take limited flood flow), and downstream (a smooth, straight channel sends flood waters rapidly to lower-lying areas)" (London Assembly 2014d: 8). In more recent times, river restoration aspires to improve the quality and operation of rivers by managing them as "thriving ecosystems" (Restore 2013:6). In 2011, London Assembly published *For a Rainy Day*, which advocates a more "natural profile to the river bank, with low-lying open areas where flood waters can harmlessly collect" (2011:43). According to the London Assembly, achieving a "natural profile" can slow down and contain "flood waters, reducing the risk of damaging flood both at the site and further downstream" (ibid). Furthermore, "Restored rivers, and green spaces with river access…[support] greater biodiversity, offering enhanced leisure use, and increasing resilience to extreme weather and climate change" (ibid). Urbanization has a degenerative impact on rivers: urban infrastructure (drainage systems, roads, industry) routes flows of watershed contaminants more directly from the surrounding land surface into the receiving river (Suddoth et al. 2011:1972). Channelling rivers places pressure on catchment areas, which in combination with excessive defoliation, intrinsically alters catchment hydrology; it also reduces interception of rainfall, in turn precipitating increased surface runoffs and erosion of the defoliated areas; leading to adverse sediment supply within river channels (Smith *et al.,* 2014). In summary:

> Common changes associated with urbanization and the urban stream syndrome include a flashier hydrological regime, changes to sediment supply, channelization and subsequent lack of both lateral and vertical connectivity, elevated levels of nutrients and pollution from runoff and sewage systems, and decreased longitudinal connectivity as a result of culverts and weirs. (ibid:250)

In February 2014, following the torrential storms of winter 2013-2014, "the London Assembly unanimously called on the Mayor to identify funding for further river restoration projects" (London Assembly 2014d:8). Apparently motivated by the realization of the limitations of scientific engineering, the London Assembly has emphasised a model of river restoration predicated upon an ecological schema, which will "greatly improve the habitat for plants and wildlife, and allow people to access the river and its nature" (ibid). This appears to indicate an environmentalist alignment of restoration and recreation; but evidence indicates toward a strategic realignment of river restoration within an international neoliberal context. Indeed, ample indication of this realignment is evident in the *London Rivers Action Plan* (2009) as referred to in *Flood Risks in London, Summary of Findings*, for it describes how:

> River restoration…could be part of wider flood alleviation schemes, especially those that use green spaces to create low-lying areas that can safely flood at times of high water flow...The London Rivers Action Plan, supported by the Mayor, aim to restore 15km of rivers by 2015: 14km have been restored to date, and there are many more potential projects for the future. (London Assembly 2014d:8)

A significant feature of the *London Rivers Action Plan* (2009) is its alignment with dictates stipulated by recent European Union Directives; it is evident that the channelling of resources necessary to meet new environmental standards has encouraged the reframing of river restoration to meet an intersecting array of domestic and international objectives. Thus, for example, in the aftermath of the 2013-2014 winter floods, the London Assembly and other prominent officials positioned river restoration as being integral to the city's flood defences. This was located within a wider political reframing, whereby restoring the physical attributes and ecosystem of rivers provides capital gains for urban community priorities. River restoration is thus operating "as part of a phase of resource reconstruction arising from changing societal priorities" (Smith *et al.,* 2014:251). It is increasingly evident that in the United Kingdom, and elsewhere in Europe, that "The restoration of ecosystems and natural capital is now viewed as an important part of the move toward a green economy" (ibid). It is thus crucial for analyses of river restoration to comprehend the national and international context of water management legislation; for this provides a basis for ascertaining the ways in which ecological movements are mobilising the international green economy's capital circulation, in order to address local environmental concerns. In this respect, the following literature review has as its focus community-based participation in river restoration, within the context of the European Union's Water Framework Directive.

River Restoration and the Political Economy Governing the Volunteering of Our Time: Literature Review

> The objective of achieving good water status should be pursued for each river basin, so that measures in respect of surface water and groundwaters belonging to the same ecological, hydrological and hydrogeological system are coordinated. (European Parliament 2000:6)

In Western Europe, river restoration is largely governed by the Water Framework Directive, and this European Union (EU) legislation directly impacts on the activities of river conservation charities. Established on 23 October 2000, the Water Framework Directive affirms a legal framework to safeguard, enhance and restore clean "inland surface waters, transitional waters, coastal waters and groundwater" across Europe, and so secure long-term sustainable use (European Parliament 2000: 8). The Water Framework Directive institutes an enterprising system of coordinated administrative arrangements for water management, sub-structured into a focus on river basins, i.e., the river's primary catchment area defined as "the area of land from which all surface run-off flows through a sequence of streams, rivers and, possibly, lakes into the sea at a single river mouth, estuary or delta" (ibid: 9). This is evident in Article 3 of the Water Framework Directive, which stipulates that Member States identify and catalogue the "river basins lying within their national territory and, for the purposes of this Directive, shall assign them to individual river basin districts" (ibid: 11). The adoption of river basins as the focus of EU water policy is guided by the belief that: "The best model for a single system of water management is management by river basin – the natural geographical and hydrological unit – instead of according to administrative or political boundaries" (European Commission 2013).

In England, the Water Framework Directive was instigated in December 2000 and inaugurated into United Kingdom (UK) law in December 2003 (Environmental Agency 2013a). The lead authority for the application, in England, of the European Commission's Water Framework Directive, is the Environment Agency. The Environment Agency has been assigned the task of applying the rules of the directive, mostly by making operational "the programmes of measures specified in the river basin management plans", which are formulated as part of the River Basin Management, of the river basin districts that are located within the European Agency's territory (European Parliament 2000: 12). Specifically, Article 13 of the Water Framework Directive states:

> Member States shall ensure that a river basin management plan is produced for each river basin lying entirely within their territory… River basin management plans shall be published at the latest nine years after the date of entry into force of this Directive. River basin management plans shall be reviewed and updated at the latest 15 years after the date of entry into force of this Directive and every six years thereafter. (ibid: 22)

The Water Framework Directive imposes a fixed series of six-year cycles for the delivery of river basin management plans; the first cycle forecloses in 2015, and will be followed by further consultation during which the river management plan will be revised and reissued (Environment Agency 2013b). The Water Framework Directive is particularly prescriptive regarding the content of river basin management plans. It stipulates that these plans must integrate environmental objectives; raise social concerns; and explore the economic factors for all bodies of water that exist within the river basin district (European Commission 2008a). An integral feature of these stipulations is public consultation, and this has provided a crucial inroad for voluntary organisations to be directly involved in the regulatory framework that increasingly governs the funding, provision and opportunities for river conservation. Thus, for example, Article 14 of the Water Framework Directive states: "Member States shall encourage the active involvement of all interested parties in the implementation of this Directive, in particular in the production, review and updating of the river basin management plans" (European Parliament 2000:23). An important feature of river basin management plans is that they incorporate procedures to report "the bodies of water not meeting the Directive's environmental targets and describe the reasons for this, such as what is causing problems and risks" (European Commission 2008a: 2). Furthermore, these plans are required to incorporate a programme of measures detailing the actions and procedures to be accomplished in order to address threats to the river basin district's aquatic ecosystem, and to enhance water quality (ibid.). Pivotal to the formulation of the river basin management plans are the river basin district liaison panels which are constituted by "representatives of businesses, planning authorities, environmental organizations, consumers, navigation, fishing and recreation bodies and central, regional and local government, all with key roles to play in implementing the plans" (Environment Agency 2013j). River conservation charities play a part in this remit in their capacity as environmental organizations. According to the Environment Agency, river basin district panels are integral to helping it "work with local stakeholders to identify the actions needed to address the main pressures on the water environment" (ibid.). Thus, public participation in the construction of river basin management plans "extends to all water users, to non-governmental organisations, such as local and national environmental groups, and to other stakeholders" (European Commission 2008a: 2).

In accordance with the Water Framework Directive's timetable for submission, the first river basin management plan for the River Thames was published and circulated in 2009 (London Assembly 2013:22). According to the Water Framework Directive's schema: "One out of 47 river waterbodies in London currently has good ecological status, 30 are of moderate status, and 16 are of poor status" (ibid). The catchment approach has, since 2011, ensured that across the region of London, non-governmental organisations (NGOs) are entering into "local catchment partnerships focused on local action plans that aim to deliver action on the ground" (ibid). Part of the incentive motivating the inclusion of NGOs in partnerships is because river restoration government funding is predominately guided by the imperatives of the Water Framework Directive (Smith *et al.*, 2014). Moreover, the directive's focus on river basins requires collaborative appreciation of a catchment area's hydrological processes. In recent years, collaborative community engagement has been a contentious issue of academic debate. In this context, academics have challenged the "co-management", "social learning" model formerly endorsed by the directive's ethos of public consultation.

In an analysis of the prospects for social learning, Lundmark and Jonsson (2014) argue that there is a disparity between the Water Framework Directive's consultative procedures and the extent to which local community stakeholders can represent their interests. In their study of the initial implementation of the EU directive, in the Lule River Basin in Sweden, Lundmark and Jonsson (ibid) agree that the directive provides procedures for the administration of forums and organisational settings for consultative dialogue. The directive is described as in line with a paradigm shift in the operation of administrative water procedures: away from a focus on rigidly managing the accelerated consumption rates towards the implementation of administrative processes that "adapt to the ecosystem" (ibid: 163). Alongside this strategy has evolved the intervention of "adaptive co-management", which implies "that bottom-up collaborative structures involving local stakeholders facilitate adaptive processes" (ibid). It is assumed that co-management with local stakeholders introduces into the administrative process local ecological knowledge, as this is believed to facilitate the implementation of environmental policy. Lundmark and Jonsson (ibid:171) in their study identified limitations to the extent to which local stakeholders can participate equally in the sharing of ecological knowledge and influencing water management policy. A central finding of their study was that the water management consultation programmes emphasised scientific knowledge as opposed to the professed basis on local ecological experience. Furthermore, the scope of the impact on local communities is often considerable making it "difficult to demarcate" and thus "many interests lack a formal channel for influence" (ibid:161). Democratic dialogue is a theme of Bernauer and Kuhn's (2010) examination: "Is there an environmental version of the Kantian peace? Insights from water pollution in Europe". The conceptual idea of the Kantian Peace refers to the supposition that nation states, with participatory democracies, and which are bound by international treaties, tend to be more environmentally responsible to their neighbours. Bernauer and Kuhn's analysis of the EU's Water Framework Directive, suggests "a positive effect of democracy on international environmental cooperation (in policy-output terms)" (ibid:98).

Smith *et al.,* (2014), in their account, *The Changing Nature of River Restoration*, also focus on participatory approaches encouraged by the EU's Water Framework Directive. Of significance to Smith *et al's* research is the positioning, in policy and governance, of river restoration as a community-based initiative, for this is in agreement with the EU's increasingly prominent ecosystem approach to water management. Smith *et al.,* argue, that in their advocacy for "citizen science" initiatives, river restoration providers coincide with their "use as a tool to fulfil legislative requirements" (ibid:249). Integral to these policy fulfilments is the framing of river restoration "as a way of improving the ecosystem services offered by rivers and this is one of the key drivers for the release of river restoration-related government funding in the UK" (ibid:252). In 2012, the Department for Environment, Food and Rural Affairs (DEFRA) established the Catchment Restoration Fund, which provides financial support for voluntary sector groups to: (1) "restore natural features in and around watercourses"; (2) "reduce the impact of man-made structures on wildlife in watercourses"; (3) "reduce the impact of diffuse pollution that arises from rural and urban land use" (DEFRA 2013a:1). Some indication of the propensity for the Water Framework Directive's policy agenda to impact on river restoration provision, is evident when one considers the constituency of the national panel that decides on Catchment Restoration Funding, for the funding body is chaired by the Environment Agency, with representatives derived from DEFRA and New England, with an advisory capacity provided by the River Restoration Centre (DEFRA 2013a:1). By the culmination of May 2012 the Catchment Restoration Fund had received 131 applications for approximately £54 million of proposed project work; and of this total, 42 projects were successfully approved, amounting to an aggregated value of £24.5 million of work (ibid). In describing the anticipated impact of the successful funding, DEFRA (2013a:1) clearly emphasises their preference for an ecosystem approach to river restoration:

> As a result of these projects, over three hundred water bodies will receive habitat improvement, improved access for fish or reductions in diffuse pollution, making significant steps towards more waters at good status as well as providing wider benefits to society and the environment.

It has been observed that early restoration programmes tended to focus on the conservation of a single species, or particular aspects of the river's fluvial morphology and physical conditions. Conversely, restoration in more recent times focuses on achieving a more concordant inter-connected approach to securing bio-diversity. The emerging ecosystem approach resonates with the EU's Water Framework Directive and its focus on catchment-scale management. In association with these trends, as Smith *et al.,* (2014:253) observe, the consultative paradigm of EU water management policy advocates collaborations between scientists and non-scientific communities. Of significance to this driver is the Water Framework Directive's insistence on conserving and improving river conditions:

> This Directive aims at maintaining and improving the aquatic environment in the Community. This purpose is primarily concerned with the quality of the waters concerned. Control of quantity is an ancillary element in securing good water quality and therefore measures on quantity, serving the objective of ensuring good quality, should also be established. (European Parliament 2000:4)

According to DEFRA (2014) "only 27% of our water-bodies in England are currently classified as being 'good status'" under the classification guidelines provided by the EU's Water Framework Directive. While this figure in and of itself is significant, it also illustrates how the regulatory framework's operationalisation of an ecological focus is premised on collaborations with the scientific community. To this end, it has been observed that the existence of local communities at the centre of river restoration has encouraged the formulation of a "citizen scientist" model of participation (Smith *et al.,* 2014: 256). Education as a feature of river restoration has been a consistent theme of environment research. Susan

Tapsell (1997) in *Rivers and River Restoration: A Child's-eye View*, attempted to gain a broader appreciation of children's perception of restoring local river environments. Central to Tapsell's (1997:62) findings is that: "children's perception often seemed to focus on the cleanliness of the river environment and the play opportunities that it could provide, rather than on its aesthetic beauty". Tapsell (ibid: 63) proposes that children learn through their lived encounters with the river environment; management of urban rivers need to integrate further experiential learning "encounters with nature and the living world, and at a time when children are being increasingly restricted in their urban environment" (ibid). Bjorkland and Pringle (2001), in their explorative study of environmental outreach education, identify the need for professional scientific communities to engage with local communities. A significant feature of the European Union's Water Framework Directive is the centrality accorded to public consultation and the opportunities for ordinary citizens to engage with the scientific community in the construction of river basin management plans. Indeed it is assumed that:

> The success of this Directive relies on close cooperation and coherent action at Community, Member State and local level as well as on information, consultation and involvement of the public, including users. (European Parliament 2000:3)

Considered as a form of environmental education, the citizen scientist model can involve "communities monitoring and collecting data that could be used both to provide important insights into post-restoration dynamics and effectiveness, and enhancing community understanding of the local environment" (Smith *et al.,* 2014:256). In this regard, third sector community-based engagement with river restoration involves making accessible the use of simple instruments for measuring and evaluating river conditions. Smith *et al.,* (2014) enthusiastically advocate the emerging citizen scientist model of river restoration, and in so doing accept, uncritically, the propensity for these interventions to render measurable territories and their populations. Smith *et al.,* (ibid:257) state that: "[T]he increasing ability to link, combine, and analyze large datasets, and the huge amount of data now available through remote-sensing techniques provide new avenues for exploring relationships between ecology and geomorphology".

It is plausible that the "intelligent combination" of large datasets will advance and further river restoration; but what is less evident is whether the integral role of community based river monitoring will be exclusively relevant to “meeting societal demands for such things as flood protection, recreation, and aesthetic value” (ibid). What this suggests is the necessity for ecological analysis to formulate a far more sophisticated conceptualisation of space in the context of river regulation and community participation. Feminist geographers have long since identified the complexity of the intersectionalities of space and the city; but the issue of space, reproductive labour, political economy and the environment is more consistently advanced by Marxist ecofeminists.

Watersheds in Marxist Ecofeminism: Theoretical Framework

Central to my theoretical framework is the feminist principle that ecological understanding of water sustainability interrogates the market form of environmental degradation. It is equally important to expand the notion of economy so as to encompass: “the multiple contributions of women’s reproductive labour to the maintenance of capital” (Salleh 2010:208). Ecofeminist ideals are not new to the lexicon of feminist struggles to preserve communities and enable equality of opportunity for all women. Global capitalism’s increasingly rapacious extension into the natural environment has raised the imperative of an ecofeminist challenge to “corporate globalization” as it relentlessly “expands and contracts, leaving no stone unturned, no body unused” (Salleh 2014:ix).

I appreciate the Marxist feminist belief that neoliberal capitalist expropriations of the natural world necessitate robust and resourceful challenge; “resourcing of women and of nature are structurally interconnected in the capitalist patriarchal system” (ibid: xi). “The stranglehold of global neoliberalism” (ibid) has indeed tightened and Marxist ecofeminism constitutes a formidable part of the feminist challenge. While in agreement with Marxist ecofeminists it is necessary to explore further the complexities of dialectical materialism; advance beyond economic deterministic ecology politics; critically engage with the governmentality of symbolic mediation; and engage reflexively with the intersectionalities of class, race and feminist environmental activism.

Marxist feminists recognise that nature “enfolds human being” (Mellor, 1997:184–5), rather than presupposing an idea of transcendence from nature, Mary Mellor (ibid:185) convincingly proclaims the imperative of assuming humans as “immanent beings”. The condition of immanence is located outside of an Archimedean timeline and thus it is recognized that: "The dynamics of the natural world exist in their own right with or without conscious human intervention…humanity cannot have a 'grand narrative' to explain its own position…Ecologically, humanity exists in a condition of radical uncertainty" (ibid.). These and other Marxist ecofeminist principles partly inform the theoretical ideas that have framed my empirical research of river restoration. The following discussion presents my theoretical framework as a confluence of Marxist ecofeminist and Foucauldian post-structuralism.

River Conservation as Water Wars

> The water endowment of a particular ecosystem depends on the region's climate, physiography, vegetation, and geology. At each of these levels, modern humans have abused the earth and destroyed its capacity to receive, absorb, and store water. (Shiva 2002:2)

Vandana Shiva (2002), in *Water Wars,* impressively argues that the economic incentives of global capitalist markets are reconstituting the proprietary rights of water. Shiva argues that profligate global markets are increasingly defining "water from being common property to private good" (ibid:19). Shiva maintains that the individualistic propensity of neoliberal capitalist markets undermine the human right to water; for it challenges regulatory frameworks so as to radicate water markets. As Shiva expresses it: "Proponents of free water trade view private property rights as the only alternative to state ownership and free markets as the only substitute to bureaucratic regulation of water resources" (ibid). Technological progress in water extraction has advanced the integration, into neoliberal capital circuits, of state owned water management. Shiva observes that contemporary water extraction technologies dispossess self-management practices and, in so doing, distance democratic engagement in the management of water. Globalization aligns with neoliberal privatization of water resources, so as to accelerate "new efforts to completely erode people's rights and replace collective ownership with corporate control" (ibid). Schmidt and Mitchell (2013) further highlight the intention of neoliberal market dynamics to include ever more water resources as part of the insatiable progression of capital accumulation. In its relentless drive to privatise the limits of water ecosystems, the neoliberal market's "drive for material abundance pits individual rights holders against one another in a perpetual competitive struggle over scarce resources" (ibid: 58). Competition between private rights holders is a continuous theme of water colonization, and Shiva (2002:53) details how imperialist intruders have invariably sought to colonize the river waterways as part of "a cultural obsession and an imperial imperative". In these historical times "Nature in general and rivers in particular, were valued for their commercial benefit and were seen as being in need of taming" (ibid:53-54). The long-term legacy of the vast water-diversion programmes established in the modern era has been to "take water from one community to another and from one ecosystem to another" (ibid:56). In so doing, capitalism ferments antagonistic relationships between individual rights holders; it shapes these competitive relations into market drivers; amended under neoliberalism, market-based competition is framed as the rightful "manipulation of social relation …itself a legitimate means of wealth protection and generation" (Schmidt and Mitchell 2013:58). Thus the ideal of neoliberal water markets transforms human relations to natural resources, into markets with respect to water as an object of private ownership (ibid). Suffice to say: "Displacement is an intrinsic aspect of wars unleashed by large water projects" (Shiva 2002:68). From this perspective, the hard engineering programmes that dominated river maintenance up until the 1960s align with a modernist capitalist vision to redirect water from natural waterways, and in so doing incorporate water drainage into capitalist property structures.

Capitalist profligate intention, within the current discourse of restoring rivers to their "natural states", is less conspicuous and complexly entangled with conflicting agenda. Shiva (ibid:69) is less reserved about capitalist intervention, for the practice of altering a river, even for a romantic notion of its natural condition, ultimately "modifies the distribution patterns of water in a basin, especially if interbasin transfers are involved" (ibid). According to Shiva (ibid) it is the tendency of capitalism that "A shift in water allocation most often generates interstate conflicts, which rapidly escalate into disputes between central governments and states". It is a principal observation of Shiva's polemic, that neither national nor international water laws are sufficiently designed to address the extreme conflicts presented by water disputes.

Shiva presents the intriguing observation that: "No legal document in contemporary law mentions the most basic law related to water – the natural law of the water cycle" (ibid:77). Shiva is consistent in the presupposition that, limitations in national water laws have propelled increasingly extravagant water extraction schemes into being, which claim rights to water in terms of their ability to conquer territory through the scheme: "the more you extract and divert water through giant projects, the more you can claim rights" (ibid:76-77). Evidently, according to Shiva (ibid), four theoretical frameworks of water rights are dominant in this area: (1) territorial sovereignty – attributes riparian nation states sovereign rights over the rivers flowing in and through their territory; (2) natural water flow theory – often described as the territorial integrity theory maintains that within a given territory the lower riparian users of a river waterway should not have their entitlement rights hindered by the water management strategies of the upper riparian owners; (3) equitable apportionment theory relates to the equitable mobilisation of international rivers as national resources; (4) a community of interest theory that "international rivers should be used by different states on an equitable basis" (ibid:78). Alongside Shiva's (2002:77-78) theories of water rights, can be positioned Ana Isla's (2005) account of *Conservation as Enclosure*. Isla (ibid) formulates a Marxist ecofeminist perspective relevant to the analysis of the integration of ecotourism into capitalist circuits of capital accumulation. Like Shiva (2002), Isla is concerned with international capitalism and the processes involved in the enclosure of the commons. Of particular concerns is the neoliberal structural framework of national debt payment that is imposed by the commercial banks and the institutional arrangements of the International Monetary Fund (IMF).

River Conservation as Gendered Enclosure

Ana Isla's (2005) *Conservation as Enclosure: An Ecofeminist Perspective on Sustainable Development and Biopiracy in Costa Rica*, advances Marxist ecofeminist concerns regarding "water wars" and international finance institutions. The case studies therein show the interrelations between the expansion of bio-prospecting, in Costa Rica, and the privatization of natural resources. Central to Isla's concerns over bio-prospecting is the manner in which capitalist markets have reformulated their traditional formulation of nature prospecting, so as to emphasise biodiversity, which in turn has been reconfigured and "reduced to the chemical compounds, genes and proteins produced by plants, insects, microorganisms and macroorganisms that can be formulated into commodities" (ibid: 49). Isla (ibid) states that capitalist ventures in bio-prospecting, involve the commodification of local proficiency regarding the attributes of the natural environment and its livestock inhabitants. Equally disconcerting is the expropriation, by pharmaceutical companies, of marketable indigenous plants and their transportation "to laboratories for modification by pharmaceutical, medical and agricultural companies" (ibid). Isla's research has relevance to a Marxist ecofeminist framework of river conservation: Isla traces inter-governmental relations between global finance capital and the rise of international commissions for sustainable development. Indeed, a key feature of Isla's thesis is that sustainable development has become a channel for the exploitative interference of capitalist bio-prospecting ventures. Much of the problem originates from a paradigm change in the financing and structuring of Third World debt. Isla (ibid:51) identifies the neoliberal capitalism of the 1980s as witnessing initial changes in the debt repayment conditions, issued by the International Monetary Fund (IMF) and the World Bank. Debt repayment schemes were "replaced by an inflow of private portfolio funds and 'debt swaps'" (ibid). "Debt-for-nature swaps" constitute a crucial example of this new regime of debt restructuring. "Debt-for-nature swaps" are a finance repayment scheme issued by international finance organisations and whereby the indebted country transfers ownership of some natural assets as part of the debt repayment restructuring. According to Isla (ibid):

> Debt-for-nature swaps are the sustainable development mechanisms of choice for the World bank, IMF, UNESCO, and large environment corporations, and they became an instrument of policy following, the United Nations Conference on Environment and Development (UNCED), headed by Maurice Strong, in 1992.

Isla documents how Costa Rica entered into a debt-for-nature swap in 1995 and this, bi-lateral agreement with Canada, provided the chance for Costa Rica to pay off a substantial amount of its national debt. As part of the bi-lateral partnership "the Canadian World Wildlife Fund (WWF-C) and the Costa Rican National Institute for Biodiversity (INBio)...became the Costa Rican government's creditor" (ibid). A central axis to the debt-for-nature swap schema is: "the creation of 'Conservation Areas" (ibid); and this has significant relevance to my analysis of river conservation.

According to Isla (ibid:52) part of Costa Rica's debt-for-nature swap repayment restructuring has involved instituting programmes of environmental conservation predicated on the establishment of sustainable development through eco-tourism. The transfer of the management of natural resources from indigenous people to public and private entities has involved the privatisation of the commons. It is possible here to map parallels with the impact of the EU's Water Framework Directive and the precipitation of public-private central management schemes. Isla identifies how the Costa Rican experience of the privatisation of the commons has severely impacted on hunting and fishing; as these are key aspects of the subsistence economy which relies upon the existence of common natural resources. According to Isla (ibid:52):

> The enclosure of this commons by the establishment of Conservation Areas closed off this source of nutrition and put an end to practices that had mutual benefit to both human and non-human sustenance and survival. In short, sustainable development destroyed sustainable economies and ... also put natural biodiversity at risk.

As a Marxist ecofeminist, Isla challenges "the devastation of the earth and her beings by the corporate warriors, and the threat of nuclear annihilation by the military warriors" (Mies and Shiva 1993:14). Furthermore, Isla's Marxist ecofeminist principles encourage a focus on the way in which women are disproportionately affected by the exploitation of natural resources. In this sense, Isla (2005), like other Marxist ecofeminist, is acutely aware "of the connection between patriarchal violence against women, other people and nature" (Mies and Shiva 1993:16). It is the central and principal belief of Marxist ecofeminists that: "The 'corporate and military warriors' aggression against the environment ... [is] perceived almost physically as an aggression against our female body" (ibid:14). In recent decades, Marxist ecofeminists have focused critically on advances in bio-engineering and biotechnology, for these developments have brought into sharp relief gender biases in scientific practices: "science's whole paradigm is characteristically patriarchal, anti-nature and colonial and aims to dispossess women of their generative capacity as it does the productive capacities of nature" (ibid:16). Isla's ecofeminist interventions focus particularly on this area of capitalist expropriation of bio-prospecting; Isla (2005:58) observes how the advance of bio-technology resourced by nature has encouraged rural Costa Rican women to apply for loans "to develop biodiversity-related micro-enterprises". According to Isla (ibid) NGOs have frequently seen biodiversity microenterprises as an opportunity to sustainably cultivate natural resources. Isla (ibid) is more hesitant about the mutual benefits of such ventures; her main concern is that biodiversity enterprises involve the commodification and privatisation of botanical knowledge and other forms of "the common knowledge of rural women" (ibid). Moreover, micro-enterprises are based on market dynamics and encourage the production of natural species that have value for the market; leading to disequilibrium in the value attributed to natural biodiversity. According to Isla (ibid:60) this has affected women whereby their targeting as ambassadors of micro-enterprises introduced external controls on forms of sustainable farming and horticulture that ordinarily would have been under local control. Isla (ibid) makes direct associations between these biodiversity micro-enterprises and what Maria Mies describes as the "housewifization of labor". Isa (ibid) defines this latter process as being of relevance to the capitalist expropriation of bio-prospecting "whereby women's autonomous subsistence work based on biodiversity is eliminated in favour of poorly paid exploitation in the capitalist market" (ibid). According to Isla (ibid): "Micro-enterprises are the first step in introducing rural women's labor into the market as a human labor commodity, making them produce for the world market instead of for their own subsistence".

Isla's general conclusions are that rural women, subjected to exploitative capitalist bio-prospecting, should form collective alliances of resistance, so as to challenge the legal dispossession from their access to common resources and curtail the biopiracy of their indigenous botanical knowledge. If we were to extrapolate Isla's argument to river conservation, one might assume that the ecotourism aspects of the EU's Water Framework Directive alongside the appropriation of local knowledges as part of the directive's focus on "public consultation", community outreach and citizen scientists, all have possibilities for exploiting a largely female community of river conservationists. Isla (ibid) cautions against adopting the conservationist rhetoric; for it is evident that this language has been appropriated by multi-national companies who, alongside national governments and in conjunction with environmental NGOs "are sacrificing the survival of forest peoples to capitalist accumulation" (ibid:60). Bio-prospecting that engages in the commodification of natural resources for capitalist gains amounts to bio-piracy: "Enclosures of common lands for the purpose of bio-prospecting liquidate the customary claims of forest ownership" (ibid:60-61).

As common resources enter into the circuits of a commodified capitalist world, they leave behind a ruinous landscape of dispossession and marginalized indigenous people. Isla observes that women are vulnerable to the privatization of the commons; this situation is exacerbated by the development programmes that seek to encourage bio-prospecting entrepreneurialism from rural women, so as to enable their economic development, through their integration into capitalist international markets. Isla (ibid:61) identifies how gender and development programmes, that are premised on entrepreneurial bio-prospecting, forcibly nudge female farmers "into capitalized biotech micro-enterprises", where they "become indebted, overextend their work time, and substitute family food production for the cultivation of medicinal plants - all for less than a minimum wage" (ibid). Capitalist biotech ventures operate internationally. These "predatory" and highly exploitative programmes that co-opt local knowledges of the natural environment, according to Isla, directly integrate local rural women, who become "tied into the world economy, not for conservation or emancipation, but to be exploited for capital" (ibid). Suffice to say, Isla's polemic is passionate and determined in its challenge to capitalized biotech conservation strategies. Isla's key concern is that conservation is being mobilized, by capitalism, to enable enclosure; and this has relevance for my analysis of river restoration. However, Isla's impassioned description and adamant Marxist ecofeminist challenge to the "predatory programs" of capitalist biotech ventures, require further elaboration of the processes involved in the commodification of the commons; for this additional element of detail might well reveal the complexity of capitalism and the possibilities available for resistance so as to extend beyond Isla's (ibid) model of "vulnerable local nature and vulnerable local women". Some indication of the complex processes involved in the cooption of the conservation of rivers, is provided in Rebecca Lave's (2012) *Fields and Streams.*

River Conservation as Neoliberal Managerial Enclosure

Rebecca Lave (2012) details the extraordinary appeal that the conservation of waterways provides for neoliberal markets, governance and environmental research. Lave focuses on stream and river restoration in America; of particular concern is "the profound restructuring of scientific production under neoliberalism" (ibid:3). Lave's framing of analysis in terms of neoliberalism is relevant to my analysis of river restoration, especially with regards to the imposition of market solutions as a part of the state's increasing retrenchment from environmental management. Markets purportedly advance service quality provision through their encouragement of competition between marketised providers, as Lave (ibid) expresses it: "A central tenet of neoliberalism is in fact epistemological: the claim that the market is the best information processor, the only entity capable of accurately comprehending the world". Lave emphasises continuities between: neoliberal market rationale, the pervasiveness of neoliberal environmental management and the increasing dominance of neoliberal science within the academy. According to Lave (ibid) a disconcerting outcome of this convergence is that neoliberal science regimes are pre-programmed to undertake a critical position against "the broader project of public science". Neoliberal environmental research, is configured by neoliberal market dynamics; in America, this has resulted in an increased emphasis on near market applied research, and has led to a constriction on the range of research "as broader, more basic questions are shelved in favour of more immediately applicable, commercializable knowledge...including the metrics used to define ecosystem service commodities" (ibid). According to Lave, within American academia, river stream restoration research is imbricated and overlapping with the neoliberal marketisation of science: researchers are apparently tailoring their empirical focus to meet the market-based criteria of government funding agencies, while also competing with the commercialised output of private producers and their "embodiment of neoliberal trends" (ibid:5). Clear parallels link Lave's (ibid) research with neoliberal market reforms of voluntary sector funding in England. Linda Milbourne's (2013) *Voluntary Sector in Transition* provides some illustration of how neoliberal market reforms of the public sector in England are affecting voluntary sector provision.

Milbourne (2013:69) documents the "organisational and ideological re-engineering" that has beset the structure of public sector funding in England. Tracing the current market reforms, to 1979, and the subsequent protracted reign of Margaret Thatcher's New Right, Milbourne describes the effects of "the political commitment to marketisation of services" (ibid). Left-wing academics tend to agree that several decades of ideologically subjecting the political economy of voluntary sector provision to market forces has undermined their responsiveness to local stakeholders and fragmented their ability to operate as a collaborative voluntary sector (Milbourne 2013:71, Kendall 2003). Voluntary sector service providers are increasingly locked into competition with other providers, and this market-based economy of dwindling service funding has "disrupted prevailing cultures of relationships between the voluntary sector and the local state, replacing less formal, often trust-based relationships with formal contractual relationships" (Milbourne 2013:71).

The issue of dismantling social democratic trust-based organizational regimes has relevance to John Clarke's (1998) account of *Doing the Right Thing? Managerialism and Social Welfare*; for it is evident that the New Right's market-based reform sought to dismantle "a thoroughly racialised and patriarchal set of norms, embedded in legislation, regulations and professional practice" (ibid:237). The New Right in their disassembling "of the political consensus on state welfare", utilised "potent public stories, 'morality tales', that were woven as a central thread in the narratives which the New Right constructed about the monstrous power of the state and how it subjugated and oppressed the people" (ibid). Clarke argues that a central motivation and outcome of these narrative devices was to challenge and expose "the claims to neutrality of bureau-professional regimes" (ibid:237). Indeed, the rise of New Right market-based reforms of the public sector accompanied a new meta-narrative of public sector management, which painted the paternalistic bureaucracies of previous regimes as subjugating and oppressive. These narratives intentionally appealed to feminist liberationist politics, (Newman 2013) and in their cooption of left wing philosophical rhetoric, appealed to an apparently progressive depersonalisation of politics and undermining of the patriarchal "machinery of domination" (Clarke 1998:238). Then and now it is evident that market-based reform of the public sector presents management as a desirable contrast to the introspective, rule-bound, dogmatically structured bureaucracies of the previous era. Neoliberal market-based reform constitutes management as "innovative, externally-oriented and dynamic"; managers are depicted as "customer-centred" they create transparent organizations, which respond effectively to the "'real world' of the marketplace"; managers, as opposed to bureaucrats, "were realists, capable of taking a strategic view and – if given the 'freedom to manage'- able to 'do the right thing'" (ibid:239). Clarke (1998) provides a means of framing patterns of reengineering that are taking place in the public sector and also third sector of state funded voluntary organisations. It is worthwhile detailing Clarke's (ibid:239-240) distinction between managerialism and managerialization:

> Managerialism is:
>
> - *an ideology* centred on expanding the right to manage in the pursuit of greater efficiency in the achievement of organizational and social objectives;
> - *a calculative framework* which orders knowledge about organizational goals and the means of their achievement, typically around an internal calculus of efficiency (inputs-outputs) and an external calculus of competitive positioning within a field of market relations;
> - *a series of overlapping discourses* which articulate different – even divergent – conceptions of how to manage and what is to be managed (Total Quality Management; Excellence; Human Resource Management; Business Process Re-engineering and so on).
>
> Managerialization is:
>
> - a process of *establishing managerial discretion/authority* over corporate resources (material, human, symbolic) and decision-making about them;
> - a process of *establishing calculative frameworks* that define the terms and conditions of decision-making, and are embedded in patterns of internal and external processes;

- a process of *creating forms of managing and types of managers*. It might be suggested that there are three forms of 'managing' visible in the restructuring of social welfare: managers, hybrids; and a 'dispersed managerial consciousnesses'.

Milbourne postulates that since its election in 2010, the Conservative and Liberal Democrat coalition in England has reinforced a competitive service contract culture, and that this has produced disillusionment within the sector, because "While VSOs expected greater funding and organisational stability to follow from formal service contracts, stability was, and has continued to be, undermined by short-term projects and policy changes" (2013:71). Clarke's (1998) framework of managerialism and managerialization provides further detailing of the re-engineering that besets the current contract culture within the voluntary sector. According to Clarke (ibid:243), neoliberal reform and its managerial discourse are instituted through the imposition of "empty rhetoric", and this is both its strength and weakness because "It cannot *fix* the meanings of the words that it deploys as a discursive strategy...[thus] it might be worth looking at some of the more localized 'tactics' that are at play in the attempt to dominate the fields of organizational meaning".

Of some relevance to my analysis of the EU's Water Framework Directive is the mobilization of managerial rhetorical devices so as to produce new "informational currency" (ibid:243), organisational liaisons and symbolic mediations. Clarke (ibid) identifies three discursive strategies and tactics as deployed in managerialist policy and practice: displacement; subordination; and co-option. Displacement has relevance to the consumer citizenship rhetoric deployed in the EU's Water Framework Directive; of particular significance is the constitution of river restoration as a right and responsibility granted to the consumers and citizens of the river territory. According to Clarke (1998:243), the notion of the customer operates rhetorically as an image:

> The customer displaces the imagery (and the implied relationships) of older ideological representations of the citizen and the client (or service-specific variants such as patient or student). The image of the customer challenges the authority of professionals or politicians to 'speak for' the needs and interests of service users. In the process, it (discursively) empowers the customer (as an active agent capable of making choices) and mangers (who are intrinsically 'customer-centred').

Clarke's (1998:243) concept of "displacement" provides a means of rendering intelligible, powerful rhetorical devices deployed within the

EU's Water Framework Directive, and in so doing, deconstructing the discursive and material constitution of the objects and subjects of river restoration. Clarke (ibid) directs our attention to the "apparatuses" of customer – provider service relations; one might alternatively refer to the technologies of the sign (Foucault 1988) through which everyday knowledge and experience are transformed into "informational currency" (Clarke 1998:243). A second rhetorical device evident in managerial discourse is "subordination", the emphasis here is to ground financial decisions in the reality of neoliberal budgetary constraints: "Where 'need' was once the product of the intersection of bureaucratic categorization and professional judgement, it is now increasingly articulated with, and disciplined by, a managerial calculus of resources and priorities" (ibid). Clarke identifies the tactic of "co-option" as a further dimension of managerial discourse. "Co-option" extends beyond the re-framing discourse; it involves the re-deployment of local knowledges as a "central mechanism for disciplining professional autonomy" (ibid:244). Of significance here are the mechanisms through which managerialism re-deploys the customer–provider service dyad, so as to discipline professional autonomy: "The systematization of quality through the production of indicators, comparable information and evaluation has sought, in Peter Jackson's delightful phrase, to 'curb the promiscuity' of quality" (ibid). Michael Apple's (2005) *Education, Markets and an Audit Culture* explores further contradictions in the managerial operation of audit indicators. Of central concern is the paradoxical tendency for neoliberal managerial marketisation to produce a centralisation of governance instituted through the vehicle of an "audit culture". The concept of "audit culture" is rooted in the observation that markets are extremely political, and that this contrasts with classical economic constructs of the market, as a realm of impersonal rational competition.

A key feature of Apple's (ibid) contribution is the valuable insight that intense conditions of market competition compel firms to locate means of exceeding the boundaries of state regulation. A consequence of which is the extension of markets into more aspects of our lives; as Apple (ibid:12) expresses it: "Increasingly, this has meant that the boundaries established to divide non-market parts of our lives must be pushed so that these spheres can be opened to commodification and profit making". One might consider here the EU's Water Framework Directive and the mechanism through which its propensity for the establishment of market relations between providers, transforms the agenda of these institutions, such that not-for-profit third sector providers become integrated into the market. Apple (ibid) operationalises a means to trace the processes through which provision becomes market orientated, and of particular significance is that: "The working conditions and outlook of the employees who work in this sector must be transformed from a model based on collective understandings and providing service to 'the public' on the one hand to working to produce profits for owners and investors and subject to market discipline on the other". An interesting observation here is that market-based reform of public sector provision invariably involves displacing a substantial amount of labour onto the consumer (ibid). Alongside the emphasis on information processing the consumer citizen, has emerged as an immensely powerful managerial tool of the audit and its embedding within work: an "audit culture" i.e., a proliferation into the labour process, work structures and culture of the organisation, technologies for the measurement of performance, service provision and consumer/client satisfaction. In Neo-liberal discourse and rhetoric auditing technologies enable the elicitation of best practice, the efficient distribution of resources and rewards for towering performance; and the ability to intervene and re-engineer underachievement.

These features of audit measurement and marketisation are, it is assumed in neoliberal rhetoric, enabling of self-governing decentralised market-based competition. However, it is a curious feature of auditing technologies that although they promise self-governance and decentralised independence, their impact is to provide for the transportation of centralised information, across organisations, to centralised institutions of governance. As Apple (ibid:15) expresses it: "The ultimate result of an auditing culture of this kind is not the promised de-centralisation that plays such a significant role rhetorically in most neoliberal self-understandings, but what seems to be a massive re-centralisation and what is best seen as a process of de-democratisation". Thus, Apple highlights endemic disjunctures between "the promised de-centralisation" of performance indicators (whereby high performing providers are motivated by the suggestion that they will achieve independence from direct supervision) and the lived actuality of auditing "massive re-centralisation" (whereby audits render the social world into measurable entities that can be aggregated to enable a centralisation of the sector's institutional governance) (ibid:15).

It is my contention that EU's Water Framework Directive is precipitating the rise of an audit culture, in which the provision and ethics of care have been transformed into a programme of objective services that can be measured against performance indicators and ranked against the competitive output of other service providers. Milbourne consistently observes that voluntary sector providers are increasingly subjected to "tighter specification of service activities" and that this increased programme of audit regulation and measurement undermines the independence of providers and reconfigures the parameters of their provision. It is a paradoxical feature of audit cultures, that they are designed and instituted to measure performance but inevitably operate as sign technologies (Foucault 1988) and inscribe reconfigurations onto the reality that they are designed to measure. Auditing is rationalised using the market-based premise that providers operating efficiently and effectively will excel in performance indicators. Milbourne (2013:95) has stated that: "Contrary to the rhetoric of establishing fairness through means of open competition, the intensification of market-driven arrangements endorses the inclusion and exclusion of different kinds of service providers and beneficiaries, exacerbating existing social inequalities". Of concern to feminist academics specialising in voluntary sector re-structuring, has been the rise of instrumental and entrepreneurial policy practices and their impact on gender, work and organisation.

During the 1990s, feminist analysis of men, the caring professions and social work, inspired a path- breaking study by Viviene Cree (1995) entitled *Why do Men Care?* Feminist research into professional care was, at that time, substantively engaging with psychological studies and the misapprehensions of Nancy Chodorow's (1978) "ethic of care" and "ethic of justice". Critical of the essentialist bias and limited political avenues presented by Nancy Chodorow's (1978) account of women's psyche and moral development, Cree's (1995:66-67) *Why do Men Care?,* sets out to explore the extent to which the caring profession is chosen by women "because caring is somehow central to their sense of self [and if so] why do men choose to become social workers?". Central to Cree's results was the existence of profound differences in why men and women choose as a career professional care. It was evident that: "Men who choose to become social workers do so in the knowledge that they are different – that they have qualities which are not stereotypically held to be male" (Cree 1995:83). In contrast, Cree's research revealed that: "Women...in becoming social workers are pursuing a career which draws on characteristics which are widely held to be feminine" (ibid:84). While these findings are interesting in their depiction of the sector during the 1990s, Cree's research has contemporary relevance in its revelatory insights into a newly emerging entrepreneurial culture within the care profession. Cree (ibid:83) identifies how, during the 1990s, men were entering the social work profession "with the confidence that their promotion prospects (should they choose to take advantage of them) are higher than those of women". Cree (ibid:84) tracked emerging trends in the occupational competencies of professional care: "Social work practice is rapidly becoming more technical, more bureaucratic, more 'masculine' in style, whether carried out by men or by women". Cree (1995) admirably avoids falling into essentialist assumptions conflating all men and ascribing to them with the competency of proficiency in technical contractual administration. Nevertheless, it is a significant feature of Cree's findings that men in the professional caring sector, gain from "the advantages of being male in a non-traditional environment" (ibid:81); "Men valued opportunities for leadership and promotion more highly than women" (ibid:78).

Apple (2005) identifies a "New Managerialism" as emerging from the audit culture in education, and the features of this new class have resonance to the forms of masculinity to which Cree (1995) refers in *Why do Men Care?*. Apple (2005:20) describes the rise of "the professional new middle class [that] gains its own mobility within the state and within the economy based on the use of technical expertise". Audit cultures involve programmatic forms of technical knowledge, calculability and instrumental reason. While members of the managerial middle class might have limited ideological commitment to the ideological basis of auditing, "as experts in efficiency, management, testing, and accountability, they provide the technical expertise to put in place the policies of conservative modernisation" (ibid). Key to Apple's (ibid:21) analysis is Pierre Bourdieu's concept of cultural capital; for it is through their habitus and social networks that members of the new middle class acquire the technical proficiency to excel within the programmatic structures of an audit culture. Apple makes interesting observations about the forms of identity offered by the spread of an audit culture:

> New identities that are centred around enhanced technical proficiency and a set of assumptions that solutions to deep-seated problems in education and the entire social sphere can be provided by enhancing efficiency...have developed over time, sponsored in part by neo-liberal discourses that have opened spaces within the state for such expertise. This enables those class fractions with technical forms of cultural capital centred around accountability and managerial efficiency to occupy these spaces and to guarantee a place for the uses of their knowledge. (ibid:20-21)

Apple's analysis has relevance for feminist research into voluntary sector funding and service provision within the context of a neoliberal market economy of river restoration. However, Apple's framing of neoliberalism in terms of a rapacious singular entity, obscures and limits a theoretical framing of the possibilities of feminist activism. Conversely, the focus on neoliberal policy as productive of spatial "heterotopias" provides a means of engaging with the complex circumstances of gender, space and power in the context of ecological praxis. It is my contention that a feminist formulation of Michel Foucault's (1986) *Of Other Spaces* provides a theoretical framework relevant to exploring river restoration in terms of the lived experience, contradictions, challenges and opportunities evident in market-based neoliberal environmental politics.

River Conservation as Ethnicity in the Context of Nationalist Neoliberal Heterotopias

> The present epoch will perhaps be above all the epoch of space. We are in the epoch of simultaneity: we are in the epoch of juxtaposition, the epoch of the near and far, of the side-by-side, of the dispersed. (Foucault 1986:22)

Neoliberal environmental governance of river conservation coupled with the organizational modernization, imposed and sustained by the European Union's (EU) Water Framework Directive engender "Other Spaces" of ecological alignment, as the location of political affinities become simultaneously public and private. The riparian landscapes of our urban cities are manifestations of political and ideological rationalities operating under the constraints of the market and saturated by the contradictions of neoliberal environmental science. Neoliberal rationalities configure river waterways as "sites", the dimensions of which emphasise "relations of proximity between points or elements" (Foucault 1986:23). Many of the modernising initiatives instituted by the EU's Water Framework Directive can also be discerned as resignifying "the space of emplacement" (ibid:22) in their application of mobile and cost effective technologies for the "storage, circulation, marking, and classification of human elements" (ibid:23). Interlacing the mobilization of these technologies is the problem of the river's proximity; its sacredness as the intersection of cyclical natural times and the lived embodiment of place; its "relations of propinquity" (ibid). Neoliberal environmental science transforms the emplacement of the space of the river into a networked matrix: a series that "takes for us the form of relations among sites" (ibid:23).

An intriguing feature of neoliberal environmental science is that having relentlessly pursued the technical appropriation of space, and despite its ability to accentuate the engineering of river waterways, the space of the river, as with other spaces in our contemporary times, "is perhaps still not entirely desanctified" (ibid). Even the most cursory perusal through recent literatures written about the River Thames provides vivid examples of the continued, sacred and profoundly spiritual reverential meaning ascribed to this river. Sanctification continues; consider my examples of the poetic conjecture relating to the appellation of River Isis as the source of the River Thames. Moreover, the creative determination of cultural histories such as Peter Ackroyd's (2007) *Thames Sacred River* suggest that we remain a long way from the complete desanctification of the space and place of our rivers. In the practice of river restoration we also find confirmation of the continuity of the sacred and sanctified; for in these encounters with the materiality of the river, we are reminded that "we do not live in a homogenous and empty space" (Foucault 1986:23). Immersion in the unmediated materiality of the riparian landscape confronts us with "a space thoroughly imbued with quantities and perhaps thoroughly fantastic" (ibid). Consider Figure I.6, thus, the photographic image appeals to:

> The space of our primary perception, the space of our dreams and that of our passions hold[ing] within themselves qualities that seem intrinsic: there is a light, ethereal, transparent space, or again a dark, rough, encumbered space; a space from above, of summits, or on the contrary a space from below, of mud; or again a space that can be flowing like sparkling water, or a space that is fixed, congealed, like stone or crystal. (ibid:23)

Foucault's poetic prose depicts the sensual and spiritual solace of space. My descriptions of cultural etymologies of the Thames, also appeal to the metaphysical virtues of space, this is particularly evident in my emphasis of: "the popular notion that the river from its source was designated Use, Ouse or Isis until at its conflux with the Tame, near Dorchester, in Oxfordshire, [where] it assumed the name Tamisis, or Tamesis" (Fearnside 1834:1); poetry depicting the River Isis as mother and progenitor of the Thames as evident here "The Isis, afterwards called Tamisis, in a long course washeth the south side of this county. The river Tame waters and fructifies the eastern parts, till at last both those rivers, with several other little streams, are received into the Isis" (Camden quoted in Fearnside 1834: 2). The internal subjective space of the River Isis is also conveyed in art; indeed the figure heads of Isis and Thamesis have been celebrated as resplendent adornments to "the keystone of Henley Bridge – Isis looking up and Thamesis down the river" (Church 1880: 1). It is notable that this sculpture was created by a woman Anne

Damer, one of the first women to have become a professional sculptor in England (ibid). Alfred Church (1880:29) in his quintessential and idyllic *Isis and Thamesis: Hours on the River from Oxford to Henley*, describes how Isis "bears a curious resemblance to 'Ouse'…It would be an interesting fact, if we could find that 'Isis' was the classical form given by the scholars of Oxford to the Ouse of the popular language".

Figure I.6., "The sound of a murmuring stream" (Robertson, *The Diary Maid*, 1784). Photographic Image, River Isis, Gloucestershire, April 2014

The internal subjective space of the River Thames, framed through the cultural etymology of the River Isis, provides part of the basis for my theoretical framework. But the spiritual image and iconography of the river and its restoration provide, in and of itself, too simple an appreciation. Indeed, as Foucault (1986:23) states: the space we inhabit, that shapes and develops our being is not only an image; for it has materiality and is heterogeneous. Furthermore "We do not live inside a void that could be colored with diverse shades of light, we live inside a set of relations that delineates sites which are irreducible to one another and absolutely not superimposable on one another" (Foucault 1986:23). Framing the heterogeneity of space, to which Foucault refers is a complex task, namely because Foucault encourages us to describe space as we live it i.e., in terms of intersecting simultaneity.

Following Foucault (1986) my research does not seek to provide a linear cartography of the River Thames' restoration, and neither does it overly indulge in the metaphysical spirituality of the river. Rather, my research is guided by the concept of the River Thames as a heterogeneous site and network of relations "that have the curious property of being in relation with all the other sites, but in such a way as to suspect, neutralize, or invert the sets of relations that they happen to designate, mirror, or reflect" (ibid:24). According to Foucault (1986), spaces that are inter-linked with other spaces, and yet are in relations of contradiction, association and subversion are of two types: utopias and heterotopias. Foucault defines utopias as: "sites with no real place" (ibid). My river travels have provided examples of Foucault's spatial utopias: In April 2014, I travelled to Gloucestershire to visit the site, just near the village of Kemble, designated as the source of the River Thames; for it is from here that cultural etymology proclaims that the River Isis springs forth. The photographic images in Figures I.7c, I.8 and I.9b provide some illustration of the utopian attributes of the Thames Head site.

Figure I.7a., "Castalian goddess, come, nor slight the class Of simplest bard; auspicious come, and prompt The flowing numbers: so may Isis lend Attentive ear well – pleas'd, nor with disdain Reject the wreath of freshest flow'rets cull'd From Pindus' hill to deck her lovely brow. Begin; what Muse to Isis shall deny The votive song? For Isis loves the muse" (Gentleman of Cambridge, *The Praises of Isis, 1755*). Photographic Image, River Isis Gloucestershire, April 2014

Figure I.7b., "Castalian goddess, come, nor slight the class Of simplest bard; auspicious come, and prompt The flowing numbers: so may Isis lend Attentive ear well – pleas'd, nor with disdain Reject the wreath of freshest flow'rets cull'd From Pindus' hill to deck her lovely brow. Begin; what Muse to Isis shall deny The votive song? For Isis loves the muse" (Gentleman of Cambridge, *The Praises of Isis, 1755*). Photographic Image, River Isis Gloucestershire, April 2014

Figure I.7c., "Castalian goddess, come, nor slight the class Of simplest bard; auspicious come, and prompt The flowing numbers: so may Isis lend Attentive ear well – pleas'd, nor with disdain Reject the wreath of freshest flow'rets cull'd From Pindus' hill to deck her lovely brow. Begin; what Muse to Isis shall deny The votive song? For Isis loves the muse" (Gentleman of Cambridge, *The Praises of Isis, 1755*). Photographic Image, Thames Head Gloucestershire, April 2014

Figure I.8., "Unnunber'd beauties rush upon the sight…Thus, Isis, while for thee I string the lyre" (Gentleman of Cambridge, *The Praises of Isis, 1755*). Photographic Image, Thames Head Gloucestershire, April 2014

Figure I.9a., "Upon that great solemnity repair'd The river gods...Unconquerable souls" (Gentleman of Cambridge, *The Praises of Isis, 1755*). Photographic Image, River Isis Gloucestershire, April 2014

Figure I.9b., "Upon that great solemnity repair'd The river gods…Unconquerable souls" (Gentleman of Cambridge, *The Praises of Isis, 1755*). Photographic Image, Thames Head Gloucestershire, April 2014

"The Arc of the Moral Universe, Although Long, is Bending Toward Justice" (Martin Luther King Jr.); Ethnicity, Britishness and the Thames

The Thames Head site in Gloucestershire is a curious utopian feature, for its enthronement in a quintessential English grassland meadow in Tewkesbury Mead presents an irreality of space "with no real place" (Foucault 1986:24). Thames Head is an "inverted analogy with the real space of Society" (ibid); for its ascription as the origin of the Thames renders it as an absolutely self-referring autotheism or deification of Thames Head as a pervasive spiritual existence. It is notable that: "The baptism of a river requires the figure of a guardian or deity" (Ackroyd 2007:24). Alongside the deification of the goddess Isis as a progenitor of the River Thames, is the magniloquent form of Old Father Thames "a water divinity…who bears a striking resemblance to the tutelary gods of the Nile and the Tiber" (ibid). Raffaelle Monti's famous statue of Father Thames, sculptured in nineteenth century, was originally commissioned for exhibition at London's Crystal Palace, and was later moved to Thames Head; and then further relocated to Lechlade, St John's Lock "the site of the first lock upon the water" (ibid:25). Raffaelle Monti's sculpture presents Father Thames as foregrounding barrels and bales of produce, signifying and paying "homage to the Thames as a river of commerce as well as sacred power" (ibid). Father Thames is also depicted yielding a spade; a denotation of the industrial labour, acumen and technique employed in the attempts to civilise the River Thames and subject it to the logic of commercial capitalism (ibid). In its depiction as the location of a river god, Thames Head and its deity Father Thames can be described as autotheistic, for they symbolise appetence "to be one with God without grace" (Nieuwenhove 2000:100). In mystic faiths autotheism refers to a heighten focus on the essence of being, which in its attempts to reach an absolute singularity from all other essences, assumes identification with God and "that the Persons will disappear into the Divinity" (ibid). Raffaelle Monti's avatar of Father Thames reverberates with a sense of Father Thames as a singular absolute deity; conqueror of all wealthy produce, having subjected nature to the rapacious power of industrial technology. John Bacon's bronze statue of Father Thames aligns signifiers of national wealth and monarchy in its dual composition of George III and Father Thames. Erected in 1780s, the sculpture is located in the forecourt of Somerset House, London (Cecil 1822). Emblazoned in bronze, King George III is depicted grasping the rudder of a sailing ship; reclining at his feet is the grandiose and prodigious figure of Father Thames depicted as a reincarnation of Neptune.

Figure I.9c., "I wander thro' each charter'd street, Near where the charter'd Thames does flow" (William Blake *London*). Photographic Image, King George III and Father Thames, forecourt of Somerset House, May 2014

Raffaelle Monti's iconographic depiction of the River Thames as Old Father Thames is in keeping with its cultural historical, quintessential Englishness. Indeed, analyses of rivers and the formation of nationalism identify the nineteenth century as a significant era in which "The growth of nationalism…created a demand for the creation and representation of national landscapes in which rivers provided significant points of reference" (Cusack 2007:101). Rivers have historically been associated with rejuvenation and renewal, thus their appropriation as a signifier of "national vitality" provides for a continuous flow of symbolic meaning across national history (ibid). Raffaelle Monti's commanding depiction of Father Thames enfolds "tales of origins and folk heroes as well as narratives of the 'homeland'" (ibid:102). In its depiction of the River Thames as a river god, Raffaelle Monti conjures up the mythical utopia of the river space and the "mythologized historical battles, perilous crossings or pioneering exploits" associated with a conquering deity. Cusack (ibid) is keen to portray collective identities as formed in and through the riverscape's mythological heroes and this requires Cusack to comprehend the collective as a direct translation "of folkloric conceptions of the 'pioneer' and the 'warrior'" (ibid). Conversely, placing more emphasis on the absolutist propensities of river god tales enables an analysis of signification and the extent to which the autothetic extremes, of river god depictions, refract their supposed singularity and create hierarchical relationality.

Baudrillard's (1968/2005) *The System of Objects*, provides further elaboration of the hierarchical relationality of absolute singularity in the analysis of "a universal code: status", in which can be discerned the operation of a sign system apparently democratising, but in reality, is insidiously productive of hierarchies. Baudrillard (ibid:212) describes how the "objects-cum-advertising system" is a code of status, which has become detached from traditional methods of determining rank; and that this is facilitated by "the social function of advertising" bringing "everyone under its sway". According to Baudrillard (ibid) the code of status is ascribed a morality; for any infringement of its principles exerts a measure of conscience. Similarly, the autotheistic tenacity of river god iconography assumes an absolute singularity, which in its "totalitarian code…no one escapes it; escaping it in a private sense cannot prevent us from participating every day in its collective development" (ibid:212-213). Tricia Cusack's (2010:6) *Riverscapes and National Identities* discusses how the belief systems and rituals associated with the veneration of river gods "have been adopted as an important part of national culture". While it is evident that in the veneration of river gods "the signs of value are entirely socialized and objectivised" (Baudrillard 1968/2005:214), one may question the extent this is solely an outcome of "any true democratization" of belief (ibid). For, it is a paradox of the insistence on a "univocal" value, that it heightens "the desire to discriminate" and precipitates "a perpetually renewed obsession with hierarchies and distinctions" (ibid). While it is evident that veneration of the river gods transcends traditional collective solidarities, the acute singularity of belief produces hierarchies of "new class or caste morality" which in their proliferation "colonize the most material and hitherto unchallengeable of spheres" (ibid). Consequently, while the veneration of river gods introduces universal platforms of signification "the code produces an illusion of transparency, an illusion of readable social relations, behind which the real structure of production and real social relationships remain illegible" (ibid). As Baudrillard (1968/2005:212-214) expresses it:

> A Universal Code: Status…What is more, the fact that a system of identification is now in place which is clearly legible to all, that the signs of value are entirely socialised and objectivised, by no means implies any true 'democratization'. On the contrary, it would appear that the insistence

> on univocal reference merely exacerbates the desire to discriminate: within the very framework of this homogenous system, a perpetually renewed obsession with hierarchies and distinctions is to be observed.

It is in this sense that one might frame theoretically how the River Thames in its progression through England is symbolically transfigured. For it is evident that the River Thames, streaming through the city of London, "acquired a 'truly national' identity only in the nineteenth century", during which time, "representations of Father Thames were adapted to an imperial iconography" (Cusack 2010:57). Clear contrasts exist, for example, between William Morris' "utopian vision of the Thames", his desire for it to "return to the purity of the river associated with a simpler, more humane and creative society" (ibid:2007:102)); as compared with Joseph Bazalgette's nineteenth century manufacturing utility model of the River Thames, as the foundation of the Victorian metropolis and waterway for the burgeoning flow of industrial capitalism. Cusack's (2010) *Riverscapes and National Identities* illustrates how as part of the River Thames' formation as a British river, the quintessential Englishness of the upper river was merged with the industrial Britishness of the lower river.

Cusack (ibid:16) states that: "while the upper Thames was associated with an image of rural Englishness and the imagined past of the nation, the London Thames represented an imperial Britishness, symbolizing its modernizing present and future". Thus, issues of nationalism, capitalist commerce and English ethnicity, distinguish Raffaelle Monti's inherently English avatar of Father Thames from John Bacon's British world commerce depiction of Father Thames. Nevertheless, both sculptures espouse the radical singularity of an autothetic "completeness of deificatio" that in its apparent intention on achieving "the unmediated union with God" (McLaughlin 1974:750), "produces an illusion of transparency", which obviates the refraction of the sign into "the real structure of production" (Baudrillard 1968:2005:214) from, which the detached migrated status of the code originally derives. Suffice to say Raffaelle Monti's iconography of Father Thames and John Bacon's sculpture both operate an objects-cum-advertising system, which is intrinsically a code of status i.e., "in its formal expression it constitutes a universal system for the identification of social rank: the code of 'status'" (ibid:212). Nevertheless the autotheistic absolutist veneration of the river god materialises in reality a signifying system, which is far removed from an "entirely socialized and objectivized", "true democratization" value system (ibid:214). Rather an effect of the absolutism of the river god is to highlight the existence of *Other* beings that exist in hierarchal relations of distinction. Even though the magnificence of the river god can suggest that "barriers of morality, social convention and language have been overturned", the spectacular apex upon which the river god resides establishes "new barriers and exclusions…new class or caste morality" which extend into and "colonize the most material and hitherto unchallengeable of spheres" (ibid:214). Consequently, the self-referring magnificent absoluteness of the river god, proclaims to produce a "universal apparatus of signification", but in actuality constitutes hierarchies of distinction into which venerating *Other* beings endlessly labour in the production of the real materiality, of the new heightened social structures and intensified social realities. Thus in contrast to Cusack's (2010) repertoire of nationalism and unity, it is my contention that the absolutism of river god iconography precipitates an enfolding of the self into the lived experience of ethnic difference. As Stuart Hall (1987:46) has observed: "this new ethnicity … isn't quite so framed by those extremities of power and aggression, violence and mobilization, as the older forms of nationalism". Rather the dialectical migration of identity from "nationalism to ethnicity" is "part of the 'decline of the west' – that immense process of historical relativization which is just beginning to make the British, at least, feel just marginally 'marginal'" (ibid). It is my contention that this observation has valuable importance as a basis for comprehending, at the level of culture and society, conflicting and contradictory ethnocultural responses directed at the European Union's Water Framework Directive.

Issues of nationalism and the territorial sovereignty of nation states are inextricably linked to the European Union's Water Framework Directive. First and foremost: "The European Union is a land of shared waters" (EC 2008b:1). It is estimated that: "60% of the EU's surface area lies in river basins that cross at least one national border, and all Member States except Cyprus and Malta contain sections of at least one international river basin district" (ibid). The European Union's Water Framework Directive establishes a legal structure to preserve and restore decent water quality across Europe and guarantee its present and future sustainable use. The official title of the Water Framework Directive is: Directive 2000/60/EC of the European Parliament and of the Council of 23 October 2000 establishing a framework for Community action in the field of water policy. The framework's natural geographical approach to water management is premised on protecting the aquatic ecosystems of river basins. The principal objective of the Water Framework Directive is "achieving a 'good status'

Figure I.9c., "I wander thro' each charter'd street, Near where the charter'd Thames does flow" (William Blake *London*). Photographic Image, King George III and Father Thames, forecourt of Somerset House, May 2014

for all of Europe's surface waters and groundwater by 2015" (EC 2008i:1). The formidable scope of this challenge is evident when one considers recent evidence that suggests "at least 40% of the EU's surface water bodies are at risk of not meeting the 2015 objective" (ibid). The classification of "good ecological and chemical status" is predicated on the control of chemical pollution to "low levels" and the maintenance of a "health ecosystem" (ibid). In order to achieve "good ecological status" member states by necessity will have to expurgate factors that are deteriorating their water eco-systems (ibid). Pollution is a primary cause of water degradation, alongside the morphological reconfigurations produced by dams; and the excessive extraction of water for irrigation and industrial manufacturing (ibid). As part of establishing evaluative criteria for limiting chemical pollution of surface water and groundwater masses across Europe, the EU has placed rigid limits on the accumulation of pollutants, which are deemed as of particular relevance to Europe; and are defined as Priority Substances and Priority Hazardous Substances. In determining controls the Water Framework Directive adopts a combined procedure: "It considers pollution in terms of what is released into the environment and the resilience of the receiving waters" (EC 2008ii:1).

In its efforts to democratise the achievement of control measures the Water Framework Directive requires EU Member States to address water quality management in their river basin management plans. The process of constructing these plans requires identifying diffuse pollution causes, the monitoring of the accumulation of priority substances and the design of apposite water quality controls. In its efforts to scientifically calibrate and programmatically monitor pollution levels the European Environment Agency is collaborating with Member States to establish a comprehensive circuitry of monitoring stations to provide expansive evaluations of water quality in Europe. The programmatic features of this process are extraordinarily calculative and intriguing in their aspiration for scientific quantification; as is exemplified in the following example: "Quantitative data on the chemical, physical, biological water quality is collected and fed into WISE [Water Information System for Europe] from more than 54,000 surface water monitoring stations and more than 51,000 groundwater stations. Member States use these stations to conduct their monitoring obligations under the Water Framework Directive, and ensure that EQS [Environmental Quality Standards] are respected" (ibid: 2). The quantitative precision of these measures, produces strict disciplinary analogues and powerfully punitive actions against Member States. On 20 January 2014, the European Commission announced: "Commission takes Denmark to Court over water management problems" (EC 2014). Denmark had apparently been accused of failing "to present plans for managing its river basins"; several years in, the lapse of the EU deadline for Member States to have constructed a river basin management plan found that "Denmark still had no river basin management plans in place…and Denmark confirmed in December 2013, the absence of management plans, noting that adoption was unlikely before mid-2014" (EC 2014a). On 16 April 2014 the European Commission announced that it was "taking Austria to Court for its failure to ensure adequate protection for the Schwarze Sulm river in Steiermark" (EU 2014b). It was the European Commission's view: "that the regional authority failed to respect the water quality requirements of the Water Framework Directive when it authorised the hydropower project in 2007" (ibid). It is not my intention here to evaluate the jurisprudence of these decisions; but rather they provide an illustration of the conditions that are leading to a resurgence of ethnic minority allegiances to national rivers. Recall Cusack's (2010:60) account of the River Thames as "constituted by a relation between discourses of an ethnocultural 'Englishness' and a modern 'Britishness'". According to Cusack (ibid:61) the Englishness of the upper region of the River Thames harkens back to "an imagined ethnic past, the age of the sturdy yeoman, and Britishness with the country's modernizing present and future".

Cusack's (ibid:67) focus on national identities induces a tendency to unify the iconography of the River Thames: "The upper Thames symbolized an ethnic Englishness with an aura of the past, and the lower Thames a modernizing and forward-looking Britishness, but urban river pageantry incorporated these different aspects of national identity". Conversely, derived from my empirical ethnographic research with river restoration charities and ecological activists engaged in river conservation, it is evident that reactions to EU water directives coupled with other social, economic and political conditions are encouraging a refraction of nationalist sentimentality. This book illustrates, with detailed empirical examples, evidence of a resurgence of transnational ethno-cultural riverscapes. Professor Stuart Hall's portfolio of writings on culture, representation and ethnicity provide a valuable basis from which to formulate a theoretical framework for the study of new ethnicities and urban rivers. Of particular relevance is Stuart Hall's (1987) *Minimal Selves*. Writing at a time during the 1980s when the children of the Windrush generation appeared to be engaged in a perplexing conjuncture of post-imperial Britishness, Hall (ibid: 44) posed "a few adjectival thoughts". Of particular concern was then, and continues to be, the intriguing centering of minority ethnic identities. Situating his own identity as a minority ethnic, Hall (ibid) quips "Now that, in the postmodern age, you all feel so dispersed, I become centered…welcome to migranthood". In his previous writings on media production and youth culture Hall (1978) had examined the disenfranchisement of young racialised masculine subjects and their positioning as part of a crisis in capitalist hegemony. In *Minimal Selves*, Hall (1987:44) documents a curious reversal in the staging of ethnic minority subjects, for these previously marginalized subjects are becoming "centered in place; without much material support". The original title of Hall's (ibid) article was "Minimal Self"; but Hall purposely pluralised this to *Minimal Selves* so as "to redefine the general feeling which more and more people seem to have about themselves – that they are all, in some way, recently migrated". Placed within the context of river conservation and European Union water directives, Hall's (ibid) observations provide valuable insights especially where he describes how: "Identity is formed at the unstable point where the 'unspeakable' stories of subjectivity meet the narratives of history, of a culture" (ibid).

One might consider here my previous discussion of the autotheism of river god iconography and the absolutist positivist metrics of the EU's Water Framework Directive. It would seem that both of these phenomena construct an angle of refraction in which the subject is compelled to confront the relationality of their position. As Hall (ibid) expresses it: "since he/she is positioned in relation to cultural narratives which have been profoundly expropriated the colonized subject is always 'somewhere else': doubly marginalized, displaced always *other* than where he or she is, or is able to speak from". Autotheistic river god iconography and the EU's positivistic quantitative water evaluation metrics both have the curious effect of producing *Others*, in the sense that their metaphysics of presence constitutes identity as a "kind of absent-present contestation with something else, with some other 'real me', which is and isn't there" (ibid:45). Hall (ibid) writes endearingly about encountering a "living through difference" and he does this by way of describing a social event. One appreciates from Hall's personal narrative a profound consciousness of the irrepressible duality of embodied and represented ethnic identity; and their refusal to be subdued. Similarly the extreme polarities of the EU's neoliberal water directive metrics collide against the ways through "which one tries to define oneself" (ibid). But as Cusack (2010:61) infers the ethnocultures of quintessential English riverscapes are not fixed but rather they are "imagined communities" as described by Benedict Anderson (1991). Indeed, as Hall (1987:45) observes "The fact is 'black' has never been just there either. It has always been an unstable identity, psychically, culturally and politically". The quintessential iconography of English riverscapes provides a basis from which to appreciate the formulation, through narratives, of racialised identity; for they are both learned through lived experience and encounters with the enduring materiality of structural inequalities. The quintessential ethnic English identities that are emerging within the context of EU water directives are similarly indexed to material disjunctures; and in this sense are embedded within a specific moment in time. Such observations run-up against the valorisation of the contemporary self as fleeting, fragmented, endlessly reconstructed and unreal; for as Hall (ibid) observes "Despite its fragmentations and displacements…'the self' does relate to a real set of histories". Hall's observation here has profound significance for conceiving of a politics of riverscape ethnic identities. A politics based on the integral reality of postmodern iconography has some scope here; but there are disconcerting limitations manifested by "The politics of infinite dispersal", for this "is the politics of no action at all" (ibid). In order to avoid the extremes of absolutist discourse and yet to retain a notion of "articulation" Hall (ibid) suggests "a politics of constituting 'unities'-in-difference"; and this concept has significance for my theoretical framing of urban riverscape ethnic identities.

A key impact of absolutist definitive totalities is displacement. Hall (ibid) observes that "more and more people now recognize themselves in the narratives of displacement". Conditions precipitating this sense of displacement include the imposition "of the arbitrary closure"; for the arbitrary construction of an edge, an absolute cessation, an utter total limit, provides for the generative basis of a new beginning (ibid). Based on my empirical research it is possible to discern that river activism is emerging from the internecine spaces at the borders of the EU's Water Framework Directive. These internecine spaces at the borders of absolute limits are generative of new ethnic identifications with river waterways; but these identities are not fleeting transitions. Hall (ibid) provides a further indication of how one might theorise the river flow of new ethnicities: "I also believe that out there other identities *do* matter … But I do think, theoretically and intellectually, it requires us to begin, not only to speak the language of dispersal, but also the language of, as it were, contingent closures of articulation". Thus the fragmentation produced by the forceful imposition of an arbitrary closure crashes through the 'frame'; but ethnic identity does not disintegrate into disassociated and fragmented miniscule selves. Rather, as this book's ethnographic research will illustrate, the imposition of absolute arbitrary closures of meaning, be it positivistic water metrics or autotheistic river god iconography, have the paradoxical tendency of unifying identity at the very aspect of fragmentation. As Hall (ibid) expresses it: "But actually you know, it isn't just the 'minimal selves' stalking out there with absolutely no relation to one another". For it is evident that arbitrary closure induces refraction, a confrontation with the lived materiality of existing conditions. The form of ethnicity that emerges from this condition "can be a constitutive element in the most viscously regressive kind of nationalism or national identity"; but it is also evident that "in our times, as an imaginary community, it is also beginning to carry some other meanings, and to define a new space for identity" (ibid:46). The ethnic identities that emerge from the closure of meaning imposed through absolutist directives and totalising iconography, collide with lived experience; refracting into the materiality of everyday struggles and real conditions; reconfiguring syncretic flows of English ethnicities from the internecine spaces of absolutist boarder controls. As Hall (ibid) observes:

> The slow contradictory movement from 'nationalism' to 'ethnicity' as a source of identities is part of a new politics.

Hall's conception of migration and the profound observation that "the general feeling which more and more people seem to have about themselves – that they are all, in some way, recently migrated"(ibid:44), has significance for my theoretical framing of the intersectionalities of class and gender in the formulation of riverscape ethnic identities. Migration is an important feature of Hall's analysis and is very distinct from a notion of mere 'movement'; this is because "Migration is a one way trip. There is no 'home' to go back to. There never was" (ibid). A principal feature of my research into European Union water directives is the effect of absolutist closures of meaning around nationhood and nationalism. Hall (ibid:45) eloquently observed that "The trouble is that the instant one learns to be 'an immigrant', one recognizes one can't be an immigrant any longer; it isn't a tenable place to be". Similarly, the European Union, in its effort to define responsibility, at the level of the nation state, for river management, motivates "another recognition of self through difference: certain clear polarities and extremities against which one tries to define oneself" (ibid). It is my contention that the European Union's conception of nationhood and its association with national identity, underestimates the complexity of ethnic identity; as Hall (ibid) observes about black identity: "It has long been thought that this is really a simple process: a recognition – a resolution of irresolutions, a coming to rest in some place which was always there waiting for one. The 'real me' at last". In actuality ethnic identity has never existed in and of itself "It has always been an unstable identity, psychically, culturally and politically" (ibid). Moreover, the historical arc of ethnic diaspora bends towards the intersectional relationality of class and gender, as Hall (ibid) insinuates: "I believe it is an immensely important gain when one recognizes that all identity is constructed across difference and begins to live with the politics of difference". When considering the intersectional relationality of class and gender, in the construction of riverscape ethnicities, one has to consider the centrality "of nation and nationalism …as one of those major poles or terrains of articulation of the self" (ibid:46). It is my contention that the European Union's water directives have a centrifugal effect on "the old discourses of nationalism or national identity" (ibid). For they strike at the centre of traditional attempts to define a nation in terms of "a culture, a language, a history" (Ibid) gender and class relations. The resulting "contradictory" progression "from 'nationalism' to 'ethnicity'" is centrifugal; for it ripples outwards, from the epicentre and in so doing unsettles the traditional hegemonic relations of the centre, such as patriarchal masculine nationalism; "It is also part of the 'decline of the west' – that immense process of historical relativization which is just beginning to make the British, at least, feel just marginally 'marginal'" (ibid).

River Conservation as Nationalist Iconography and the Relationality of Masculine Identity

Theologians and ecclesiastics have ceaselessly philosophised over the puzzling and at worst seemingly heretical tendency for erudite mystical scholars to exercise language that implies autotheistic connotations "i.e., language that bespeaks a strict identification of oneself with God" (Wiseman 1990:230). The extensive history of the eminent monotheistic religions - Judaism, Christianity and Islam – is beset with conflicts precipitated by the use of autotheistic language that assumes oneness with God without intermediation or otherness (ibid). The central concern raised by autotheistic language is eloquently reflected in Ambroise Gardeil's statement: "God cannot unite Himself to the substance of the soul to modify it in its very being in the way He modifies it through grace; otherwise the soul would be God" (quoted in Wiseman 1990:234). In their religiosity to preserve the transcendence of God, theologians have tended to be hostile to any text which undermines that transcendence. Thus the consistent insistence in the Judeo-Christian tradition "that God's ways are not our ways" and extensive scriptural chronicles "of creation and redemption by a God 'whose throne is in heaven' have tended to make any claims to identity with God suspect" (ibid:236). These contemplations are not restricted to theological disciplines but extend into the social sciences; for it is evident that the formulation of love in psychiatric discourse can pronounce human love as readiness for fusion, i.e., "the merging of the self with another person or ideal, creating a fused identity" (Gaylin 1986 quoted in Wiseman 1990:242).

Figure I.9c., "I wander thro' each charter'd street, Near where the charter'd Thames does flow" (William Blake *London*). Photographic Image, King George III and Father Thames, forecourt of Somerset House, May 2014

Figure I.9d., "I wander thro' each charter'd street, Near where the charter'd Thames does flow" (William Blake *London*). Photographic Image, King George III and Father Thames, forecourt of Somerset House, May 2014

Figure I.9e., "I wander thro' each charter'd street, Near where the charter'd Thames does flow" (William Blake *London*). Photographic Image, King George III and Father Thames, forecourt of Somerset House, May 2014

In Sigmund Freud's (1930/2010:10-11) *Civilization and its Discontents* the realisation of love absorbs into itself all boundaries: "At its height the state of being in love threatens to obliterate the boundaries between ego and object. Against all the evidence of his senses the man in love declares that he and his beloved are one, and is prepared to behave as if it were a fact". Freud's pronouncements on the fusion of identity in the union of love provide for an autotheistic text, this is because the conjugation of lovers in the entirety of their union expresses an unmediated "ontological identity" (Wiseman 1990:243). Thus the "capacity for fusion" in psychoanalytical prescriptions of the "desire to lose oneself utterly in another, to the exclusion of everyone and everything else" suggests a condition of self-referentiality in which the union is the "only source of gratification" (ibid:248-249). The autotheism in a union of totality therefore raises concerns in both its "claiming personal identity with God" and the pursuit of a "capacity for fusion" which in its totality produces a "kind of addictive attachment to a sensation that not only lessens but practically obliterates concern for anything else in one's environment" (ibid:249). And this has direct relevance for my analysis of Thames Head as a utopian space; for in its manifestation as an apex of origin is evoked a metaphysics of presence that alludes to the possibilities "to be God with God" (ibid:250) and in so doing suggest a heresy that obviates the relational bond and embeddings of space in place. Relationality therefore provides for a countenance to the extreme individuality of autotheistic statements. According to Wiseman (1990:250) relationality is a restorative procedure, in that it is applied to "The intensity of the union they experience with God"; and thus Wiseman (ibid) states that: "The crucial question is whether one remains fixated in that phase of love or allows the love itself to move outward to others" (ibid). Conversely, it is my contention that the insistence on a totality of union is an apparition achieved through the sublimation of relationality.

As with any theocratic autotheism the irreducibility of the crypto-trinity to one, and of the particular to the whole, propels into being reclamation of the supremacy of the whole, for one can never make a god entirely unto one's self. The presence of Isis in the River Thames' cultural etymology ameliorates the heresy of a self-referring origin of the River Thames, aspiring to be entirely identical with God. Isis restores the relationality of the particular to the whole; the cultural etymology of River Isis ameliorates the transgressive autotheism of self-referring signs, and re-establishes their relationality. It is in this sense that Alexander Pope's "The winding Isis, and the fruitful Tame", Benjamin Martin's account of "The

River Isis [as] remarkable for being the Head of the Thames, and [giving] one Half of the Name to that famous River Tamisis" and William Camden's "Isis…chief of the British rivers; of which we may truly say, as the ancients did of Euphrates in the East, that it both plants and waters Britain", in these and many other instances, Thames Head is depicted as a utopian space of "inverted analogy" (Foucault 1986:24), in which autotheism surrenders to the relationality of the sign-object and its "real space of Society" (ibid:24).

Inscribed on a monument that marks Thames Head (see Figure I.9f) is the statement: "The Conservators of the River Thames 1857-1974. This stone was placed here to mark the source of the River Thames". In this monument is presented "society itself in a perfected form, or else society turned upside down" (Foucault 1986:24). For Thames Head is the signified source of the River Thames that streams triumphantly through England's capital city and yet springs so humbly from a rivulet of the River Isis. In his enchanting account of René Magritte's *Ceci n'est pas une pipe*, Michel Foucault sets about deconstructing *The Unraveled Calligram*; it is my belief that the techniques Foucault employs have relevance to deciphering the intriguing sign system deployed on the monument of Thames Head. Foucault's (1983/2008) concerns about the perplexities of René Magritte's pipe have relevance to the strangeness of the Thames Head, monument's inscription. Foucault (ibid:19) identifies how in René Magritte's pipe the relation between text and image initially appears as a "simple demonstrative"; for the text is a label: *Ceci n'est pas une pipe*. Just as in René Magritte's pipe, the inscription on the monument at Thames Head might be conceived as a simple label. But similarly to René Magritte's pipe the initial mirror effect of text and image, soon gives way to a propositional contradiction, so the Thames Head's statement that "This stone was placed here to mark the source of the River Thames" is logically challenged because as Foucault remarks "its 'referent' [obviously a monument] does not verify it" (ibid). The monument enthroned in a quintessential meadow in Tewkesbury appears consistently as intersecting grid coordinates on ordnance survey maps; but the mark of the stone monument as "the source of the River Thames" is perplexing. Referring to René Magritte's *Ceci n'est pas une pipe* Foucault (ibid:20) says that: "The operation is a calligram that Magritte has secretly constructed…Each element of the figure, their reciprocal position and their relationship derive from this process, annulled as soon as it has been accomplished". Foucault (ibid) states: "[T]he calligram has a triple role: to augment the alphabet, to repeat something without the aid of rhetoric, to trap things in a double cipher".

Returning to the text on the monument at Thames Head, it is evident that the primary aspect of the calligram involves the construction of the image in words: through the lines delimiting the text of the statement on the monument; for the sequencing of the lettering "lodges statements in the space of a shape, and makes the text *say* what the drawing *represents*" (ibid:21). Having visited the monument and spent time reading the inscription, set in the context of the monument's enthronement in a relatively sparse meadow, one senses the eerie possibility that it is the *source of the River Thames*, because it says it is. That is to say: "The calligram is thus tautological" and as with René Magritte's *Ceci n'est pas une pipe,* the calligram effectively utilises the "capacity of letters to signify both as linear elements that can be arranged in space and as signs that must unroll according to a unique chain of sound" (ibid).

Foucault (ibid:29) raises further relevant issues regarding the formal layout of René Magritte's *Ceci n'est pas une pipe:* for the image of the pipe and its corresponding text have features common to curatorial orthodoxy: "A painting 'shows' a drawing that 'shows' the form of a pipe; a text written by a zealous instructor 'shows' that a pipe is really what is meant". But in actuality René Magritte's *Ceci n'est pas une pipe* breaks with the methodological conventions of orthodox curatorial traditions and its effort to "stabilize a unique space" (ibid:30). Although the picture's label is positioned conventionally, it usurps the orthodoxy of the curator's epithet; for it declares about the picture: "This is not a pipe" and reveals about itself: "This is not a pipe". Foucault (ibid) observes: "the painting, written sentence, drawing of a pipe – all this is not a pipe". Thus the label produces an oppositional meaning in which "the drawing of the pipe and the text presumed to name it find nowhere to meet and be superimposed, as the calligrapher so presumptuously had attempted to bring about" (ibid:30-31). Foucault's (ibid) reflections here have relevance to my theoretical framing regarding the inscription on the monument at Thames Head. The monument's letter inscription operates simultaneously as a sign and a shape: in the former the inscription represents that *This stone was placed here to mark the source of the River Thames*; but in the latter meaning, the calligram's sign as "letter permits us to fix words; as line, it lets us give shape to things" (ibid). The calligram is literally written in stone; the inscription is etched out of the very rock of the stone; it is the mark to which it refers. The inscriptive sign does not merely represent; for it is the very thing that it is supposed to be depicting; the sign has materiality, it ontologically signifies; and in so doing produces an oppositional meaning, namely that the inscription on the monument at Thames Head should be objectively indicating, rather than intrinsically shaping. Foucault (ibid:21) describes how: "the calligram aspires playfully to efface the oldest oppositions of our alphabetical civilization: to show and to name; to shape and to say; to reproduce and to articulate; to imitate and to signify; to look and to read". It is my contention that the inscription on the monument at Thames Head is an alluring calligram: for it operates to intriguingly compel relationality and thus an oppositional reading to the monument's autotheistic divination. The calligram inscription on the monument at Thames Head coupled with its location, and cultural–political relevance, constitute this space as an intriguing utopia. My theoretical framework emphasises the significance of symbolic mediations as part of the construction of the river space as a utopian space; it is necessary to frame the heterogeneous utopian space of the river. Foucault's (1986) account of utopias as *Other Spaces* provides a valuable means of empirically framing the symbolic mediations of river restoration.

Figure I.9f., "To rise from nature up to nature's God" (Gentleman of Cambridge, *The Praises of Isis, 1755*). Photographic Image, Thames Head Gloucestershire, April 2014

The concept of symbolic mediation and its relationship to cultural politics are detailed by Ulrich Beck (1995). According Beck's (ibid: 45) *Ecological Politics in an Age of Risk*, the action of "Destruction and protest are symbolically mediated". Cultural political protest involves an engagement with a nature that is at one and the same time real and dislocated in a real place. Beck (ibid: 48) observes how nature is constructed through discourses, ideals and representations. A consequence of these symbolic mediations is the fact that "ecological protest is a matter not of natural, but of cultural, fact; a phenomenon of cultural sensibility and of the attentiveness of institutions". My theoretical framework conceives of river restoration as having been constructed through representations which engender the river space as utopian idealism. The concept of nature as "pampered" infers that the "social integration of nature" through environmental conservation involves cosseting nature, indulging nature, mothering nature. Indeed, Beck (1995:52) states that part of the integration of nature as a realm of calculative social action produces dangers which "even where they remain impossible to calculate, involve greatly increased labour for housewives and mothers". Beck (ibid: 39) argues that the integration of nature into the social, transforms nature as it becomes "a kind of anchor by whose means the ship of civilization, sailing over the open sea, conjures up, cultivates, its contrary: dry land, the harbour, the approaching reef". In other words the symbolically mediated nature of ecological activism is a transformed malleable resource; a mystifying inversion of a naturalistic terrain. Beck (ibid: 46) is keen not to render ecological activism as mere apparition, enfolded into the symbolic mediation of nature; for "it is not some mysterious sort of alchemy that gives the protest over certain poisons social resonance". While it is evident that ecological activism and the social integration of nature share a symbolic mediation of nature these acts "are isolated from one another by the cultural readiness to accept despoliation" (ibid). Alongside the formulation of the River Thames as a utopian site, my theoretical framework has as its additional central focus the river space as a "heterotopia". According to Foucault (1986:24) heterotopias are:

> [R]eal places –places that do exist and that are formed in the very founding of society – which are something like counter-sites, a kind of effectively enacted utopia in which the real sites, all the other real sites that can be found within the culture, are simultaneously represented, contested, and inverted. Places of this kind are outside of all places, even though it may be possible to indicate their locality in reality.

It is evident that through intervening in the existence of the river space as a utopia and its actuality as a heterotopia, river spaces can assume the character of a sort of hybrid of heterotopia and utopia: a mirror reflection. Indeed, Foucault (1986:24) states that:

> In the mirror, I see myself there where I am not, in an unreal, virtual space that opens up behind the surface; I am over there, there where I am not, a sort of shadow that gives my own visibility to myself, that enables me to see myself there where I am absent: such is the utopia of the mirror.

Mirror reflections are a recurring theme of river restoration, and this is largely a condition and consequence of neoliberal policies, which shift responsibility for the conservation and restoration of rivers, onto local voluntary organisations as part of Big Society entrepreneurial initiatives. Part of my theoretical framework involves examining the modelling of rivers in the discourse and practice of community-outreach programmes. It is my belief that in these programmes can be found the mirror reflections of neoliberal subjects, self-disciplined to engage in entrepreneurial activities so as to address environmental issues that increasingly present immediate risks to local communities. Their status as mirror reflections is caused by the neoliberal entrepreneur of environmental consumer citizenship "reflects the present power of ideology to frame the world and how people ought to act in it" (West and Carrier 2004:485-486). Neoliberal community-outreach environmental initiatives involve powerful sign technologies, which inscribe and signify the subjects and objects of neoliberal governance. At the same time however, these sign technologies present simulacra that can be culturally and politically mobilised "that gives my own visibility to myself, that enables me to see myself there where I am absent" (Foucault 1986:24). West and Carrier's (2004) *Ecotourism and Authenticity: Getting Away from it All?* provides an account of the economic dimensions of neoliberal ecotourism. Central to West and Carrier's (2004) observations of the economic side of neoliberal ecotourism development programmes, is that it demands "the reduction of state financial power and fosters market-based models of political-economic and even social life" (ibid: 484). Thus it is my contention that river restoration operates partly as a mirror image of the neoliberal Big Society civil community involving activities and policies which galvanise individual subjects to take responsibility for the maintenance of local rivers. But the mirror reflection can also empower cultural political activity as it is mobilised by subjects to achieve ecological gains.

West and Carrier (2004: 488) describe how, in their study of ecotourism, "the neoliberal orientation of the context in which environmental managers operated obliged them to embrace ecotourism in order to fund their parks". My empirical research into river restoration identifies systematic processes through which, individuals faced with the local implications of global environmental crisis, are obliged to invest social capital in networking into river restoration charity organisations. In so doing, their solutions to local environmental concerns were shaped by the policies of the river restoration organisations to which they had networked. Thus while the subjects of neoliberal environmental programmes, such as the EU Water Framework Directive, might initially present as mirror images, their position in the river space is at one and the same time located within heterotopias:

> [I]n so far as the mirror does exist in reality, where it exerts a sort of counteraction on the position that I occupy. From the standpoint of the mirror I discover my absence from the place where I am since I see myself over there. Starting from this gaze that is, as it were, directed toward me, from the ground of this virtual space that is on the other side of the glass. I come back toward myself; I begin again to direct my eyes toward myself and to reconstitute myself there where I am. The mirror functions as a heterotopia in this respect: it makes this place that I occupy at the moment when I look at myself in the glass at once absolutely real, connected with all the space that surrounds it, and absolutely unreal, since in order to be perceived it has to pass through this virtual point which is over there. (Foucault 1986:24).

Foucault (1986:24) defines as "heterotopology", the "simultaneously mythic and real contestation of the space in which we live". Foucault's dimensions of heterotopology have relevance to my formulation of a theoretical framework for the feminist critical analysis of river restoration. Following Foucault (ibid), the river space as a heterotopia can be classified restoratively as "privileged or sacred or forbidden places, reserved for individuals who are, in relation to society and to the human environment in which they live in a state of crisis". The conception of the river space in terms of "crisis" might initially appear melodramatic, but it has compelling resonances with the mobilisation of river restoration in late modernity.

Rivers are intriguing heterotopias, and river restoration interpolates these "Other Spaces" in fascinatingly complex channels. Streaming through the city of London, the River Thames navigates an astonishing environmental history and disconcerting environmental present. In 1197, conservancy of the River Thames was bestowed upon the Mayor and Corporation of London, in a charter instituting, among other stipulations: “that all weirs that are in the Thames be removed” (quoted in Herbert 1966:98). Chicanery machinations about weirs are further evident in the Magna Carta (1215): “All weirs from henceforth shall be utterly put down, by Thames Medway, except only (nisi) by the sea coasts” (quoted in Herbert 1966:98). Governing the application of river technologies, increasingly connected commerce and politics in historic endeavours to harness river flow navigation and configure river contours. Hydraulic forms of river engineering such as weirs and aqueducts advanced commercial imperatives; they harnessed the powerful energy of rivers so as to rotate immense waterwheels and dredging operations deepened the transport capacity of river waterways. Nevertheless, legislative acts and governance of river waterways suggest that the conservancy of the River Thames has become infinitely entangled with the emergence of the “risk society”.

Ulrich Beck (2007:5) describes how the reverse side consequence of industrial capitalism has been the emergence of the “risk society”, whereby, the excesses of capitalist science, its industrial manufacturing, and its rapacious accumulation, produce risks which exceed the steering and protective capacities of social, political and economic institutions. According to Beck (ibid) this ‘reverse side’ to “the obsolescence of the industrial society” and the rise of “the risk society” have occurred in two stages. In the initial stage “the effects and self-threats are systematically produced, but do not become public issues or the centre of political conflicts” (ibid). As modernity enters into its later stages: “a completely different situation arises when the dangers of industrial society begin to dominate public, political and private debates and conflicts” (ibid). River restoration increasingly enters into this later stage of the risk society; it is increasingly politically mobilised as part of government defences against the hazards of climate change. My theoretical framework is particularly concerned with the formulation of river space in terms of our local experiences of global risks. For Beck (ibid) “Risk society is not an option that one can choose or reject…It arises in the continuity of autonomized modernization processes which are blind and deaf to their own effects and threats”. In the wake of industrial society, we are assured of “reflexive modernity” and with this, an era of self-confrontation in which increasingly localised risks present “opportunities, threats, ambivalences”, which were once “previously possible to overcome in a family group, in the village community or by recourse to a social class or group, must increasingly be perceived, interpreted and handled by individuals themselves" (ibid:8).

The river space conceived as a heterotopia brings the issue of self-confrontation into sharp relief. This is because it "is capable of juxtaposing in a single real place, several spaces or sites that are in themselves incompatible" (Foucault 1986:25). Urban river spaces are intersected by many other spaces that align with difficulty, and this situation is accentuated by, as opposed to being resolved by, the EU's Water Framework Directive; the market-based re-structuring of voluntary sector funding; local governance and community-outreach environmentalism. It is the crisscrossing of the public and private spheres through the river's heterotopia that provides opportunities and constraints for feminist praxis. This is because the heterotopia presents "a sort of absolute break with … traditional time" (ibid: 26). The intersecting places of the river space heterotopia engender a unique heterochrony in which the *longue durée* of embodied social relations intersects with "time in its most fleeting, transitory, precarious aspect" (ibid) in the current rhythm of capitalist accumulation. In the former heterochronies, time resonates with the time of reproductive labour; for "time never stops building up and topping its own summit" (ibid). In reproductive labour, time is relationally configured (Odih 2007). The capacity of the river space's heterotopia to intersect public and private spheres, through river restoration, constitutes a space for reproductive labour that is at one and the same time "outside of time and inaccessible to its ravages", while also constituting "a sort of perpetual and indefinite accumulation of time in an immobile place" (Foucault 1986:26). It is my belief that the intersecting heterochronies of the river space confirm this *Other Space* as a heterotopia *par excellence*.

Methodology and Method

"Interpretive structuralism" informs the methodology of my empirical research. According to Raymond Morrow (1994:24), Interpretive Structuralism is committed to the principle: "social relations and social analysis always have an interpretive (hermeneutic) dimension". To this end organisational ethnography has been applied to understanding the culture and the values of river conservation charities. Organisational structures, systems, processes and actors are recognised as engaged in complex dialectics of discursive construction. Thus critical discourse analysis was utilised to comprehend the manifold ways language operates to connect social practices with "social relations of power and domination"

(Fairclough 2001: 230). Critical discourse analysis also ascribes semiotic dimensions to social practice as Fairclough (ibid: 234) expresses; "semiosis figures in representations. Social actors within any practice produce representations of other practices, as well as ('reflexive') representations of their own practice". A further feature of interpretive structuralism is the transformative capacity of agency, for it is argued that "structures may be species-specific or historically constituted, and sometimes consciously transformed even if they have a kind of objective facticity that appears independent of immediate actors" (Morrow 1994:24). To this end, my research was disinclined to approach respondents with prescripted questions; rather, interviews were subject-centred and guided by the direction of the flow of responses to my thematic unstructured questions. Immense care was invested in balancing the "thematic" and "dynamic" properties of these qualitative interviews (Kvale 2009). Attention to the interview dynamic required that "thematic" questions were introduced as and when they coincided with the dynamic interchange of the interview. This was particularly useful during my interviews along the foreshore of the River Thames as part of my participation in, and ethnographic observation of, river restoration events. Indeed, attention to the dynamic interchange of field work is axiomatic to a further feature of interpretative structuralism: "meaning and structures constantly are reproduced (statically) and produced (dynamically) across space and time" (Morrow 1994: 24). Commitment to the time and space location of agents within the empirical field, guided my development of *live streaming* ethnographic research. The Classical Greek philosopher Heraclitus (circa535 - circa475 BC.), espoused the idea that the universe is in perpetual fluctuation; continuity and motion epitomise the ebb and flow of social life. Flux is an apposite visual metaphor for the field of ethnographic research. Fluctuation also highlights the uniqueness of insight provided by entering into the stream of actions, and being embedded within the moment of the live event. The metaphor of the stream resonates with Heraclitus' famous aphorism that we never step into the same river twice: "The river where you set your foot just now is gone – those waters giving way to this, now this" (Heraclitus 2003:27). *Live streaming* ethnography is about engaging in the moment of the movement; observing, listening and learning while embedded in the rivulet of the event; flowing with the rhythm of meaningful engagement. *Live streaming* empirical ethnography requires an appreciation of social identities as constructed in and through time and space; an idea incisively advanced by Heraclitus (ibid:51): "Just as the river where I step is not the same, and is, so I am as I am not". Methodologically, my formulation of *live streaming* interpretative structuralism involves the design of ethnographic techniques that enable observations and interviews to be embedded in the ebb and flow of meaning making. In total, over thirty empirical interviews were carried out during August/September 2013–February/March 2014.

The extensive qualitative interviews, involving three river conservation charities, were carried out either in their offices; in situ during river restoration events; or in bookshop, cafés in South London. In addition to the qualitative interviews with members of the river conservation charities, the empirical research involved over twenty brief moments of discussion with volunteers as they participated in river restoration events along the foreshore of the River Thames.

Organisation of Book

On 28th July 2010, the United Nations General Assembly agreed a landmark resolution affirming "the right to safe and clean drinking water and sanitation as a human right that is essential for the full enjoyment of life and all human rights" (United Nations 2010:2). Prior to this resolution, the United Nations Millennium Declaration had pledged to halve, by the year 2015, the proportion of the world's population who have limited access to safe drinking water (ibid). Corresponding with these initiatives has been a deepening concern about gender and inequality associated with water provisioning; water quality management and the subsistence agricultural labour carried out by rural women in the Third World. It is a disconcerting feature of gender inequality that while women constitute 43 percent of rural farmers, in developing countries, women only represent 5 percent of agricultural land holders in North Africa and South-west Asia (FAO 2012: 19). These limitations, on access to agricultural assets, exacerbate time poverty and water poverty experienced by rural women in the Third World. Securing clean and reliable water in these conditions becomes even more labour intensive in the context of competition for water advanced by water-scarce countries and profligate capitalist manufacturing industries. In recognition of gender inequality in the provisioning of and access to reliably safe, clean water, it is the central objective of this book to formulate an ecofeminist geopolitics of water.

> [I]n sub-Saharan Africa … women and girls in low-income countries spend 40 billion hours a year collecting water – the equivalent of a year's worth of labour by the entire work force in France. (United Nations Department of Economic and Social Affairs UNDESA 2013)

Ecofeminists perceive the exploitation of nature to be bound up with a patriarchal ideology of domesticity which has exploited and continues to exploit the reproductive labour of women. For women subsistence farmers in the Third World, patriarchal ideology combines with capitalism in the production of gender stereotypes that shape the design and planning of irrigation programmes. An outcome of the patriarchal ideology of domesticity is that "women are primarily housewives and mothers, while men are farmers and irrigators" and thus, irrigation programmes neglect to be informed by the agricultural needs and priorities of female subsistence farmers (FAO United Nations 2012:4). It is no wonder that from this observation emerges the rallying assertion that social oppression and the exploitative oppression of nature are commensurate. For ecofeminists, environmental degradation inter-connects with "patriarchal violence against women, other people and nature" (Mies and Shiva 1993:14). In the context of water ecology in the Third World "Experience has shown that where irrigation design fails to recognize that women are water users and farmers in their own right, risks are high that women lose existing access to land or the products of their own labour" (FAO United Nations 2012:4-5). And this raises important issues concerning "corporate and military warriors" and their "aggression against the environment" for ecofeminists have perceived this "almost physically as an aggression against our female body" (Mies and Shiva 1993:14). This book aligns with the ecofeminist materialist perspective; it formulates a theory of environmental degradation based on the materiality of capitalist exploitation with regards to the female body and the contours of reproductive labour. The central objective of this book is to theoretically formulate and empirically engage the ecofeminist geopolitics of water ecology. My sociological research problem is the voluntary activity of provisioning (restoring) river waterways and the relation of this activity to Marxist ecofeminist critique of "meta-industrial labour". My research question: **Is the provisioning of the river foreshore, as part of river restoration conservation programmes, part of the "meta-industrial" domestication of the environmental degeneration of natural river ecosystems?**

Chapter One: Time for an Ecofeminist Marxist Water Ecology. It is the central contention of this chapter that Marx's ecology provides the foundations of an ecofeminist Marxist ecology of water. Consequently, the chapter identifies key conceptual ideas evident in Marx's ecology, and sets about formulating their basis for a feminist Marxist geopolitics of water. To this end, the central theme of this chapter is "meta-industrial labour" (Salleh 2003) and the provisioning of water quality; and this is themed under the sub-title: “Meta-Industrial Labour and Water Quality”. In September 2000, world leaders at the United Nations Millennium Summit adopted the *Millennium Declaration*. Key to this declaration is the commitment to a series of Millennium Development Goals (MDGs) dedicated to the eradication of the extreme conditions of poverty. As part of the MDGs’ intention to safeguard environmental sustainability, a target was established to halve, by the date 2015, the numbers of the world’s population without sustainable access to uncontaminated drinking water and elemental sanitation (UN Water 2013j:2). Facilitated under the patronage of UN-Water, the “Universal access to water, sanitation and hygiene (WASH) is a development imperative” and integral to “a holistic approach to all aspects of the water cycle [which] will contribute to the achievement of multiple development goals related to education, health, food and energy, and towards reducing inequality, boosting employment and empowering women” (ibid). The conditions favouring the establishment of sustainable approaches to water quality management are predicated upon commercial viability and profit incentives; this raises disconcerting issues of limitations to capitalist investment in the development and innovation of water quality. Further to this issue, the UN Water (2013j:4) report identifies that limitation on universal access to clean water, sanitation and hygiene “costs 1.5% to 4.3% from GDP, stunts childhood growth, drains women’s time and energy, empties school chairs, forces needless risks, and denies human dignity”. Women and girls are disproportionately burdened with the responsibility of water provisioning; and it is the contention of this chapter that capitalist accumulation benefits from the unpaid “meta-industrial” water labour of women in the Third World.

Chapter Two: Dispossession: “Isolated, with no before and after” (T.S. Eliot East Coker). This chapter focuses on the theme of water time poverty and primitive accumulation. In 1978 the Water and Sanitation Program (WSP) was established through a multi-donor affiliation superintended by the World Bank “to support poor people in obtaining affordable, safe, and sustainable access to water and sanitation services” (WSP 2010). A central contention of WSP is that women and girls, in developing countries, are invariably responsible for the provisioning of water and the management of household hygiene. In the absence of a domestic water supply, female members of the household are forced to travel arduous distances to collect water.

Figure I.10b., 'Time present and time past, Are both perhaps present in time future' (T.S. Eliot *Burnt Norton*). Photographic Image, South London, River Thames, September 2013

Figure I.10c., 'Time present and time past, Are both perhaps present in time future' (T.S. Eliot *Burnt Norton*). Photographic Image, South London, River Thames, September 2013

Time poverty is clearly a gendered dimension of water poverty. Indeed, according to WSP (2010:9), "Cumulatively, one estimate suggests that some 40 billion hours a year, are spent, collecting water in sub-Saharan Africa – equal to a year's labor for the entire workforce of France" (ibid). According to the UN Water (2014) report *Water and Energy*: "In developing countries, women and girls bear most of the work burden associated with managing water and energy scarcity, fetching water for the 780 million people lacking access to improved sources of drinking water". WSP (2010:9) advocates intervention by the water and sanitation sector so as to alleviate the time poverty experienced by women in the provisioning of water. They identify "time poverty" as "the situation where women's time is inflexible, consumed by routine and non-productive tasks perpetuating their absence from decision making and other profitable pursuits" (ibid). While concurring with this description of time, gender and water poverty, further conceptual development is required to appreciate the scope and depth of its circumstances.

The basis for a geo-political formulation of water, gender and time poverty is evident in Marx's account of primitive accumulation and ecofeminist Marxist dialogue. In the postcolonial ecofeminist writings of Ana Isla (2009), vociferous critiques of the modern day occurrence of primitive accumulation are put forward. As a Marxist ecofeminist, Isla concurs with Marx's observations regarding the historical conditions of primitive accumulation. However, as with other Marxist ecofeminists, Isla disagrees with the contention that primitive accumulation is a mere precursor to the establishment of industrial capitalism. According to Isla (2006b:61): "We cannot look at the ecological crisis as separate from the complex of social crises produced by economic relations". Referring to the work of Maria Mies, Isla further states that: "ecofeminists can no longer be satisfied with the Marxist argument that the "first contradiction'' is capital versus wages which, once resolved, would create conditions necessary to resolve secondary contradictions" (ibid). Chapter Two, illustrates these debates within the context of an online ethnographic empirical case study of a river restoration cleanup carried out along the foreshore of the Rio Virilla in Costa Rica in 2010.

In March 2010, Costa Rica's Rio Virilla became the focus of environmental campaigns motivated to address sewage pollution. It has been suggested that a major source of river water contamination in Latin America derives from improperly discarded industrial waste products: "synthetic chemicals, and hazardous wastes" that seep into the groundwater

(San Martin 2002:21). Rio Virilla's contaminated water ecology is partly due to pollution from "agro-industrial activities and metropolitan development" (UNEP1999/2009:127). In addition to industrial waste and agrochemicals, Rio Virilla is subject to sewage pollution: "approximately 250,000 cubic meters of waste pours into the Rio Virilla every day" (Mitchell and Pentzer 2008:186). Industrial agro-chemical discharge coupled with the sewage production of impoverished metropolitan development is suggested to be polluting the water ecology of Rio Virilla (ibid). In March 2010, hundreds of volunteers were assembled into a series of Rio Virilla restoration clean-up events, and successfully extracted approximately 30,000 plastic bottles from the banks of the river (Ministra de Salud Costa Rica 2010). The organizers, Mountain of Hope, enrolled public and private sector agencies to participate and sponsor the event; among these were recycling companies which became the recipients of the extracted plastic bottles. Commercial investment in the event can partly be explained as Corporate Social Responsibility (CSR); less evident are the motivations of non-corporate volunteers. This case study advances and progresses beyond Marxist ecofeminist analysis of the incorporation of Third World women's work into the International Division of Labour. In this sense, Chapter Two, provides the theoretical basis for an analysis of the conservation of river waterways in terms of sub-politics, and thus advances beyond the Marxist ecofeminist percept that provisioning environmentally degenerated natural habitats involves exploitative meta-industrial labour i.e., housewifization.

Figure I.11., "And the pool was filled with water out of sunlight" (T.S. Eliot *Burnt Norton*). Photographic Image, South London, River Thames, September 2013

Chapter Three: Watersheds in Marxist Ecofeminist Analysis. Chapter Three engages with a recent empirical theme, namely *Securing London's Water Future*. To this end, the chapter identifies key reports, published by the office of the Mayor of London, wherein each of these publications is critically detailed as part of an analysis of water time poverty issues experienced by First World women. Reports that are critically discussed include: *The Greatest City on Earth: Ambitions for London by Boris Johnson* (Greater London Authority2013) and *Securing London's Water Future* (Johnson 2011). In 2011, the Greater London Authority published a strategy for securing a "world class" (ibid: 6) and environmentally resilient infra-structure for the River Thames. Central to this document is the proclamation that levels of the city's water are unsustainable in the context of rapid population growth in the numbers of Londoners, as well as climatic decreases in the summer rainfall. In order to meet this demand for water, "some water companies are currently abstracting more water from the environment than it can sustain" (ibid: 13) resulting, in recourse, to an increased use of London's reservoirs and desalination plants. According to *Securing London's Water Future* (Johnson 2011) a significant reason why London's water supply is unsustainable is because of the poor stewardship demonstrated by domestic households in their consumption of water. Thus the report states that:

> Londoners use more water than the national average (167 litres per person per day in 2009-10 compared to 146 litres per person per day), largely because we live in small households, which are less water efficient. Many Londoners have little incentive to save water – only one in four homes has a water meter. We also lose enough water in leakage from our water mains network to fill more than 238 Olympic-sized swimming pools every day. (ibid:13)

A central conclusion of the report is that London's demand for water will increase exponentially with its population, alongside depleting summer rainfall and the popularity for smaller dwellings. "This means that London's already tenuous supply-demand balance will become increasingly unsustainable – we therefore need to act to balance supply and demand" (ibid). An integral feature of the proposed action plan is "managing water use" and instituting efficiency in the "disposal of waste water". Both of these features encourage Marxist engagement, for they echo the concerns that Marx expressed in the nineteenth century about environmental degradation, consequent of accelerated capitalist production. While it is necessary to factor classical Marxist relflections into modern day analysis, the persistent coupling, in documents and dictates produced by government agencies, of water ecology and household consumption makes imperative a time for ecofeminist Marxist analysis of water sustainability. Consequently, this chapter is organised as follows: *Securing London's Water Future* provides for an empirical basis for a classical Marxist ecological engagement with the accelerated time of capitalist exploitative production; each aspect of the classical Marxist ecological debate is interpolated with Marxist ecofeminist critical reflections so as to highlight and address the deficiencies in classical Marxist ecology. The resulting tapestry provides a theoretical basis and conceptual formulation for a Marxist, ecofeminist geopolitics of water sustainability.

In the 2011 report, Greater London Authority's *Securing London's Water Future* a programme of initiatives, policies and ideas was instigated establishing the city's water resources; managing rainfall; dismantling "the hydrophobic policies of the 1960s – which saw natural rivers and waterways encased in concrete"; and committing the environmental governance of London to working "in harmony with water in our landscape to ease the consequences of heavy rainfall and beautify our city at the same time" (ibid: 6). *Securing London's Water Future* proclaims its dedication to environmental ecology and the securing long-term strategies of water sustainability. While it is evident that these ecological principles are a dedicated feature of the water strategy, it is also evident that the contradictions and conflicts of capitalist production encumber many of the well intentioned policy initiatives. This chapter is concerned with highlighting how the general conditions of capitalising sustainable water supply involves an endeavour to achieve transcendence from an ecocentric metabolic exchange with natural water supplies, while also exploiting the national and transnational meta-industrial labour of people who are marginalized from "economic sectors". Meta-industrial labour is expropriated through the extraction, without remuneration, of provisioning and reproductive labour as material acquisition and inputs to capital. Marxist ecofeminist analysis of water sustainability mobilises activists to focus on meta-industrial labour as human energies expended, and exploited, in the capitalization of water resources. This chapter aims to provide a further theoretical basis for analyzing the time water poverty experience of First World women in relation to Marxist, ecofeminist geopolitics of water sustainability.

Figure I.12., "And every moment is a new and shocking valuation of all we have been". (T.S. Eliot *East Coker*). Photographic Image, South London, River Thames, September 2013

Chapter Four: Organizational Values of River Conservation Charities. Chapter Four is based on ethnographic narrative interviews. This chapter explores the organizational values and culture of three River Thames waterways charities. Each of these charities is dedicated to River Thames water ecology, and this involves working with local communities, schools, and tourism leisure industries. Environmental education is a consistent feature of the citizenship orthodoxy narrated by these organizations. My research findings identify river conservation charity community-outreach programmes as encouraging an educational advocacy engendered by creative, transformative pedagogies. Framed within the philosophy of education espoused by Paulo Freire's (1996) *Pedagogy of the Oppressed*, this chapter describes and engages with River Thames waterways charities as being informational, actor networked and democratic agents.

Chapter Five: Volunteering Our Time in the Sub-political Geographies of River Conservation. The denaturing of environmental sustainability is observably accelerating, and continuously bringing into question the proficiency of capitalism, to superintend the proliferating risks of global climate change. Industrial modernization involves exploitative subjugation, and the capitalization of the natural environment is an integral feature of this capitalist dynamic. It is evident that preceding modes of industrial capitalism modernized in disregard of issues of sustainability in the expenditure of renewable resources. The "risk society" is the unintended outcome, or reverse-side consequences, of industrial capitalist opulence. The "risk society" refers to a second order of modernity in which risks precipitated by a previous order of modernity exceed existing regulative mechanisms: it "designates a developmental phase of modern society in which the social, political economic and individual risks increasingly tend to escape the institutions for monitoring and protection in industrial society" (Beck 2007:5). According to Beck (ibid), the conditions of risk society precipitate changes at the level of individual apprehension. In the risk society, insecurities erupt through the disjunctures, contradictions and conflicts that capillary from industrial society into subsequent orders of modernity. The current milieu of privatized securities is eroding collective solidarities, individuating identities, and undermining social defense against local experiences of global risks. Nowhere is this clearer than in the paralyzing grip of an environmental crisis. Overwhelmed by unprecedented magnitudes of environmental risk, political institutions no longer profess to guarantee present and future wellbeing. Faced with the local immediacy of international environmental risks, individuals are compelled into defensive action. Beck (ibid) describes "sub-politics" as confrontational and immediate, a direct political engagement indexed to the particular locality of disjuncture. It is a politics of the moment, in the moment. Chapter Five explores river conservation in terms of Beck's (2007) concept of "sub-politics".

Chapter Six: Gender and the Geopolitics of Care in River Conservation. Marx conceived that freedom, at its most basic level, resides with rationally mastering the realm of necessity. In the writings of Marx, the life of the species consists materially in the fact that the human being (as with animals) "lives on inorganic nature" (1844/2007:74). Humans are least like animals, when their creativity of action loosens them from the shackles of immediate necessity, providence, and happenchance, in the material conditions of existence. According to Marx (ibid) cultivating the natural world engages humans in inspiriting activities, advancing character and developing our creative attributes. Thus, from Marx's ecology one can try to discern a sense of the spirit which mobilizes volunteers to invest themselves in the provisioning of their local river waterways. However, attempting to impose a classical Marxist ecological framework immediately reveals limitations, for river waterways provisioning can be discerned as a domestication of environmental issues. Maria Mies' (1998) concept of "housewifization" appears to present an apposite Marxist feminist lens to empirically and theoretically formulate an ecofeminist ecological analysis of the geopolitics of river waterways provisioning. According to Mies (1998:126), "The housewife is the optimal labour force for capital at this juncture and not the 'free proletarian', both in the underdeveloped and overdeveloped countries". It is Mies' observation (ibid) that: "Whereas the consumer-housewife in the West has to do more and more unpaid work in order to lower the costs for the realization of capital, the producer-housewife in the colonies has to do more and more unpaid work in order to lower the production costs". Mies observes that both women are subjected to powerful ideologies that impose patriarchal assumption about the actions and practices of ideal womanhood. Chapter Six critically engages with Mies' conception "housewifization". The chapter further extends the question of domestication, for it deconstructs the feminine ideal care ethic of river conservation, and innovatively evaluates the ecological activism that is taking place on the river foreshore. Primarily informed by my interviews, *in situ* with volunteers cleaning the foreshore of the River Thames, this chapter develops a post-structural analysis of gender and ecological activism.

Michel Foucault (1988:18) identifies inter-relationships between sign technologies (signs, meanings, signification), technologies of production (transformation of material things), technologies of power (objectivising categorization), and technologies of the self (self-regulating abilities of subjects to impact discourse on the cultivation of body and soul). Formulating an account of gender and ecological activism therefore requires examining "the application of objective capacities" in terms of the "relationships of communication" that circulate in and through the discursive constructs of gender identity. Symbolic mediations are inextricably bound to the objectivizing capacity of technologies of power ("whether they consist of obligatory tasks" or processes of visibility). Symbolic mediations are meaningful communications, for they "imply finalized activities (even if only the correct putting into operation of elements of meaning), and by virtue of modifying the field of information between partners, produce effects of power" (Foucault 1982:218). Consequently, symbolic mediations:

> can scarcely be dissociated from activities brought to their final term, be they those which permit the exercise of this power (such as training techniques, processes of domination, the means by which obedience is obtained) or those which in order to develop their potential call upon relations of power (the division of labor and the hierarchy of tasks). (ibid:218)

In this analysis of the subject of gender and ecological activism, it needs to be appreciated that the "transformation of the real" (technologies of production); "signs, communication, reciprocity, and the production of meaning" (technologies of the sign); and "domination of the means of constraint" (technologies of power) are interrelated features of symbolic mediations (ibid). Nevertheless, their application is never uniform and invariant; this is because "in a given society there is no general type of equilibrium between finalized activities, systems of communication, and power relations" (ibid). Part of Foucault's thinking here derives from his advocacy of a definition of "government" that extends beyond the formal institutions and structures of politics and the state. Rather, "government" refers to a shifting assemblage of formal and informal agencies, practices and institutions that variously and differentially align the self-regulating ability of subjects with the design, objectives and scope of a regime of governance. As Foucault expresses it:

> The exercise of power consists in guiding the possibility of conduct and putting in order the possible outcome. Basically power is less a confrontation between two adversaries or the linking of one to the other than a question of government. This word must be allowed the very broad meaning which it had in the sixteenth century. "Government" did not refer only to political structures or to the management of states; rather it designated the way in which the conduct of individuals or of groups might be directed: the government of children, of souls, of communities, of families, of the sick. (ibid:221)

Symbolic mediations are inextricably linked to the exercise of power in government i.e., they are involved in the designation of "the way in which the conduct of individuals or of groups might be directed" (ibid). A consistent feature of my research into the feminine ideal, as part of a symbolic mediation of river restoration, has involved focusing on governance both in terms of conventional political structures and "also modes of action, more or less considered and calculated, which were destined to act upon the possibilities of action of other people" (ibid). Thus, for example, I have empirically researched the impact of the European Union's *Water Framework Directive*, on river conservation charities. My interviews with three river conservation charities produced primary data which has been triangulated with data derived from my interviews along the foreshore of the River Thames. Both sets of data have trace elements of government in the sense of "political structures or … the managements of states" (ibid); but the data also illustrates "the exercise of power as a mode of action upon the actions of others" (ibid). Emerging from this triangulation of data is the suggestion that the coordination of technologies of production, signs, and power, constitutes a shifting and irregular field of practice. Symbolic mediations of ecological activism tend not to easily slot into the formal apparatuses of political institutions and environmental organisations. Rather, as Foucault observes:

> [T]here are diverse forms, diverse places, diverse circumstances or occasions in which these interrelationships establish themselves according to a specific model. But there are also 'blocks' in which the adjustment of abilities, the resources of communication, and power relations constitute regulated and concerted systems. (ibid:218)

The symbolic mediation of ecological activism is constituted through a complex milieu of moderated communications (environmental citizenship programmes, advertising signs and scheduled events) and, less formally "by the means of a whole series of power processes (enclosure, surveillance …)" (ibid:219). Further to the operation of assemblages of actions upon actions, it is evident that these symbolic mediations of ecological activism, in terms of the feminine ideal, are not prescriptive or directly imposed on the subject. Rather, symbolic mediations of ecological activism exemplify the manifold way in which the exercise of power opens up a field of possibilities for the cultivation of gender identities.

Figure I.13a., "What might have been and what has been, Point to one end, which is always present" (T.S. Eliot *Burnt Norton*). Photographic Image, South London, River Thames, September 2013

Figure I.13b., "What might have been and what has been, Point to one end, which is always present" (T.S. Eliot *Burnt Norton*). Photographic Image, South London, River Thames, September 2013

Figure I.13c., "What might have been and what has been, Point to one end, which is always present" (T.S. Eliot *Burnt Norton*). Photographic Image, South London, River Thames, September 2013

Figure I.13d., "What might have been and what has been, Point to one end, which is always present" (T.S. Eliot *Burnt Norton*). Photographic Image, Scuth London, River Thames, September 2013

PART ONE:

MARXIST ECOLOGY AND ECOFEMINIST WATERSHEDS

Figure P.1: "Time past and time future" (T.S. Eliot *Burnt Norton*). Photographic Image, South London, River Thames 2013

CHAPTER ONE

TIME FOR AN ECOFEMINIST MARXIST WATER ECOLOGY

> For the sake of completeness, it should be noted that what we understand here by land also includes water, etc. insofar as this has an owner and appears as an accessory to the land. (Marx *Capital Vol., III* 1894/1981:752)

There is "a colossal wastage in the capitalist economy" (Marx 1894/1981:195). The capitalization of processes, by which the waste outputs of production are reintegrated into the cycle of production, are predicated on social labour; thus, they are mediated by existing structural inequalities between capital and labour. Capitalism's colossal output of waste is an intractable consequence of its expansive assemblages of production and thus "results only when labour is carried out on a large scale" (ibid:196). Nevertheless, this imposing scale of waste output presents the avaricious enumerations of capitalism with the attractive possibility of capitalising on waste (Sandler 1994); an issue originally raised by Marx in *Capital Vol., III* and defined as the "Utilization of the Refuse of Production" (ibid:1894/1981:195). Capitalism generates immense waste while at the same time seeking to intricate waste into commodity capitalism through its transformation as an exchange-value. Marx (ibid) observed paradoxical contradictions in the capitalization of waste, and he appeared particularly frustrated by the ineffective capitalist environmental management of the River Thames, as is evident in the statement: "In London, for example, they can do nothing better with the [natural human waste] produced by 4 ½ million people than pollute the Thames with it, at monstrous expense" (ibid). Over a hundred years later the ecology of the River Thames continues to be contaminated by organic and inorganic pollutants. In June 2013, Thames Water Utilities LTD was fined £41,000 for polluting the River Ash in Ashford; contravening the obligations of its environment permit (Environment Agency 2013). It has been estimated that 4,300 fish perished as a result of the water pollution incident at the River Ash (ibid). Earlier, in May 2013, Thames Water Utilities LTD was fined £27,779 for the contamination of the Engine River,

Figure 1.1. "Investing form with lucid stillness, Turning shadow into transient beauty", (T.S. Eliot *Burnt Norton*). Photographic Image, South London, River Thames, September 2013

in the vicinity of Walton Lane, Weybridge (Environment Agency:2013a). In February 2013, Thames Water Utilities LTD was mandated to pay an estimated £14,000 fine, for contaminating Faringdon Stream in Faringdon (Environment Agency: 2013c). It is often claimed that part of the problem is due to London's archaic Victorian sewer system, which channels domestic waste and rainwater into subterranean networks. Polluting discharge often occurs when the sewer overflows, and subsequently dispenses untreated waste into the River Thames through the sewage network's 57 Combined Sewer Overflows (CSOs). Approximately 39 million tonnes of raw sewage are discharged into the tidal Thames annually. The emission of untreated sewage discharge takes place an estimated "60 times a year, after as little as two millimetres of rain" (Environment Agency 2013d). According to a 2013 Science and Technology Committee report entitled *Water Quality: Priority Substances*, inorganic chemical pollutants are entering into English waterways. Scientific studies identify these chemical pollutants as responsible for poisoning aquatic organisms, agglomerating in the eco-system, as being detrimental to habitats, and of significant harm to human health (House of Commons 2013:3). European Commission Directives for water quality stipulate that chemicals which endanger the aquatic environment are regulated at the European Union (EU) level under the auspices of the *Water Framework Directive*, and are defined as "priority substances". Member states of the EU are duty bound to control the levels of these priority substances in order to ensure that they do not exceed the standards stipulated by the European Commission. Of particular significance have been recent additions to the list of priority substances namely: "The inclusion of two oestrogen-based pharmaceutical products" (ibid: 3). The oestrogen-based pharmaceuticals designated for additions to the European Commission's list of priority substances are: "ethinyl oestradiol (EE2), used in oral contraceptives, oestradiol (E2), used in hormone replacement therapy, and diclofenac, an anti-inflammatory" (ibid:7). The designation of oestrogen-based pharmaceuticals as presenting risks to the aquatic environment has provoked immense controversy. Of particular concern is the claim that the increased presence of ethinyl oestradiol and oestradiol in the aquatic environment can: "affect the health of aquatic animals through the feminisation of male fish" (ibid:3). Further to this claim is the stipulation that ethinyl oestradiol and oestradiol should be subject to European Union control under the *Water Framework Directive*; and thus designated as priority substances. In support of this, ruling research evidence provided by the Environment Agency is often cited. This is because "In 2004, the Environment Agency found that 86% of male fish sampled at 51 sites around the UK could be classified as intersex, as a result of displaying these characteristics" (ibid:7-8). One interesting feature of the controversy precipitated by this revelation that counter arguments tend to converge around is the issue of cost. It has been observed that: "the cost of treating wastewater to remove pharmaceutical products has been one of the main objections to these proposals" (ibid:9). Furthermore, it is reported that the water industry has expressed reservations that the European Commission's proposals will require wastewater to be treated in order to achieve the same standard as the production of drinking water; this is presumed to be a prohibitively expensive requirement. Indeed, resistance to the European Commission's stipulations have suggested that "upgrading wastewater treatment plants to remove the ethinyl oestradiol alone would cost £27-31 billion over the next twenty years in England and Wales" (ibid:9-10). While debating the financial cost incurred to improve the wastewater treatment by the removal of oestrogen-based chemicals, it is important to critically frame our comprehension of this issue.

In an article entitled *Drowning in a Sea of Estrogens*, Celia Roberts (2003) directs attention to a heightened social concern, that across the globe, biospheres are being biologically transformed by the presence of endocrine disrupting, estrogens in the aquatic environment. For Roberts (ibid:196), biological denaturing as a consequence of these estrogens, raises intriguing "questions about the 'nature' of sex, gender and reproduction, and their relations to each other". Elsewhere Non-governmental Organisations (NGOs) have accumulated internationally comparative data documenting the role of women in the management of water consumption and sanitation. The United Nations Department of Economic and Social Affairs (UNDESA) recently published its plans for the International Decade for Action "Water for Life" 2005-2015. This resolution "calls for women's participation and involvement in water-related development efforts" (UNDESA 2013). It is evident that women, in developing countries, have a disproportionate responsibility for the provision and care of water (ibid). According to UNDESA (2013), "In Africa, 90% of the work with respect of gathering water and wood, for the household and for food preparation, is done by women". Inadequate sanitation also betrays gender differentials; for it is evident that incommensurate planning, design and provision of hygienic private sanitation amenities can be a source of embarrassment and physical unpleasantness for millions of women across the globe (ibid). In 2013 the United Nations Human Settlements Programme (UN-Habitat) also published revelatory information about women's unequal access to water sanitation facilities and the provision of water supplies. In a report entitled *State of Women in Cities 2012-2013*, UN-Habitat (ibid) identifies that "women experience time poverty due to inadequate urban services". It is an intriguing fact that urban dwelling intensifies the destitution of poverty and this has particular implications for women's disproportionate responsibility for provisioning household services. According to UN-Habitat (ibid:xii): "The urban poor face specific circumstances which can exacerbate poverty such as spending more on food and services, especially water, sanitation, housing, energy and transport". Women are inordinately affected by these circumstances because of their culturally ascribed unpaid duties in the domestic division of household labour. The UN-Habitat (ibid) report also documents worrying correlations linking the inadequate provision of private sanitation amenities with the occurrence of violence against women (ibid:xiii); along with physical bodily strain where domestic water supplies are meagre and women have to trek distances carrying heavy vessels of water (ibid:30). The UN-Habitat (2013:35) report concludes: "Without adequate attention to the need for a gendered approach to urban planning, these processes can undermine urban prosperity". The inference here is that gender differences in the management, provision and access to water impinge on basic human rights. Indeed, it is notable that:

> Access to safe drinking water and sanitation were established as human rights by the UN Human Rights Council in 2010, adding impetus to MDG [Millennium Development Goals] 7 Target C, which is to 'reduce by half the proportion of the population without sustainable access to safe drinking water and basic sanitation by 2015'. (UN- Habitat 2013:40)

Evidence of the need to incorporate feminist geopolitics into the design and planning of actions geared towards water rights is particularly pertinent in the above resolution. This is because the rights to water, as decreed by the UN Human Rights Council in 2010, do not incorporate the right to water for bathing, cleaning and washing (ibid). Scarcity of this kind "not only impacts women's and girls' workloads, but can also seriously compromise their dignity and self-respect" (ibid: 40). In these, as well as other examples, is evidence of the imperative need for feminist research into the formulation of a geo-politics with respect of women and water.

It is the central contention of this chapter that Marx's ecology can provide the foundations of an ecofeminist Marxist ecology of water. Consequently, the chapter identifies key conceptual ideas evident in

Marx's ecology and sets about formulating their basis for a feminist, Marxist geopolitics of water. To this end, the central theme of this chapter is meta-industrial labour (Salleh 2003) and the provisioning of water quality; this is discussed in the section entitled *Meta-Industrial Labour and Water Quality*. In September 2000, world leaders at the United Nations Millennium summit adopted the Millennium Declaration. Key to this declaration is the commitment to a series of Millennium Development Goals (MDGs) dedicated to the eradication of the extreme conditions of poverty. As part of the MDGs' intention to safeguard environmental sustainability, a target was entrenched to halve, by the year 2015, the numbers of the world's population who do not have sustainable access to uncontaminated drinking water and elemental sanitation (UN Water 2013j:2). Coming under the patronage of UN-Water, the "Universal access to water, sanitation and hygiene (WASH) is a development imperative" and integral to "a holistic approach to all aspects of the water cycle [which] will contribute to the achievement of multiple development goals related to education, health, food and energy, and towards reducing inequality, boosting employment and empowering women" (ibid). The conditions favouring the establishment of sustainable approaches to water quality management are predicated on commercial viability and profit incentives (ibid); and this raises disconcerting issues of limitations to capitalist investment in the development and innovation of water quality. According to UN Water (2013j:4) women and girls are disproportionately burdened with the responsibility of water provisioning. It is the contention of this chapter that capitalist accumulation benefits from the unpaid "meta-industrial" (Salleh 1995, 2001, 2003) water labour of women. That is to say, capitalism relies upon the existence of non-capitalised water resources. The forms of labour that are associated with the provisioning and maintenance of these resources are particularly lucrative when they reside outside of the wage-labour contract, and are carried out as non-economic reproductive labour, or through an impoverished, precarious and exploitative wage. Such observations are often amiss in analysis of consumerism and water consumption, as is evident in Arjen Hoekstra's (2013) *The Water Footprint of Modern Consumer Society*. Hoekstra's analysis of water consumption appears to neglect gender inequality; thus, inequalities in the domestic division of household management of water both locally, in the Western world, and globally, in the developing world, are disregarded features of Hoekstra's analysis. Conversely, this study of water and consumer society recognises the imperative to place women and gender at the centre of analysis; as is appropriate with the Marxist ecofeminist tradition.

Meta-Industrial Labour and Water Quality

> Safe drinking water and basic sanitation are intrinsic to human survival, well-being and dignity. Without a serious advance in implementing the water and sanitation agenda, there is little prospect of achieving development for all (United Nations Secretary General Ban Ki-moon 2008 quoted in WWDR4 2012:403)

> 783 million people do not have access to clean water and almost 2.5 billion do not have access to adequate sanitation (UN Water 2013i)

> 6 to 8 million people die annually from the consequences of disasters and water-related diseases (UN Water 2013i)

> Treatment of wastewater requires significant amounts of energy, and demand for energy to do this is expected to increase globally by 44% between 2006 and 2030...especially in non-OECD countries where wastewater currently receives little or no treatment (UN Water 2013i)

> Pollution knows no borders...Up to 90% of wastewater in developing countries flows untreated into rivers, lakes and highly productive coastal zones, threatening health, food security and access to safe drinking and bathing water. (UN Water 2013i)

According to UN Water (2013): "approximately 3.5 million people die each year due to inadequate water supply, sanitation and hygiene"; "80% of sewage in developing countries is discharged, untreated, directly into water bodies"; "industry dumps an estimated 300 – 400 MT of polluted waste in waters every year"; and "Nitrates from agriculture is the most common chemical contaminant in the world's groundwater aquifers". For centuries, the River Thames has been integral to the water ecosystem and sustainability of London. In the 19th century, heightened perturbation over pollution of the River Thames was ameliorated by Joseph Bazalgette's innovative underground wastewater disposal network. The River Thames at this time was in an exceptionally foul state; rainwater cascading over the industrialising metropolis was impracticably streaming into the river; the Metropolitan Board of Works outfalls were directly discharging sewage into the River Thames; noxious gases escaping from negligent riverside manufactories and industrial chemical discharge were contaminating the River Thames (Parliamentary Papers 1884, 1857). Karl Marx observed the dire condition of the River Thames and raised thought provoking questions concerning the "utilization of the refuse of production" (*Capital Vol., III*: 195). Indeed Marx observed that: "As the capitalist mode of production extends, so also does the utilization of the refuse left behind by production and consumption" (ibid). *Capital Volume III* details contradictions which beset capitalism's endeavour to reduce waste production and conserve capital investment through the recycling of refuse.

Industrialisation precipitated formidable and revolutionary developments in the capitalist mode of production. Marx observed parallels linking acceleration in the forces of production with increased incidence of river pollution. As capitalist industrialisation expanded, so also did the intractable problem of waste disposal; and much of this problem resulted from the capitalist system itself. Of central concern is capitalism's market economy of competing agents. Marx observed capitalist manufacturing to be, "marked by the waste of much material" and because companies compete against each other, the production of waste accumulates: "since nothing is done according to a social plan, but rather depends on the infinitely varied circumstances, means, etc. with which the individual capitalist acts. This gives rise to a major waste of productive forces" (*Capital Vol., II* 1885/1978:252). Marx observed the paradoxes and contradictions that also blighted capitalism's endeavour to recycle the refuse of production, "its so-called waste products, back into new elements of production, either in the same branch of industry or in others" (*Capital Vol., III.,* 1894/1981:172). The Industrial Revolution's insidious accumulation through dispossession, and its obstinate centralisation of manufacturing, had corralled wage-labourers into cities, provoking disturbances in "the circulation of matter between man and the soil, i.e., prevents the return to the soil of its elements consumed by man in the form of food and clothing; it therefore violates the conditions necessary to lasting fertility of the soil" (*Capital Vol., 1.,* 1887/2003:474). In the sphere of industrial farming, "all progress in capitalist agriculture is a progress in the art, not only of robbing the labourer, but of robbing the soil; all progress in increasing the fertility of the soil for a given time, is a progress towards ruining the lasting sources of that fertility" (ibid:474-475).

Marx's materialist dialectic of nature includes soil and water as resources that are needlessly depleted through capitalism's negligent metabolic exchange with the surrounding environment. Marx details in *Capital Vol., 1*, dialectical oscillations involving "absolute" and "relative" surplus value extraction: "surplus-value produced by prolongation of the working-day" is defined as "absolute surplus-value"; whereas "surplus-value arising from the curtailment of the necessary labour-time, and from the corresponding alteration in the respective lengths of the two components of the working-day" Marx describes as "relative surplus-value" (*Capital*

Vol., 1., 1887/2003:299). The more advanced the capital, the more predicated it is upon the continuous expansion of market relations, and the more it strives for the accelerated appropriation of surplus value through the "greater annihilation of space by time" (Marx, *Grundrisse* 1857/1973: 539). But in relative surplus value production, the accelerated pace of surplus value creation requires both a quantitative (i.e., in terms of actual numbers) and qualitatively subjective expansion (i.e., in terms of commensurate values and desires) of existing consumer markets. As Marx expresses it: "relative surplus value…requires the production of new consumption [through]…quantitative expansion of existing consumption; secondly: creation of new needs by propagating existing ones in a wide circle" (ibid:408). This is because the creation, by capital, of absolute surplus value involves "making deductions from a given quantity of [value-creating] labour" (ibid:554). In these circumstances, capital is continually maximizing the expropriation of objectified labour through extended working hours, and this in turn enables "capital to renew this profitable bargain,....on a more enlarged scale" (ibid:550). Conversely, in relative surplus value, an increase in the productivity of work is accompanied by the ability to expropriate more surplus value in a reduced number of hours worked. This leads to an accumulation of undifferentiated commodities and the need for the "discovery and creation of new use values" (ibid:408). This suggests the creation of values which exceed the notion of useful values (e.g., commodities that have a clear functionality) and encompass a spectrum of qualitative differences (e.g., commodity fetishism). To this extent "the surplus labour gained does not remain a merely quantitative surplus, but rather constantly increases the circle of qualitative differences within labour" (ibid:408). It is, therefore, apparent that the production of relative surplus value must be augmented by the expropriation of a "non-capitalist social strata as a market for its surplus value" (Luxemburg, 1971: 368). And this non-capital reservoir does not entirely refer to the quantitative resources of the productive economy. Marx provides some foundation for this mode of analysis, when he states:

> The value of the old industry is preserved by the creation of the fund for a new one in which the relation of capital and labour posits itself in a new form. Hence exploration of all of nature in order to discover new useful qualities in things; universal exchange of the products of all alien climates and lands; new (artificial) preparation of natural objects, by which they are given new use values. The exploration of the earth in all directions, to discover new things of use as well as new useful qualities of the old; such as new qualities of them as raw materials etc.; the development, hence, of the natural sciences to their highest point; likewise the discovery, creation

> and satisfaction of new needs arising from society itself; the cultivation of all the qualities of the social human being… This creation of new branches of production, i.e., of qualitatively new surplus time, is not merely the division of labour, but is rather the creation, separate from a given production, of labour with a new use value; the development of a constantly expanding and more comprehensive system of different kinds of labour, different kinds of production, to which a constantly expanding and constantly enriched system of needs corresponds. Thus, just as production founded on capital creates universal industriousness on one side – i.e., surplus labour, value-creating labour – so does it create on the other side a system of general exploitation of the natural and human qualities, a system of general utility, utilising science itself just as much as all the physical and mental qualities, while there appears nothing higher in itself, nothing legitimate for itself, outside this circle of social production and exchange. (*Grundrisse*, 1973: 409)

It is evident that capital accumulation requires a reservoir of non-economic resources as the basis for the production of relative surplus value. In contemporary regimes of capitalist accumulation this translates as "the universal appropriation of nature as well as of the social bond itself by the members of society" (ibid). Just as production founded on absolute surplus value expropriates the objectified time of "value-creating labour", so the production of relative surplus value entails "the exploitation and exchange of natural and mental forces" (ibid: 410). Consequently, Marx affirmed that:

> [C]apital creates the bourgeois society, and the universal appropriation of nature as well as of the social bond itself by the members of society. Hence the great civilizing influence of capital; its production of a stage of society in comparison to which all earlier ones appear as mere local developments of humanity and as nature-idolatry. For the first time, nature becomes purely an object for humankind, purely a matter of utility; ceases to be recognized as a power for itself; and the theoretical discovery of its autonomous laws appears merely as a ruse so as to subjugate it under human needs, whether as an object of consumption or as a means of production. In accord with this tendency, capital drives beyond national barriers and prejudices as much as beyond nature worship, as well as all traditional, confined, complacent, encrusted satisfactions of present needs, and reproductions of old ways of life. It is destructive towards all of this, and constantly revolutionizes it, tearing down all barriers which hem in the development of the forces of production, the expansion of needs the all-sided development of production, and the exploitation and exchange of natural and mental forces. (*Grundrisse*, 1973:409-410)

Writing in the 19th century, Marx observed the *fons et origo* of industrialised agricultural production. Enclosure Acts had already dictated the dispossession of peasant farmers from common pastures; thus paving the way for large-scale capitalist farming to annihilate "the peasant, that bulwark of the old society, and replaces him by the wage-labourer" (Marx *Capital* Vol., I 1887/2003:474). Technological advances in geology, alongside capitalism's preference for scientific farming machinery, challenged the metabolic ecology of pre-industrial farming. Marx observed how the revolutionary effect of the new scientific discoveries was accentuating the disaggregation of agriculture from manufacturing: "Capitalist production completely tears asunder the old bond of union which held together agriculture and manufacture in their infancy" (ibid). Marx observed that over time, the dichotomised fields (of industry and agriculture) were converging as a consequence of synergies in their uptake of scientific technologies. But this "higher synthesis" of agriculture and manufacture produces what Marx describes as a "metabolic rift", i.e., the disruption between humans and the natural metabolism of the earth. Metabolic rift frequently typifies the ecological instabilities instantiated by the migration of rural labour to capitalism's industrial towns; for "it disturbs the circulation of matter between man and the soil" and the application of short-term intensive farming to rural land "prevents the return to the soil of its elements … therefore violating the conditions necessary to lasting fertility of the soil" (ibid). "But while upsetting the naturally grown conditions for the maintenance of that circulation of matter, it imperiously calls for its restoration as a system, as a regulating law of social production, and under a form appropriate to the full development of the human race" (ibid). Marx observed how capitalism's endeavour to reconcile the "lasting" metabolic disjuncture consequence of industrial agriculture, was persistently curtailed. Capitalism's creation of "restorative systems" was predicated on the marketised exchange relations of capital and labour. In the "higher synthesis" of agriculture and manufacturing, rural and factory labour were now subject to the exploitative and dispiriting temporality of the capitalist labour-process, in which "the instrument of labour becomes the means of enslaving, exploiting and impoverishing the labourer" (ibid). It was Marx's belief that:

> In modern agriculture, as in the urban industries, the increased productiveness and quantity of the labour set in motion are bought at the cost of laying waste and consuming by disease labour-power itself. Moreover, all progress in capitalistic agriculture is a progress in the art, not only of robbing the labourer, but of robbing the soil; all progress in

> increasing the fertility of the soil for a given time, is a progress towards ruining the lasting sources of that fertility. (Marx *Capital Vol., I,* 1887/2003:474-475)

Marx's observations of metabolic rifts between humans and nature are as relevant now as they were in the inception of industrial agricultural production in the 19th century. Insofar as capitalist profit propels "restorative systems" of sustainable production into market relations of exchange, its quest for "progress in increasing the fertility of the soil" will persistently succumb to metabolic rift. As Marx expresses it: "Capitalist production, therefore, develops technology, and the combining together of various processes into a social whole, only by sapping the original sources of all wealth – the soil and the labourer" (ibid:475). Evidence of capitalism's continued exploitation of natural resources forms the basis of recent United Nations water reports. In 2012 the UN's World Water Assessment Programme (WWAP) published the fourth edition of its *UN World Water Development Report* (WWDR4). WWDR4 (2012a) identifies three main drivers affecting water quality. The first is the hydrological cycle: natural conversions in which atmospheric transportation can affect the quality of water by sedimenting atmospheric pollution from one place to another. For example sulphur emission derived from fossil fuels can be carried over expansive distances and precipitate in the form of acid rain (ibid:403). In rivers and lakes with limited resilience, sulphur emissions can cause acidification in the ecosystem (ibid). International regulation of sulphur emission have, to a considerable extent, moderated this source of acidification in the ecosystem; but a disconcerting number of manufacturing plants remain insufficiently equipped, and are inadequately provisioned to meet international standards of treatment methods. Further, disruptions to the natural processes of the hydrological cycle derive from climate change. Of concern is the projected rise of the global temperature in excess of 2°C – 3°C by the year 2050 (ibid:404). Increased global temperatures can precipitate drought conditions of acute low flows in water systems, and in so doing reduce the ecosystem's competency to absorb and convert contaminated water (ibid). A second factor of some concern is the "social drivers" in water quality management. Across the globe an increasing advocacy for biofuels, both in the West and in the developing countries, is placing additional pressures on intensive farming production techniques and heightening the propensity for management systems to contend with the chemical waste deposits derived from intensive agricultural production (ibid). Population growth is commonly cited as a key determining factor of water quality management. It is estimated that: "by 2025, 1.8 billion people will live in countries or regions with absolute water scarcity, and two-thirds of the world population could be affected by water stress conditions" (ibid:541). The third factor driving "water quality risks" globally relates to "economic drivers" (ibid:405). Amongst these "the impacts of economic growth on urban settlements, industrial development and food production" which "translate directly into increasing risks and emerging water quality issues" (ibid).

Demographic forecasts continue to identify urbanization as a huge pressure on water quality management. More and more people are now dwelling in urban areas and in increasingly extreme circumstances are occupying informal peri-urban areas that are ill-equipped to accommodate water consumption and sanitation needs: "Every second the urban population grows by 2 people"; some "827.6 million people live in slums, often lacking adequate water and sanitation services"; there was an "[i]ncrease in the overall slum population between 2000 and 2010: 6 million people a year. On these trends, the world slum population should reach 888 million by 2020" (UN Water 2013g). It is also evident that: "An estimated 2.6 billion people worldwide live without adequate sanitation facilities …and the majority of wastewater in developing countries is discharged, untreated, into receiving water bodies" (ibid). Suffice to say that disruptions to the natural hydrological cycle, coupled with population growth and urban migration, are placing an excessive strain on water quality management. Further areas of water quality management risks relate to "economic drivers". Agriculture constitutes approximately 70% of global water consumption, thus the conceivable hazards of water quality impacts which derive from agricultural return flows are of real significance (UN Water 2012:405). Agricultural applications: "cause nutrient contamination and the sector is the major driver of eutrophication, except in areas with high urban concentration" (ibid). It has been established that nutrient enrichment and pesticide application are steadily becoming one of the planet's most epidemic water quality problems (ibid). Industrial development is associated with these hazardous water polluting forces, and technological progression provides the conditions for the increased manufacture of consumer goods, thereby increasing incidents for the further introduction of toxic and dangerous chemicals into the water ecosystem (ibid:406). It is evident that in the developing world, insufficient capital accumulation undermines the building of the regulatory infrastructure necessary to cultivate sustainable industrial development. In these circumstances, the imposition of international environmental specifications is particularly challenging, making it difficult to prevent the discharge of untreated industrial waste directly into freshwater systems (ibid). Alongside these sources of contamination are also: excessive leakages of hydrocarbons; hazardously acidic mining drainage waters; exorbitant deforestation logging and the proliferating practice of clearing natural forests so as to establish oil palm plantations dedicated to the cultivation of biofuels (ibid:407); "Industry is responsible for dumping an estimated 300 – 400 million tonnes of heavy metals, solvents, toxic sludge and other waste into waters each year" (UN Water 2011a:2). Nitrates derived from agriculture have been identified, across the world, as the most common source of chemical contamination in groundwater aquifers (ibid). Statistical evidence confirms that: "In the United States of America, manure and fertilizer run-off from agriculture is the single greatest source of water pollution, with croplands alone accounting for nearly 40 per cent of the nitrogen pollution and 30 per cent of the phosphorus pollution" (ibid). Water salinization also impedes effective water quality management. In the vast majority of countries with expansive areas of land salinization, water salinization is an associated problem. To this effect: "Major problems have been reported in Argentina, China, India, Sudan and many countries in Central Asia, where more than 16 million hectares of irrigated land are salinized" (ibid). The encroachment of saline seawater into fresh water aquifers causes further deterioration of water quality, as it causes the salinization of water in seaboard areas, especially in locations where groundwater extraction is immoderate (ibid).

Ecosystems dangerously impaired by degraded water can directly affect human health, well-being and livelihood. It is noted that: "Every year 1.8 million people die from diarrhoeal disease attributable to unsafe water or poor sanitation and hygiene" (UN Water 2011:3); "every 20 seconds, a child dies as a result of poor sanitation" (UN Water 2013c); "2.5 billion people still lack improved sanitation. 1.1 billion people still practice open defecation, that's 15% of the world population" (ibid). Managing water quality is thus an immediate and crucial priority for vast numbers of the world's population.

Of particular concern is the fundamental role that women perform in water management and sanitation; this has been internationally recognized ever since the 1977 United Nations Water Conference; the inauguration of the Drinking Water and Sanitation Decade (1981-1990); and the 1992 International Conference on Water and the Environment (UN Water 2006:1). In February 2003, the Interagency Task Force on Gender and Water (GWTF) was established; an outgrowth initiative stemming from the Interagency Network on Women and Gender Equality (IANWGE). In

2004, IANWGE was assimilated into UN Water, it became one of UN Water's taskforces; and was designated the duty "to take responsibility for the gender component of the International Water for Life Decade 2005-2015" (UN Water 2013k). Gender sensitive water and sanitation policies are now key programme areas within UN Water. Indeed, each of the aforementioned international events paid explicit recognition to the crucial role of women in the provisioning and safeguarding of water (UN Water 2006). More recently, the International Decade for Action "Water For Life" 2005-2015 has stated that: "It has become increasingly accepted that women should play an important role in water management, and that this role could be enhanced through the strategy of gender mainstreaming" (UN Org 2013). Gender mainstreaming refers to: "a strategy for making women's, as well as men's, concerns and experiences an integral dimension of the design, implementation, monitoring and evaluation of policies and programmes in all spheres so that women and men benefit equally" (ibid). Studies corroborate the idea that women's involvement in the planning of water quality management strategies can increase their effectiveness (UN Water 2006). To this end there has been an increase in the number of women appointed to international policy forums that have water quality management as their focus: e.g., female membership of the African Ministerial Council on Water (AMCOW); the African Ministers Initiative on WASH (Water Sanitation and Hygiene); and the Water Supply and Sanitation Collaborative Council (WSSCC) (ibid:2). UN Water (2006:2) states: "These leaders constitute the critical mass needed to have gender integrated into water and sanitation policies and programmes. Working closely with these dynamic women leaders is important for advancing a gender perspective at global and national levels during the "Water for Life" Decade, and developing a network on gender and water" (ibid).

It is difficult to challenge women's "primary responsibility for water management" as a basis for the involvement of women in international water quality management initiatives. Nevertheless, it is imperative that such forums critically reflect and explore the wider picture of why women in the developing world are ascribed the responsibility for the provision of water quality; furthermore, it is necessary to ask economic sociological questions about global capitalism and its expropriation of the labour expended in the provisioning of water. It can be argued that the international focus on women and water sustainability has tended towards a first generation feminist response: emphasizing equal access to the policy platforms of international organisations. From a second wave feminist perspective, the inclusion of women into the decision making arenas of international organisations, falls far short of challenging the material structures that systematically position women as responsible for water provisioning, management and sanitation. Indeed, Marxist ecofeminists have problematised the equal opportunities institutional approaches to eco-politics; for it arises "from the modernist industrial division of labor and its inevitable knowledge fragmentation" (Salleh 2003:61). Steeped in modernism's segmentation of daily life, it is easy to marginalise the tacit experiential knowledge of capitalist accumulation, and the environmental degradation of water ecology. Marxist ecofeminists can provide an important initial approach to the development of Marx's ecology, so as to conceptually formulate an account of feminism, global capitalism and the environmental degradation of water.

In *Patriarchy and Accumulation on a World Scale: Women in the International Division of Labour*, Maria Mies (1998:xi) uses a highly illuminating metaphor to explain the material limitations that women experience in capitalist regimes of accumulation. Mies (ibid) argues that capitalism is very different from the gender indifferent abstractions of Marxism, in which their bourgeoisie versus the proletariat divide simplifies the inter-sectionality of gender, race and class in capitalist exploitation. Capital accumulation and its insatiable pursuit of global economic domination are only "possible as long as huge areas of human and non-human production" are colonized (ibid). Axiomatic to this hegemonic economic agenda has been the extension of colonial pursuits to "Women, nature and the people and lands of Africa, Asia and Latin America" (ibid). It is evident that the colonisation of these terrains was not only inter-connected but also formed an integral feature of capitalist accumulation: "the invisible underground foundation for this accumulation process" (ibid). Mies (ibid) adopts the metaphor of the iceberg to illustrate the relation of paid labour to unpaid labour in capitalism's international division of labour. The tip of the iceberg is supported by multitudes of colonised human and non-human productive resources. As Mies (ibid) expresses it: "We used the metaphor of an iceberg, where capital and waged labour form the visible economy, 'above the water', counted in the GDP, where waged labour is protected by a labour contract, and where housework, work in the informal sector, work in the colonies and nature's production form the underwater part of this economy". In this formulation, can be conceived an initial approach to theorising economically a feminist account of women and water provisioning, along with capitalist accumulation. Central to this formulation is Ariel Salleh's (2003) sociological conception of "meta-industrial labour".

In an article entitled "Ecofeminism as Sociology", Salleh (2003:63) challenges classical Marxist ecology. Of initial concern is Marx's reference to nature as "inorganic". In the context of a feminist geopolitical ecology of water, Marx's depiction of nature as "inorganic" suggests that an analysis of capitalism and water sustainability from a Marxist perspective, will presuppose water as inorganic. Thus a first step in a feminist geopolitics of water, involves revitalising the material status of nature in Marxist ecology; this involves adopting an eco-centric approach to ecology. Further to this initial step, a Marxist ecofeminist approach to water ecology needs to critically engage with Marx's "division between productive and reproductive labour" (ibid: 2001:2). Of particular concern is the neglect, in classical Marxist analysis, of reproductive labour as a feature of the dialectical tensions that are evident "between forces of production versus external nature" (2003:63). In the 1970s, second wave feminists formidably critiqued the emphasis on paid labour in Marx's labour theory of value. It was argued that reproductive labour is integral to the conditions enabling the expropriation of surplus value from paid labour. Ecofeminist praxis extends this earlier feminist critique "by integrating ecopolitical concerns – equality, cultural diversity, and sustainability" (ibid:64). To this end, and following the domestic labour debates of the 1970s, materialist ecofeminists have identified how gender inequality and women's reproductive labour are inextricably correlated with the forms of subsistence labour that are integral to the provisioning of natural environments (Salleh 2010:208). As may have already been discerned, a feminist Marxist ecology of water is premised on the assumption that women's domestic labour in the provisioning and management of water is inextricably affected by water quality. Indeed, it is my contention that the water provisioning carried out by women, forms part of an unpaid substratum of water management integral to capitalist profit. Some basis for this contention is provided in Salleh's (ibid) conception of "meta-industrial labour". Salleh (ibid) has formulated the concept of "meta-industrial labour" to accentuate forms of labour that are one and the same, time external and "indispensible intermediaries of ecosystemic flows" necessary to the maintenance of capitalism. (ibid:211). As Salleh states: "The term meta-industrial labor denotes workers, nominally outside of capitalism, whose labor catalyzes metabolic transformations – be they peasants, gatherers, or parents" (ibid:212). The concept of meta-industrial labour provides a highly valuable lens through which to re-examine and critically comprehend the actualities of the water provisioning that is carried out by women, and which is widely documented in the water reports of international organisations. Consider for example the following statements:

> Pollution of our rivers places billions of thirsty, hungry urban families that live downstream in danger. The dialogue on wastewater management and water quality confirmed that dilution of pollution was no longer a solution. Parties discussed how prevention, reduction, or removal of pollution to be both possible and profitable. They reconsidered the use, reuse, value and even meaning of "waste" water. As more than half of humanity lives in cities, contaminated discharge and surface runoff spread water-borne disease among billions. While real solutions yield high returns, the urban poor who need them most generate almost no taxes or influence; meanwhile politicians rarely invest if direct costs are immediate and indirect benefits appear only after they leave office or help those at a distance. Downstream, the impacted billions who depend on food from marine ecosystems could look upstream to slow, stop, or reverse nitrogen and phosphorous loads in the 90% untreated urban wastewater. Rewards could go beyond resilience to boost growth, jobs and business certainty. But scale matters. Reuse options must relate to local circumstances, cultural norms, safety of use, awareness, and capacity. (UN Water 2013j:5)

The above statement derives from a recent UN Water (2013j) report entitled *The Post 2015 Water Thematic Consultation*. The report links water quality initiatives to political conditions when it states that: "To address impacts and build capacity requires political will. Alas, politicians rarely act to provide wastewater infrastructure because often benefits are felt after their 'political time' and are most visible downstream" (UN Water 2013j:19). Economic conditions are also highlighted as being of significance to the initiation of water quality management programmes:

> Water pollution and poverty often go together. This often means that those people who are most affected by bad waste-water management generate almost no taxes and even less influence. Poverty traps the poor in a vicious circle of degradation and disease. Yet urban areas represent an opportunity in disguise, if leaders can turn liabilities to assets, and introduce sustainable approaches with judicious investments. (UN Water 2013j:19)

The conditions favouring the establishment of sustainable approaches to water quality management are predicated on commercial viability and profit incentives; this in turn raises disconcerting issues of limitations in capitalist investment in both the development and innovation of water quality. Further to this issue, the UN Water (2013j:4) report identifies that limitation in the universal access to clean water, sanitation and hygiene "costs 1.5% to 4.3% of GDP, stunts childhood growth, drains women's time and energy, empties school chairs, forces needless risks, and denies human dignity". Women and girls are disproportionately burdened with the responsibility of water provisioning, and it is the central contention of this chapter that capitalist accumulation benefits from the unpaid "meta-industrial" (Salleh 2010) water labour of women. The theoretical basis for this contention is informed by Rosa Luxemburg's (1913/2003) *The Accumulation of Capital*, and more specifically the ecofeminist developments expressed in the work of Ariel Salleh, Mary Mellor and Maria Mies who describe various accounts of "meta-industrial" labour. All three of these ecofeminists consistently advance the view that capitalist production produces a condition that "undermines its own social metabolism" and in so doing precipitates "a metabolic rift" between itself and the surrounding natural environment (Salleh 2010:205). Indications of metabolic rift are clearly evident in water scarcity; degradation of water; and water pollution: "By 2025 1800 million people will be living in countries and regions with absolute water scarcity, and two-thirds of the world population could be under stress conditions"; (UN Water 2013f); "Approximately 3.5 million people die each year due to inadequate water supply, sanitation and hygiene" (UN Water 2013). Marxist ecofeminists convincingly argue that, in capitalist regimes of accumulation, metabolic rift is traversed by expropriating unpaid labour. In an article entitled "From Metabolic Rift to Metabolic Value", Salleh (2010) identifies those engaged in reproductive labour as essential "intermediaries of ecosystemic flows and human aspirations – a gestation that is received globally in denial, a 'symptomatic silence' to re-coin an old 1970s phrase" (ibid:211). Reproductive labour is transformative, thus Salleh (ibid) ascribes the prefix *meta* to denote upscaling processes, which in turn create alterations in substantive materials. Meta processes applied to organic and inorganic matter metabolise productive change and thus create value. Salleh (ibid) ingeniously situates the concept of "metabolic value" as a coetaneous determinant of labour value, whereby: "The term metabolic value denotes the value sustained and enhanced" by workers who are ostensibly marginal to capitalist economic relations of work, and yet at the same time integral "in supporting ecological integrity and the social metabolism" (ibid:212). Salleh conceives this form of work as resonant with the organic productive rhythms of the metabolic exchange with nature. It is often suggested that provisioning water, as part of reproductive labour, assumes a temporality resonant with the natural processes of the metabolic exchange with nature. Indeed, the UN's World Water Development Report WWDR4 (2012:109) describes how: "Rural women often rely upon common water resources such as small water bodies, ponds and streams to meet their water needs, but in many regions these sources have been eroded or have disappeared due to changes in land use, or have been appropriated by the state or industry development needs, or to supply water to urban areas". Of significance is the supposition that the meta-industrial labour of rural women involved in the provisioning of water, is technically elementary; often assuming candid, intuitive, embodied skills. Emblematic of the meta-industrial labour to which Salleh refers is the primitive management of water quality through the provisioning of river water supplies. Indeed, the study by UN Water (2006) *Gender, Water and Sanitation: A Policy in Brief* describes how:

> Indigenous people possess traditional knowledge and skills concerning the sensing/locating of water and protection of the source. Water sources on indigenous lands are often considered a sacred, and element, and indigenous women may be the holders of 'water knowledge'. Their traditional land management skills often provide the most effective method of water resource management in their settlement areas. However, indigenous people are seriously affected by their uncompensated and unsustainable loss of water to farming and other industries introduced from outside of their communities. (UN Water 2006:8)

When imagining the improvised, unsophisticated requisites demanded of this form of venture it is appreciable that:

> Meta-industrial workers translate principles learned hands-on in the material world – a vernacular epistemology, cognizant of cross cutting synergies in living processes, and replicating these thermodynamic circuits of nature. This labor is relational, flow orientated, and regenerative of biotic chains. Its unique rationality is a capacity for economic provisioning in a way that preserves metabolic value as it goes. Unlike the extractive mode of production with its inevitable metabolic rift, this economy is eco-sufficient without externalized costs in social, ecological, or embodied debt. (Salleh 2010:212)

Salleh's presuppositions regarding the processual "relationality" of meta-industrial labour imply a task orientation conception of time in which labour resonates with the "living processes" rhythms and the "biotic" duration of harmonious metabolic transformations (2010:212). Suffice to say that the quantitative chronometric time of privatised water supply is experientially in discordance with the processual nature of meta-industrial labour, and yet these two forms of time co-exist. This is because primitive accumulation is a condition of global capitalist accumulation. Consequently, the reproductive labour of the "nonmonetized domestic sphere" and "indigenous cultivators" of the developing world's meta-industrial labourers "supplies metabolic value to capital" through their capacity to undertake metabolic transformations that are crucial to capitalist accumulation (ibid). As Salleh expresses it:

> Meta-industrial labor exists concurrently with capitalism and even so called dematerialized production cannot do without it. (ibid:212)

Mies (1998) contends that the international division of labour, as the basis for capitalist accumulation, depends upon the exploitation of meta-industrial labour. Mies commences from the premise that the international division of labour, in the underdeveloped nations has not consistently enabled these countries to control the established industries. Rather, it is evident that the international division of labour has actively exploited labour markets, in the Third World, alongside their primary resources, in order to advance the relentless expansion of global capitalism. Mies (ibid) introduces into this debate the convincing contention that the progression of the international division of labour has been towards "the integration of Third World women's work within the global market economy" (ibid:114). Within the field of agriculture, the integration of women's work into the world market encompasses "family labour", in which women's unpaid subsistence labour subsidises the production of goods for market. Mies (ibid:115) provides examples, amongst which include: "'family labour' in small peasant units which produce independently or on a contract basis for agro-business firms"; "women working as unpaid 'family labour' within cooperatives which produce for export"; "women working as casual labour in commercial agriculture". Mies' (ibid) observations and examples have significance to the formulation of ecofeminist geopolitics of water ecology; for it appears that the water provisioning responsibilities, disproportionately ascribed to women, constitute a proportion of the unpaid Third World women's labour that is increasingly incorporated into the world economy. Indeed, consider the following account of women and subsistence farming:

> In the developing world, men tend to focus on market-oriented or cash crop production, whereas women work with subsistence agriculture, minor crops and vegetable gardens, and often grow a wider diversity of crops...Consequently, the use and management of irrigation water is likely to be different for men and women farmers...For example, in gender-based farming systems in sub-Saharan Africa, where men and women often

> cultivate separate fields…this reality has often been ignored in irrigation projects. As a result, while men could irrigate their cash crops, women did not have access to irrigation systems for vegetable gardens and subsistence crops. (WWDR4 2012:748)

According to Mies (1998:118), an integral component of modernization in the developing world has involved revaluating labour in the domestic sphere; and defining this activity within the ideology of the "housewife". As part of this process of universalising the housewife doctrine, the subsistence labour conducted by women, in the third world, becomes amalgamated with the salaried income of the head of the household; and is thus valued as a subsidiary to that income. As Mies (ibid) states: in "the model of the nuclear family as signs of progress, it is also possible to define all the work women do – whether in the formal or informal sectors – as supplementary work, her income as supplementary income to that of the so-called main 'breadwinner', the husband". Disconcerting outcomes of this nuclear family model can be observed in the above stated extract provided by WWDR4 (2012:748). The extract asserts that subsistence farming carried out by women is neglected when designing agricultural irrigation infrastructure. Thus, women involved in subsistence farming have to trek further than commercial farmers when collecting water for farm irrigation. Capitalism and its global market economy further exploits these arrangements through the construct of the family wage; whereby the unpaid food, farming and water provisioning carried out by women forms the basis for capitalism's calculation of the family wage, along with women's rights as stakeholders in land reform. As Mies (1998:119) describes: "The capitalist sexual division of labour … is the strategic principle which is responsible for the fact that women …are not defined and paid as wage-workers…in land reform provision they are not given independent and legal ownership of land, that they do not get access to other productive property, that in cooperatives they are often mere appendices". Suffice to say, unpaid subsistence work carried out by women in the Third World has increasingly gained relevance in capitalism's exploitative calculations of the family wage. Defining women's water provisioning, along with subsistence production and reproductive labour as housewife income supporting activities, denies many women in the Third World rights as workers. It is precisely this deeming of women's unpaid subsistence work which makes primitive accumulation a co-existent feature of capitalist accumulation, for it undermines women's land rights and precludes challenges to the privatisation of communal resources. Consequently, water privatisation, or its acquisition through enclosure, severely impacts upon time dedicated to the collecting of water; It has thus been observed that "women in sub-Saharan Africa spend 40 billion hours per year collecting water" (WWDR4 2012:584); this situation is partly a condition and consequence of a situation in which "women hold title to less than 2 per cent of the world's private land" (UN Water 2006:4); capitalist accumulation gains from time impoverished provisioning of water because it is unpaid labour irrespective of its duration. In the following section, these concerns are developed further.

Timely Reflections: The Nature of Capitalism's Ecological Disjuncture

In *Capital Vol.1.,* Marx presents the evolution in the labour process as signifying, not a metabolic equilibrium with nature, but an ecological disjuncture with the natural environment. This is particularly evident in capitalist commodity production, a system in which labour becomes a commodity i.e., "the capitalist buys labour-power in order to use it, and labour power in use is labour itself" (*Capital Vol. 1,* 2003:173). The intrinsic worth of labour is thus objectified into the fetish of the equivalences of commodity exchange. But it is also evident that "in order that his labour may re-appear in a commodity, he must, before all things, expend it on something useful" (ibid). Consequently the labourer is compelled to convert his or her natural propensities into a mere instrumental apparatus of material survival i.e., "in degrading spontaneous activity, free activity, into a 'means', estranged labour makes man's life a means to his physical existence" (1844/2007:76). The full potential of human beings resides in emancipating their natural productive propensities to creatively transform the world through self-enriching alienation. Central to the realisation of this freedom, is the development of material conditions in which associated producers govern metabolism with nature in a rational way. According to Marx, it is only in these circumstances of contiguity with nature that humans actualise their species' being:

> [S]ince for the socialist man the entire *so-called history of the world* is nothing but the begetting of man through human labour, nothing but the coming-to-be of nature for man...Since the real existence of man and nature has become practical, sensuous and perceptible – since man has become for man as the being of nature, and nature for man as the being of man – the question about an alien being, about a being above nature and man – a question which implies the admission of the inessentiality of nature and of man – has become impossible in practice. (Marx 1844/2007:113-114)

Conversely, in capitalist regimes of accumulation, workers are unable to satisfy their natural needs directly because the mode of production is privately owned. Consequently the labourer is forced to satisfy their natural needs through the mediation of the waged labour market. "Expenditure of wages on goods means that they are positing their needs as necessary – 'necessary needs' –within the alienated system of capital" (Fraser 2000:152). When one considers necessary needs to be "the form that the satisfaction of 'natural needs' takes in capital", it is evident that their need-form is formulated exponentially with the value of the worker's labour power (ibid). What is significant here is that, in capitalism, the satisfaction of needs forces individuals into exploitative mediations with the object-form of nature.

Ecofeminists perceive the exploitation of nature to be bound-up with a patriarchal ideology of domesticity, which has exploited and continues to exploit the reproductive labour of women. For women subsistence farmers in the Third World, patriarchal ideology combines with capitalism in the production of gender stereotypes that shape the design and planning of irrigation programmes. One outcome of the patriarchal ideology of domesticity is that "women are primarily housewives and mothers, while men are farmers and irrigators"; and thus irrigation programmes neglect to be informed by the agricultural needs and priorities of female subsistence farmers (FAO 2012: 4). It is no wonder that from this observation emerges the rallying revelation: that social oppression and the exploitative oppression of nature are commensurate. For ecofeminists, environmental degradation interconnects with "patriarchal violence against women, other people and nature" (Mies and Shiva 1993:14). In the context of water ecology in the Third World: "Experience has shown that where irrigation design fails to recognize that women are water users, and farmers in their own right, risks are high that women lose existing access to land or the products of their own labour" (FAO 2012:4-5). And this raises important issues concerning "corporate and military warriors" and their "aggression against the environment"; for ecofeminists have perceived this "almost physically as an aggression against our female body" (Mies and Shiva 1993:14). This book aligns with the ecofeminist materialist perspective, for it formulates a theory of environmental degradation based on the materiality of capitalist exploitation with regards to the female body, and the contours of reproductive labour. From these convictions arises a distinct trajectory of ecofeminist geopolitical analyses of water ecology; each dedicated to challenging the conventional Marxist proposition that the history of capitalist development is equivalent to the dialectics of class struggle.

Figure 1.1a., "Investing form with lucid stillness, Turning shadow into transient beauty", (T.S. Eliot *Burnt Norton*). Photographic Image, South London, River Thames, September 2013

Figure 1.1b., "Investing form with lucid stillness, Turning shadow into transient beauty", (T.S. Eliot *Burnt Norton*). Photographic Image, South London, River Thames, September 2013

Figure 1.2a., "Ere young Ideas spring; to local Thought" (Ann Yearsley, *On Genius Unimproved*, 1787). Photographic Image, South London, River Thames, September 2013

Figure 1.2b., "Ere young Ideas spring; to local Thought" (Ann Yearsley, *On Genius Unimproved*, 1787). Photographic Image, South London, River Thames, September 2013

CHAPTER TWO

DISPOSSESSION: "ISOLATED WITH NO BEFORE AND AFTER" (T.S. ELIOT *EAST COKER*)

> In developing countries, women and girls bear most of the work burden associated with managing water and energy scarcity, fetching water for the 780 million people lacking access to improved sources of drinking water (WHO/UNICEF, 2012) and collecting firewood for the 2.7 billion depending on traditional biomass for cooking…This adds to their time and work burdens and seriously compromises their educational and employment opportunities, perpetuating the intergenerational transfer of poverty and disempowerment. Available statistics often fail to recognize or measure their real contributions to their economies and communities as unpaid water and energy providers. (United Nations Water 2014:21)

In 1978, the Water and Sanitation Program (WSP) was established through a multi-donor affiliation superintended by the World Bank "to support poor people in obtaining affordable, safe, and sustainable access to water and sanitation services" (WSP 2010a). A central contention of the WSP is that women and girls, in developing countries, are invariably responsible for the provisioning of water and the management of household hygiene. In the absence of domestic water supply, female members of the household are forced to travel arduous distances to collect water. Time poverty, therefore, is clearly a gendered dimension of water poverty. Indeed, according to WSP (2010:9): "Cumulatively, one estimate suggests that some 40 billion hours a year, are spent, collecting water in sub-Saharan Africa – equal to a year's labor for the entire workforce of France". WSP (ibid) advocates intervention by the water and sanitation sector so as to alleviate the time poverty experienced by women in the provisioning of water. They identify "time poverty" as "the situation where women's time is inflexible, consumed by routine and non-productive tasks, and perpetuating their absence from decision making and other profitable pursuits" (ibid). While in agreement with this description of time, gender and water poverty, further conceptual development is required to appreciate the scope and depth of its circumstances. The basis

Figure 2.1a., “Emptying the sensual with deprivation, cleansing affection from the temporal. Neither plenitude nor vacancy”, (T.S. Eliot *Burnt Norton*). Photographic Image, South London, River Thames, September 2013

for a geo-political formulation of water, gender and time poverty is evident in Marx's account of primitive accumulation and ecofeminist Marxist dialogue. In the postcolonial ecofeminist writings of Ana Isla (2009), can be found vociferous critiques of the modern day occurrence of primitive accumulation. Isla (ibid) convincingly identifies the enclosure of previously communal water resources as disproportionately affecting women. Focusing mostly on Latin America, Isla identifies gender segregations in the domestic provisioning of river water; for example, Isla's work concerning the Peruvian Amazon Basin identifies how "men gather at the river's edge before sunrise to fish and hunt, and women congregate there to wash the clothes, collect water for cooking, take baths, and bathe their young children" (ibid:25). Isla identifies how women are disproportionately affected by primitive accumulation in the capitalist exploitation of natural resources. As a Marxist ecofeminist, Isla concurs with Marx's observations regarding the historical conditions of primitive accumulation. However, like other Marxist ecofeminists, Isla disagrees with the contention that primitive accumulation is a mere precursor to the establishment of industrial capitalism. According to Isla (2006:61), "We cannot look at the ecological crisis as separate from the complex of social crises produced by economic relations". Referring to the work of Maria Mies, Isla further states that: "ecofeminists can no longer be satisfied with the Marxist argument that the 'first contradiction' is capital versus wages, which once resolved, would create conditions necessary to resolve secondary contradictions" (ibid). In the final section of this chapter, it is argued that the water and time poverty experienced by women, in the developing world, can be theorised through an ecofeminist advancement of Marx's account of primitive accumulation.

Water Time Poverty and Primitive Accumulation

According to World Water Development Report (WDR4 2012:34) the increasing shift towards the commercial acquisition of land can facilitate significant advances in the development of infrastructural resources. Commercial providers are increasingly attracted to landmasses with relatively abundant water resources. While water consumption can drive this commercial acquisitioning, the manufacture of water-intensive products has become a significant feature of the "virtual water trade" (ibid). However, there are also causes for concern regarding the "equitable distribution of benefits to the host country and its people" (ibid). A disconcerting feature relates to the fact that: "water is typically not explicitly mentioned in disclosed land deals. In the few cases where water is referred to, the amount of water withdrawals to be permitted is not specified" (ibid). Consequently, the indigenous rural inhabitants of lands where previously communal access to natural water resources was the tradition, are increasingly competing against corporate commercial companies for water use. These circumstances relate to Marx's concept of primitive accumulation.

It is a fundamental premise of Marxist ecology that primitive accumulation desecrated the rural economy of pre-industrial Western Europe; thus by the 19th century, the traces of an affinity between the agricultural labourer and communal investment in farmland had dissipated, as Marx passionately recounts: "[T]he robbery of the common lands, the usurpation of feudal and clan property, and its transformation into modern private property under circumstances of reckless terrorism, were just so many idyllic methods of primitive accumulation. They conquered the field for capitalistic agriculture, made the soil part and parcel of capital and created for the town industries the necessary supply of a 'free' and outlawed proletariat" (Marx *Capital Vol., 1* 1887/2003: 685). Accumulation through the dispossession of free farming proprietors enabled the transformation of this agricultural population into: "material elements of variable capital. The peasant, expropriated and cast adrift, must buy their value in the form of wages, from his new master, the industrial capitalist" (ibid:697). As capitalist agriculture begins to take hold of the mode of production, so it is that the "parcelling of the soil" gained universal application and accelerated the transformation of the previously scattered means of production into concentrated farming enterprises: "the pigmy property of the many into the huge property of the few, the expropriation of the great mass of the people from the soil, from the means of subsistence, and from the means of labour, this fearful and painful expropriation of the mass of the people forms the prelude to the history of capital" (ibid:714). It is significant that Marx identifies primitive accumulation as a prerequisite condition and thus a prelude to the establishment of industrial capitalism, as Marx expresses it: "A certain accumulation of capital, in the hands of individual producers of commodities, forms therefore the necessary preliminary of the specifically capitalistic mode of production…It may be called primitive accumulation, because it is the historic basis, instead of the historic result of specifically capitalist production" (ibid: 585). One gains here an insight into Marx's conception of the time of nature; for the acquisition of natural resources and their privation into enclosure are presupposed as a point in the linear dialectic of historical materialism. The relation of primitive accumulation to the advent of industrial capitalism is a paradoxical feature of Marx's ecology; for it is clearly evident that primitive accumulation is a continuous and co-existent condition of capitalist accumulation. In the context of water ecology and urbanisation, it is apparent that parallels linking restrictions on the communal access to water and privatisation of the commons are aspects of primitive accumulation. Numerous United Nations water ecology reports, published in recent times, substantiate this claim.

In 2011, the United Nations Human Settlements Programme (UNHSP), published its *Final Report World Water Day 2011: Water and Urbanization.* In a section entitled *Groundwater Management in Urban Areas,* the report outlines challenges and threats to urban ground water management – a crucial source of drinking water. Central to the report's findings is: "the vulnerability of groundwater resources in urban areas to over-exploitation, pollution and different types of flooding" (ibid:68). Groundwater is connected with elements of urban water systems, for it is an integral part of the hydrological and urban water cycles. As UNHSP (ibid:69) expresses it: "Groundwater plays an important role in sustainable water management in cities by providing a vital source of water supply for different uses and sustaining certain types of aquatic and terrestrial eco-systems". Groundwater depletion is also an indicator of water scarcity, as UN Water (2007:4) describes: "Symptoms of water scarcity include severe environmental degradation (including river desiccation and pollution), declining groundwater levels, and increasing problems of water allocation where some groups win at the expense of others". Water scarcity further exacerbates the disproportionate responsibility that women are ascribed in the provisioning and management of water. Consequently, water scarcity has been associated with increased time poverty as women find themselves having to spend arduous hours trekking longer distances to collect water. It has been suggested that: "Many women suffer permanent skeletal damage from carrying heavy loads of water over long distances day after day" (ibid:20). According to UNDESA (2013), "In Africa, 90% of the work of gathering water and wood, for the household and for food preparation, is done by women". Limits on access and the ownership of land, by women, appear to be significant determinants and conditions of the time water poverty experienced by women in the developing world. UN Water (2006:4) in its *Gender, Water and Sanitation: A Policy Brief* identifies that: "women hold title to less than 2 per cent of the world's private land. Moreover, even where women do have a legal right to land, customs often prevent them from taking de facto control of said land and natural resources, such as in Burkina Faso, Cameroon and Zimbabwe" .

Postcolonial ecofeminists embrace the fundamental interrogation of capitalism and nature in the context of the Global South. As Steady (2009:56) expresses it: "Feminists, primarily from the Global South, have argued for a more fundamental transformation in relations of production, to eliminate the gross social and gender inequalities that economic exploitation produces". An issue of compelling pertinence to this debate relates to the structural reforms instituted by global financial institutions as part of the aid provided to developing countries. Global markets for the international importation of water have relevance to postcolonial ecofeminist concerns regarding "conservation as enclosure"; "debt-for-nature" bio-piracy, and the expropriation of the meta-industrial labour that is traditionally carried out by women. For example, according to WWDR4 (2012), the economic solvency of many Latin American countries relies on the export of their natural resources; water has become a key to the global demand for natural resources. Part of this demand has been propelled by the development of competing markets for water, and the growth of global markets for the export of "virtual water" (ibid: 9). This term generically refers to the fact that water is an integral element of all natural foods and inorganic commodities, and specifically relates "to the volume of water used in the production of a good or service" (ibid: 33). Increases in the production of food and goods have resulted in an increasing expenditure of embedded "virtual water". Climate change and the excessive exploitation of sustainable water management have led to a situation whereby capitalism is scouting for water resources globally; in essence, whereby capitalism is chasing water across the surface of the globe. Trade in virtual water involves countries in the commercial transaction of water that is embedded in products and food: "As a result, billions of tonnes of food and other products that require water to produce are traded globally" (ibid:33). Countries that are water scarce, including numerous nations in the Middle East, are steadily becoming "net importers of virtual water, relying on the importation of agricultural commodities to meet the food needs of their growing populations" (ibid). Additionally, Western consumer capitalist countries have also become net importers of virtual water that is embedded in the production of consumer goods (ibid). For these net importers of water, the trade is economically profitable and cost effective. However, in global capitalism, equitable exchange relations are the first casualty of profitability. Indeed, it has been observed that:

> Although global water trade can lead to significant national water savings, through virtually transporting water from a place of relative abundance to a place of relative scarcity, trade alone cannot guarantee sustainable water management at the source. Indeed, as virtual water-exporting nations increasingly respond to global demand, the dimensions of responsibility for sustainable water management are subsequently elevated to a more complex and supranational relationship between consumer and producer. (WWDR4 2012:33)

The virtual water trade clearly raises concerns regarding the inevitability of inequitable relations arising between water abundant territories/producers, and cash rich consumer territories; for the latter countries can gain from these exchanges in circumstances where "they can afford the infrastructure to harness the water and there are no artificial barriers in international trade" (WWDR 2012:34). It is evident that these circumstances are exacerbated by the globalization of neoliberal development programmes that require Third World countries to liberalise their trade markets as part of their debt repayment conditions (Odih 2007). Postcolonial Marxist ecofeminists extend this argument so as to frame the exploitation of natural resources. A key contribution to this debate is Ana Isla's (2009, 2005) analyses of the debilitating consequences of the World Trade Organization's imposition, onto the Third World, of structural adjustment programmes which obligate partnerships between private sector financiers, indebted nations and environmental conservation agencies. Conditions that have precipitated these arrangements converge towards the late 1980s, when loans ordinarily provided by the International Monetary Fund (IMF) and the World Bank began to be substituted for listings of commercial portfolio funds and "debt swaps" (Isla 2006:51). The latter paved the way for the establishment of debt-for-nature swaps. These deals, popular with the World Bank, the IMF, UNESCO and international environmental development organisations, have become "an instrument of policy following the United Nations Conference on Environment and Development (UNCED) headed by Maurice Strong, in 1992" (ibid). The finance aide mechanism of debt-for-nature swaps operates by rewarding participating countries for conserving their natural environment. These rewards invariably entail writing–off a proportion of the country's debts in exchange for their demonstrated commitment to designated sustainability programmes. Concern arises, however, where debt-for-nature swaps enrol debt ridden developing countries into iniquitous bio-diversity exchange relationships. Isla (ibid:52) provides detailed accounts of Latin American countries that have been required to capitalize on their bio-diversity so as to establish ecotourism as part of sustainable development programmes;

and more sinisterly debt-for-nature swaps have provided commercial avenues for "non-renewable resource extraction, such as mining" (ibid). Primitive accumulative enclosures of the land are a further disconcerting feature of debt-for-nature swaps; in these, programmes areas are closed-off to form Conservation Areas, resulting in the cessation of the bio-diverse metabolism between humans and nature, as well as the sequestering of natural resources (ibid). It is evident that these enclosures impact on access to water resources and the sustainability of aquatic eco-systems. Isla (ibid:55) identifies how, in the wake of debt-for-nature swaps in Costa Rica, it became illegal to hunt and fish in previously communally owned areas. The resulting confrontation involved: "Hunters and fishermen strongly [defending] their rights to the land and water resources on the grounds that they have been doing this forever without harm to the land" (ibid).

Isla (2009) identifies how the enclosure of previously communal water resources disproportionately affects women as part of traditional gender segregations in the domestic river orientated tasks of rural villagers in Latin America. For example, Isla's work concerning the Peruvian Amazon Basin identifies how "men gather at the river's edge before sunrise to fish and hunt, and women congregate there to wash the clothes, collect water for cooking, take baths, and bathe their young children" (ibid:25). Latin American waterways are also the focus of my empirical research into gender and waterways provisioning. The following provides a case study of a river cleanup in Costa Rica's Rio Virilla. The case study highlights international features of the domestication of environmental water ecology issues.

Case Study: Waterways in Latin America

> Este pásado 13 de marzo, varios voluntarios tuvimos la oportunidad de darle un respiro al cañón del Río Virilla recogiendo aproximadamente 30 mil botellas de plástico que se encuentran a las orillas de este río, uno de los ríos más contaminados de Costa Rica y de Centroamérica. Gracias a los organizadores de "Montaña de esperanza" la ayuda y apoyo de la Muncipalidad de Mora, la empresa recicladora Servicios Ecologicos, la asistencia de la Cruz Roja y grupos como los scouts de la zona, un equipo de montañismo y todos los demás voluntarios que asistimos al evento, se logró realizar un trabajo de limpieza significativo en la zona de Ciudad Colón donde se encuentra este aglomerado de botellas plásticas en el río Virilla. Esfuerzos como estos, que aunque parezcan muy pequeños comparados con la cantidad de basura que se acumula en muchos ríos alrededor de Costa Rica, demuestra la voluntad de muchas personas y

organizaciones por ayudar a combatir el problema de la basura y educar a la sociedad con respecto a la importancia que tiene el conocer sobre lo que es basura y lo que no es basura, o sea, lo que se puede reutilizar y reciclar. Luego de la recolección de miles de botellas de plástico, la empresa Servicios Ecológicos las lleva a la planta de procesamiento en donde las botellas llevan su debido proceso de limpieza y reciclaje. Este es apenas el primer día de recolección de botellas de plástico en el río Virilla, quedan miles más por recoger. (Reciclando En Espiral Costa Rica 2010)

Figure 2.2. "Volunteers Remove Plastic Bottles from the Banks of the Polluted Waters of Virilla River in Ciudad Colón , about 55 kilometers South of San Jose on April 10, 2010" (Getty Images). Photographic Image by: Yuri Cortez/AFP/Getty Images, Created 10 April 2010, Editorial Image: 98394301. Permission of Getty Images.

This past March 13, volunteers had the opportunity to relieve Virilla River canyon of about 30,000 plastic bottles that were found on the banks of this river, one of the most polluted rivers in Costa Rica and Central America. Thanks to the organizers of 'Mountain of Hope', and with the help and support of the Muncipalidad of Mora, the recycling company Ecological Services, the assistance of the Red Cross and groups such as scouts in the area, a team of mountaineers and all the other volunteers who attended the event, we were able to undertake a significant cleanup operation in the area of Ciudad Colón where [there was a] cluster of plastic bottles in the Virilla River. Efforts such as these, that even if they seem very small compared to the amount of trash that accumulates in many rivers around Costa Rica, demonstrates the willingness of many people and organizations to help combat the problem of garbage, and educate society about the importance of knowing what is trash and what is not trash, that is, what can be reused and recycled. After the collection of thousands of plastic bottles, the Ecological Services Company takes them to the processing plant where the bottles undergo the due process [of] cleaning and recycling. This is just the first day of collecting plastic bottles in the Virilla River; with thousands more still to be collected. (Reciclando en Espiral, Costa Rica 2010)

Figure 2.3., "According to Environmentalists the Virilla River of Costa Rica is One of the Most Polluted in Central America" (Getty Images). Photographic Image by: Yuri Cortez/AFP/Getty Images, Created 10 April 2010, Editorial Image: 98394314. Permission of Getty Images.

Campana De Limpieza De Rios Costa Rica Te Quiero Limpia. El Ministerio de Salud está apoyando una campaña de Limpieza de ríos denominada Costa Rica te quiero limpia, que va por capitulos o misiones. La primera misíon es la limpieza del Río Virilla en diferentes tramos del cauce, especialmente en aquellos donde hay mayor presencia de residuos. Ya se han realizado dos vistas a sus márgenes, la primera en Atenas el pasado 20 de marzo y el sábado 24 de abril en la Represa Brasil en Santa Ana. Está siendo organizada por Teletica con el apoyo de la Asociación ambientalista TerraNostra y cuenta con el patrocinio de empresas como Banco Nacional, FEMSA Coca Cola, WPP Y ostras empresas que donan la

> alimentación, guantes, bolsas, el transporte al sitio, entre ontros. (Ministerio De Salud Costa Rica 2010a:1)
>
> Cleaning Campaign Costa Rica Rios Love You Clean. The Ministry of Health is supporting a campaign called Clean Ríos Costa Rica.... The first cleaning mission is Río Virilla in different sections of the channel, especially where there is a high proportion of waste. (Ministerio De Salud Costa Rica 2010a:1)

In March 2010, Costa Rica's Minister of Health, Dra. María Luisa Ávila Agüero, participated in a river waterways cleanup event situated at Rio Virilla in San José; estimates suggest that approximately 2 tons of solid waste were extracted as part of the three day event (Ministerio de Salud Costa Rica 2010:1). Rio Virilla streams through Costa Rica's Central Valley, and drains into Rio Tárcoles, forming a confluence into the Pacific Ocean. Unfortunately, the Rio Virilla is the recipient of much of the Central Valley's urban and industrial waste; consequently, it is one of the most contaminated rivers in Latin America (UNEP 2000a:38). In its assessment of the *Eastern Equatorial Pacific*, UNEP (2006:35) have stated that: "Agricultural run-off is the main source of chemical pollution, caused by the widespread use of pesticides in agriculture, especially for export production. Pesticide application (2 kg/person) is one of the highest in Latin America ... An estimated 250 000 m³ of liquid residuals are deposited daily in the Virilla River (Grande de Tárcoles upper basin) which take only 24 hours to reach the Gulf of Nicoya". The Rio Virilla river cleanup event was conducted in March 2010 at the place where Rio Tárcoles intersects with Rio Virilla. Over 200 volunteers were involved in the initiative, and volunteers came from a multitude of communities, ranging from private enterprise to the public transport sector. The event was comprehensively coordinated, for it commenced at 7:30am and finished 6 hours later at 13:30pm with operational communications provided by attendants and patient medical care led by the Costa Rican Red Cross (Ministerio de Salud Costa Rica 2010:1). Private finance enterprises, such as the National Bank of Costa Rica, featured significantly amongst the volunteers and gave financial sponsorship. Sanitizers and anti-bacterial gel were donated by commercial companies such as Red Dot and Zepol laboratories (ibid:2). The arrival of the volunteers was facilitated by the vehicle donations of transportation companies; and in conjunction with the volunteers' own vehicles were used to transport the extracted waste materials away from the foreshore of Rio Virilla to be recycled. Reports of the clean-up event agree that it was a quantitative success: volunteers are estimated to have extracted the equivalent of 145 garden bags of refuse; vast quantities of recyclable materials were also extracted and have reportedly been donated to workshops for adults with learning disabilities; alliances with recycling companies provided lucrative avenues for the donation of the colossal numbers of plastic bottles extracted from the river. According to the Ministerio de Salud Costa Rica: "Se pretende realizar esta iniciativa al menos una vez al mes, con el objetivo de eliminar la gran cantidad de residuos sólidos que se encuentran, actualmente, en este rio y sus alrededores" ("the initiative is intended to be performed at least once a month, with the goal of eliminating the large amount of solid waste that is currently in this river and its surroundings") (ibid:2). Elsewhere, Matamoros et al (2012), in *Heroes Del Reciclaje* have described the promotional imagery employed in the Rio Virilla cleanup, in addition to other river conservation events in Costa Rica. Matamoros et al's (ibid) focus on the comedic effect of satirical advertising aesthetic has merit, but their analysis is uncritical of the structured inequalities through which river conservation can be expropriated and transformed into meta-industrial labour. Moreover, the social-psychological basis of Matamoros et al's (ibid) consumer behaviour analysis focuses excessively on agency at the expense of social structure, and in so doing decontextualizes sign-making from the social-political, economic, and social structural context in which actor networks (e.g., advertisers) are constantly engaged in complex and labyrinthine processes of meaning transfer. Furthermore, Matamoros et al's (2012) highly descriptive focus on advertising parody is far removed from a systematic critical analysis of advertising as technologies of the sign and the effect of advertising as powerful inscriptive devices that render the social world knowable, and thus amenable, to governmental regulation. Conversely, the case study presented here provides an analysis of the Rio Virilla in terms of gendered meta-industrial labour, media communication and the geopolitics of river conservation.

Social Mediated Collective Action

Social media featured significantly as a means of rendering meaningful the cultural political issue of Rio Virilla's rapidly degenerating water ecology, and social media was prominent in mobilizing a local communal response to this environmental crisis. Of prominence were the multimedia advertising postings provided by the ecotourism company Wild Water Tours. During its period of operation, Wild Water Tours Costa Rica branded itself as an eco-tourism provider that was dedicated to environmental issues; it would communicate these issues via educational campaigns and collaborative ventures with non-governmental agencies. In March 2010, the eco-tourism company Wild Water Tours Costa Rica published a multimedia documentary entitled: *Montaña de Esperanza Rescate del Rio Virilla* (Mountain of Hope's Rescue of the Virilla River). Written by Monica JF, the account of Montaña De Esperanza's crusade to rejuvenate the Rio Virilla's natural ecology is as convincing as it is intriguing. Consistent themes running through the multimedia documentary appeal to humanity in general, and emphasize the imperative for human beings to mobilize agency in the unquestionable endeavour of addressing the perils of environmental degradation. Rhetorical strategies are utilized effectively so as to suggest that, elsewhere, the culture industry's incorporation of ecological messages dilute their gravitas: "Saving the planet under the 'recycle, reuse and reduce' slogan is very familiar to everybody but in many cases that is just a simple catching phrase since unfortunately many people don't practice it" (Monica JF 2010). The rhetorical organisation of the documentary attempts to further illustrate this issue with reference to comedic advertising aesthetics, designed to use humorous visual metaphors in their lighthearted appeal; but these apparently were less than successful. Thus the multimedia text critically states that: "Despite multiple environmental conservation campaigns and other movements that used funny characters like 'Jacinto Basurilla' and 'Yoyito', to raise awareness and create a different attitude in people towards garbage disposal, the problem still persist" (Monica JF 2010). Elsewhere the rhetorical organisation of the text builds into a series of self-affirming claims pronouncing the inadequacies of previous attempts to ally culture with environmental issues, as is evident in the following extracts:

> Even though many organisations have tried to solve the problem, their efforts have been in vain, as Costa Ricans keep throwing garbage everywhere. The dimensions of the pollution problem were witnessed by Montaña de Esperanza, an organization that exposed the litter 'crisis' [that] the Rio Virilla course is experiencing in Rodeo area in Ciudad Colón through a news report broadcast on the TV programme '7 Días'. This group of young entrepreneurs and nature lovers who identified with the environment decided to propose a serious solution, achievable and integral to the litter problem in that area. In conjunction with the Mora Municipality, the recycling company Ecological Services, [the] Red Cross, some private companies such as Wild Water Tours and volunteers in general, the Montaña de Esperanza coordinators drafted a plan for solid waste collection consisting of at least four different dates, three of them aimed for adults and one for children. The main objective of this activity is to create awareness among the Costa Rican population. 'The solution to

> this problem is a shared responsibility; we understand this is an issue of awareness, advocacy and education. To educate the population, the government and private companies, must first integrate all the actors and prompt an action where everyone is committed', said Felipe Castro, coordinator of the Montaña de Esperanza project …About 100 volunteers were able to collect approximately 30,000 plastic bottles. These bottles were taken to the Ecological Services recycling plant to be recycled. (Monica JF featured in Wild Water Tours 2010)

The above account of the Rio Virilla cleanup event is intriguing in its rhetorical composition, for it appears on the one hand to be merely challenging the inhumanity of the pollution and degradation of the Rio Virilla; but on the other hand its engagement with the use of parody and comedic aesthetic in environmental campaigns does not challenge the market for commercial eco-tourism travel to Rio Virilla. Further examples of inter-relationships between commerce, advertising and environmental campaigning for the conservation of Rio Virilla are evident in the many blogs that were posted about the cleanup. Indeed, social media has provided a platform for the computer mediated expression and meaningful articulation of on-going projects and strategies devised to sustain the spirit of the Rio Virilla cleanup campaign, as exemplified the following posting:

> Costa Rica te Quiero Limpia Rescatemos el Rio Virilla
>
> El pasado 15 de mayo del presente año se realizó la Campaña "Costa Rica te quiero Limpia Rescatemos el rìo Virilla" en playa Guacalillo ubicada en el Pacífico Central, en donde toneladas de basura proveniente de la GAM es depositada por el arrastre del río Grande de Tárcoles y el Virilla, constituyendo la pérdida irreversible de ecosistemas, el aumento de especies en peligro de extinción, entre ellas el cocodrilo (Crocodylus acutus), y además de la formación de criaderos de microorganismos e insectos vectores de enfermedades dañinos para la salud.
>
> Ante esta gran problemática se está promoviendo una conciencia ambiental responsable en la solución de la contaminación por desechos sólidos, hacia un manejo adecuado de residuos y a la erradicación de enfermedades generadas por la acumulación de los mismos, de ahí esta importante actividad, logrando la integración de las comunidades, instituciones gubernamentales, ONG, empresas privadas y la sociedad civil en general, así como la excelente organización de Terra Nostra, Canal 7 y el personal del Ministerio de Salud de Garabito, Orotina y San Mateo. Esta actividad presentó un gran éxito al recolectar 8 toneladas de residuos sólidos de playa Guacalillo, siendo esto apenas un granito de arena aportado por los que participamos en esta campaña y deseando que estas actividades se han solo el principio de una lucha sin fin por nuestro querido Planeta Tierra.
>
> (Campañusea Costa Rica Te Quiero Limpia Jueves 03 de Junio de 2010)

> On May 15 of this year was the campaign "I love Costa Rica Rescuing the Rìo Virilla" cleanup took place on Guacalillo beach located on the Central Pacific, where tons of trash from the GAM is deposited by the drag of the Rio Grande de Tárcoles and the Virilla, constituting the irreversible loss of ecosystems, increasingly endangering species, including the crocodile (Crocodylus acutus), and in addition aiding the formation of areas of breeding of microorganisms and insect vectors of diseases harmful to health.
>
> Given this large problem [it is necessary to encourage] promoting responsible environmental awareness, in solving the solid waste pollution, to [enable] a proper waste management system and the eradication of diseases caused by the accumulation of garbage. Hence this important activity, achieving the integration of communities the government, NGOs, private companies and civil society in general as well as showing the excellent organization of Terra Nostra, Channel 7 and the staff of the Ministry of Health of Garabito, Heredia and San Mateo. This activity was a great success and collected 8 tons of solid waste from Guacalillo beach, but this was just a tiny grain of sand contributed by participation in this campaign and we hope that these activities are only the beginning of an endless struggle for our beloved Planet Earth.
>
> (Campaña Costa Rica Te Quiero Limpia Jueves 3 June 2010)

In March 2010, Costa Rica's Virilla River also became the focus of environmental campaigns motivated to address pollution from industrial bio-chemicals and human sewage. A major source of river water contamination in Latin America derives from improperly discarded industrial waste products: "synthetic chemicals, and hazardous wastes" that seep into the groundwater (San Martin 2002:21). The Virilla River's poor water ecology is partly due to pollution from "agro-industrial activities and metropolitan development" (UNEP 1999/2007:127). In addition to industrial waste and agrochemicals, the Virilla River is subject to sewage pollution: "approximately 250,000 cubic meters of waste pours into the Rio Virilla every day" (Mitchell and Pentzer, 2008: 186). Industrial agro-chemical discharges, coupled with the sewage population of the impoverished metropolitan development, provided the conditions whereby the Virilla River was unmistakably polluted with debris and its water quality contaminated. The organizers, "Mountain of Hope", enrolled public and private sector agencies to participate and sponsor the event; among these were recycling companies that became the recipients of the extracted plastic bottles. Commercial investment in the event can be explained as corporate social responsibility (CSR); less evident are the values and motivations of non-corporate volunteers. Marxist ecofeminist analysis provides a way to begin to theorize these events.

Rio Virilla Clean-up: Marxist Ecofeminist Reflections

Marx conceived that freedom, at its most basic level, was based on rationally mastering the realm of necessity. In the writings of Marx, the life of the species consists materially in the fact that the human being (like animals) "lives on inorganic nature" (Marx, 1844/2007: 74). Humans are least like animals, when their creativity of action frees them from the shackles of immediate necessity, providence, and happenchance, in the material conditions of existence. Axiomatic with the concept of immediate necessity is that of "natural needs". Marx understood "natural needs" to include nutrition, clothing, shelter, etc. Marx treated natural needs as a "universal concept and general abstraction respectively" (Fraser, 2000: 148). Natural needs exist across human societies and their satisfaction constitutes the "first premise of all human existence", that human beings "must be in a position to live in order to be able to 'make history'" (Marx and Engels, 1846/2007: 48). Basic to existence, "the first historical act" involves "the production of the means to satisfy these [natural] needs, the production of material life itself" (ibid.). While it might seem that the satisfaction of natural needs relates to drives of bygone years, Marx is keen to point out that this is both "an historical act, a fundamental condition of all history" and also "must daily and hourly be fulfilled merely in order to sustain human life" (ibid.). Marx continuously explains the changing composition of natural needs and dialectical shifts in the form of their satisfaction, as part of his broader analysis of "the contradictory movement of human beings in the shaping and making of their world" (Fraser, 2000: 146). Given that the form by which natural needs are gratified has divergent structural properties, it is evident that the socio-historical analysis of need satisfaction can render transparent the dialectical infrastructure of a given society (ibid: 147). It is, therefore, unsurprising that Marx's analysis progresses from the "general to the determinant abstraction" so as to interrogate the composition and economic infrastructure of need satisfaction in a given society and at a given historical interval. Continuous with this presupposition, is Marx's treatise that, in capitalist modes of production, the particular manifestation of the satisfaction of natural needs "leads to a complex and overbearing system of need satisfaction" (ibid: 148). Central to Marx's conceptual and procedural strategy is the discovery of "inner connections" of need-forms (ibid.). Thus, from Marx's ecology one can discern the sense of motivation and spirit which mobilized the volunteers to invest tremendous effort in the Virilla River clean-up events. However, attempting to impose a classical Marxist ecological framework immediately reveals its limitations; for the river provisioning conducted by the Virilla River volunteers constituted unpaid labour and thus was not subject to the laws of Marx's labour theory of value. Maria Mies' (1998) concept of "housewifization" appears to present an apposite Marxist feminist lens with which to formulate a theoretical analysis of the Virilla River campaign and the river clean-up event.

According to Mies (1998), the post-1970s era of capitalist production is defined by systemic transformations in the operational strategy of the international division of labour (IDL). In the previous IDL, western countries produced raw materials in (ex-)colonial territories and transported these raw materials to the West to transform them into industrial products suitable for western industrialised markets. In the deregulatory neoliberal marketplace of post-1970s capitalist production, the labour markets of indebted developing countries are the sought-after locations for Western capitalist labour-intensive production (ibid.: 113). Integral to this transition has been the "integration of women as housewives" in capitalism's new IDL (ibid: 114). Maria Mies convincingly argues that in conjunction with the imposition of the nuclear family model, the integration of the housewife's role involves the expropriation "of Third World Women's work" (ibid.), whereby capitalism actively exploits practices that are integral to reproductive domestic labour. Mies' proposition has significant implications for an analysis of river clean-up activities: was the Virilla River's clean-up an example of capitalism's expropriation of domestic labour? Mies (ibid: 117) argues that "a very large proportion of female labour in Third World countries is engaged in production of goods for the market in the rich countries". To this can presumably be added the meta-industrial labour involved in provisioning natural resources that presently constitute reserves for capitalist exploitation or are the exhausted remnants of capitalist production. River clean-up appears resonant with this form of labour, but the complex contours of enrolment, agency and gender identity suggest that volunteers involved in the Virilla River clean-up campaign were not merely passively mobilized into providing capitalism with valuable meta-industrial labour. Mies' conception of "housewifization" presents insights, but also limits appreciation of agency. In an article entitled, *Women's Community Work Challenges Market Citizenship*, Collins *et al.* (2011) provide conceptual resources relevant to theorizing agency and community work in the context of marketized environmentalism.

Collins *et al.*'s (2011) research focus is women's community work within the context of the neoliberal restructuring of public sector services. In recent years this sector has particularly struggled in the face of reductions to state-sponsored programmes of social welfare and individualizing notions of market citizenship. Issues of this nature may appear distant from the conditions of river clean-up campaigns, nevertheless similarities in social relations suggest important theoretical relevancies. In Collins *et al.*'s research, women working in the voluntary sector "find themselves limited, squeezed and regulated into focusing their services in ways that are consistent with a reconfigured citizenship in which individuals are responsible for their own material provision" (ibid: 297). Collins *et al.*, identify forms of exploitation that have resonances with Maria Mies' (1998) concept of "housewifization". Collins *et al.*, state: "A central dilemma facing women's organizations is that even as women challenge the erosion of social citizenship, their provisioning work can become a substitute for public resources and a tool for assigning individual responsibility" (2011: 297–8). Nevertheless, Collins *et al.*, (ibid: 298) identify practices, employed by women, to challenge the individualizing marketized citizenship of neoliberal governance. Provisioning is defined by Collins *et al.*, "as the work of securing resources and providing the necessities of life to those for whom one has relationship of responsibility". In order to analyse women's work conducted in community organizations, Collins *et al.*, develop the concept of collective provisioning that "includes all the work women do collectively in organizations to provide for themselves and their households". Collective provisioning involves assuring the minimum of resources and exigencies for those for whom the collective has responsibility. Similarities exist here with the collective provisioning involved in the Virilla River clean-up event, for both activities involved individuals having to "negotiate across the boundaries of paid, unpaid and volunteer work" (ibid.). Provisioning in community work and river clean-up activities also reveal "the ways in which women's work is embedded in relationships" (ibid). Of central concern to Collins *et al.*, (ibid) is that: "Provisioning also captures the complex context of women's work and the ways in which they negotiate across the boundaries of paid, unpaid and volunteer work".

Central to the conceptual findings provided by Collins *et al.*, (2011) is an analysis of agency and volunteering in the collective provisioning of river waterways. Of initial interest is the proposition: "the voluntary sector [is] a new arena for demonstrating worthiness" (ibid: 303). By implication, provisioning river waterways might be discerned as the deployment of agency so as to actively demonstrate self-value. According to Collins *et al.*, this relationship retains the neoliberal market's conception of exchange: "There is an 'exchange' … and the exchange is individualised" (ibid.). If it were the case that the volunteers involved in the Virilla River cleanup event had been paid, one could easily identify the operation as an exchange system, for "In this context, volunteering has a utilitarian function of reinforcing the idea that material needs should be met through the exchange of the market". An additional concept introduced by Collins *et al.*, (ibid.) and of interest to an analysis of provisioning river waterways is the idea that voluntary sector participation has become relevant to employability, the sustainability of voluntary organizations and a means for individuals to access services. Thus, Collins *et al.*, refer to a sense among the volunteers interviewed that if it were not for their participation, the organization would cease to operate. By implication, it might be argued that volunteers participating in river waterways clean-up campaigns assume a sense that the organizing institution is beholden, for its existence, to voluntary membership. Significantly Collins *et al.*, identify how in a neoliberal era of individual responsibility and marketized employability skills, there has emerged "a focus on volunteering as the new requirement for access to services and to employment" (ibid). From this perspective, voluntarily participating in river waterways clean-up can be perceived as deploying agency in the acquisitioning of employability transferable skills. As Collins *et al.*, observe: "Volunteer experience is required on a resume by many employers to demonstrate community involvement or an ability to work as part of a team. Additionally, volunteering may serve as an entry point to a paid position" (ibid.). Alongside these direct responses to neoliberal market citizenship, Collins *et al.*, identify "the solidarity of a shared struggle" (ibid). Altruism defines this form of solidarity because:

> Helping others because you have been helped takes place within a collective space where those who have been helped in turn help others who share the same needs. And it is not an exchange so much as a paying it forward, a line of assistance that extends from one person to another to another. (ibid: 304)

One senses that collective provisioning involves long-term time horizons involving generations, past, present and future, and this is enhanced by collectively working with others who are experiencing the same situation. Communities who are collectively exposed to contaminated waterways will feel a sense of solidarity and resolution to actively address the situation. Collins *et al.*, describe solidarity achieved through community work as "about finding social support or a support network. It is about finding a place where your citizenship is restored because you are not deemed unworthy by virtue of being in need or belonging to a marginalized group" (ibid: 305). Concern for the river waterways in one's neighbourhood is validated through collective provisioning. In this activity a space is cultivated to achieve recognition in the validity of caring about environmental issues; and opportunity is provided to participate in the resolution of environment issues. This latter issue is particularly relevant to understanding river waterways provisioning in terms of agency: Collins *et al.*, argue that: "For these women the connection to advocacy begins with their participation in a collective space that challenges an individualized explanation for their situation and recasts the notion of social citizenship." Similarly it may be seen that, for those individuals engaged in river waterways, provisioning collectively working with volunteers to address the environmental degeneration of shared waterways is the first step towards advocating agency. The women in Collins *et al.*'s research began "to build new collective identities and to reject the status assigned to them as less than ideal citizens, who are not meeting all of their needs via the marketplace" (ibid: 306). Similarly, it is conceivable that volunteering to collectively provision river waterways involves agency as individuals actively recast marketized citizenship and cultivate new spaces for embodied material engagement with the environment in order to be practically involved in challenging the environmental degeneration of river waterways.

Timely Reflections

Marxist ecofeminists emphasize the inclusion, in Marxist ecology, of reproductive labour as part of the complex matrix through which capitalist environmental degradation operates. Referring to Bennholdt-Thomsen and Mies, Isla aligns herself with the Marxist ecofeminist use of the metaphor of the iceberg:

> [W]here capital and waged labor form the visible above-water economy, which is counted in GDP and frequently protected by a labor contract. The unseen 'feminine' support of this exposed tip of the global capitalist patriarchal iceberg is women household workers in both the global North and South, along with peasants, indigenous peoples, and nature. (Isla, 2006: 60)

The iceberg metaphor is apposite in its analogy of an unseen multitude that resides underneath the apex of capital labour relations and affably transported internationally traded water. UN Water reports consistently advance our appreciation of the vast scope and scale of women's unpaid labour invested in the provisioning of water and management of water quality. To these accounts needs to be added UNICEF (2012) reports highlighting the extent to which children's lives are blighted by the inequities of poverty and water provisioning responsibilities:

> An analysis of data from 25 countries in sub-Saharan Africa, representing 48 percent of the region's population, reveals that women and girls bear primary responsibility for water collection. (UNICEF, 2012: 31)
>
> In 71 per cent of all households without water on the premises, women or girls are mainly responsible for water collection. In 29 per cent of households, men or boys assume this task. (ibid)
>
> In these 25 countries, it is estimated that women spend a combined total of at least 16 million hours each day collecting drinking water; men spend 6 million hours; and children, 4 million hours. (ibid)

From these and other statistics it is evident that the base of the water quality management iceberg is anchored by "externalised economic sectors", who are not merely exploited through low wages but also through their unpaid reproductive or meta-industrial labour, which functions to "provision, 'free of charge' [the] service and material inputs to capital" (Isla, 2006b: 60). It is clear that the enclosure of water common resources disproportionately affects women. Indeed, Marxist ecofeminists have observed that the meta-industrial and reproductive activities that are carried out by women are vitally important to capitalism, but capitalism obfuscates, demeans or derides these workers, subsuming their labour as housework so as to deny remuneration. In the unpaid water provisioning of women in the Third World, it is this labour that capitalism seizes through acquisitive dispossession: "these products are taken from them with no or very poor compensation, through capitalist patriarchal violence" (ibid: 60). It is my contention that where the international importation of water is considered a part of a programme of water sustainability, it is imperative that discussions of global footprints and environmental degradation include consideration of postcolonial ecofeminist analysis. For the postcolonial ecofeminist tradition interrogates the intersection of gender, class and colonialism in order to critically challenge the "ever greater enclosure, parasitism, and pimping by capitalists" (ibid: 63). It is my contention that transnational programmes of water sustainability render crucial the development of postcolonial Marxist ecofeminist geopolitics of water sustainability. The aim of this chapter has been to formulate the theoretical foundations of this new and immensely significant progression in eco-feminism.

In *Eco-Feminism and Eco-Socialism: Dilemmas of Essentialism and Materialism, Capitalism, Nature, Socialism*, Mary Mellor (1992: 51) attempts to formulate an ecofeminist account of the material bases of economic human society. According to Mellor, there are three interconnected bases to the material existence of economic life: "the forces/relations of production, the forces/relations of reproduction and the relations between human society and nature" (ibid.). Mellor rebukes classical Marxist ecology for its disjunctive partitioning of economic productive labour from non-monetized reproductive labour and the sphere of nature. In this exposition of Mellor's embodied materialism it is evident that patriarchy is constituted as the protagonist linking the exploitation of women and the exploitation of nature: "the sphere of "production" does not just represent the interests of capital, it represents the interests of men" (ibid.). Such conclusions are relevant as the basis of a critical riposte to international organizations that emphasize "gender mainstreaming" as the central strategy through which to address women's disproportionate role in water provisioning. Consider this suggestion in light of the following WWDR4 observation:

> Water policies based on broad, generalized perspectives are more likely to miss out local knowledge, and social and gender dimensions and their implications. Recognizing the various purposes for which these local water resources are used by different groups of men and women in the community would help successfully integrate gender considerations not only in water resource management, but also in sectors such as urban water supply, agriculture, industry and energy that depend upon water resources, and which are often in conflict over water allocations and their demand for freshwater resources. By working together in partnership with these sectors, synergies and trade-offs in providing access to different groups of men and women in local communities can be understood and addressed by decision-makers in government bodies, private sectors and civil society. This would help in anticipating risks and uncertainties and planning for safe-guards to be put in place for the most vulnerable groups in the community. (World Water Development Report 4, 2012: 744)

In as much as international organizations resolve to address gender inequality through an emphasis on equal opportunity in the access to the decision-making arena, the wider issues of capitalism and meta-industrial labour remain unchallenged. It is thus necessary to highlight and challenge the economic benefits that capitalism accrues from women's unpaid provisioning of water. Indeed, it may be argued that in the absence of a critical Marxist approach, an emphasis on equal opportunities partakes in a patriarchical logic "separating off production from reproduction and from nature" and thus colluding with patriarchal capitalism's cultivation of "a sphere of 'false' freedom that ignores biological and ecological parameters" (Mellor, 1992: 51). Focusing on institutional participation and policy initiatives emerging from these decision-making forums neglects a second contradiction of capitalist exploitation "the way in which the male/capitalist sphere of production is materially dependent upon women and nature" (ibid.). As Mellor further reveals:

> It is in women's lives that the relationship between the social/biological is constructed and it is this under-laboring work that women do that is not incorporated into the 'material' world of men as represented in the theoretical framework of historical materialism. (ibid: 52)

Consequently, merely inviting women into the decision-making of international organizations "will not resolve the "second contradiction" of the conditions of production for either women or nature" (ibid: 52). It is therefore part of the advocacy of Marxist ecofeminist praxis to actively engage with environmental issues and to do so from a feminist standpoint. The case study featured in this chapter provides an example of the possibilities suggested by Marxist environmental water ecology activism; the case study also opens up debates about environmentalist alliances with government agencies, manufacturing industries, social media and the gender of river clean-up activities.

Figure 2.1b., "Emptying the sensual with deprivation, cleansing affection from the temporal. Neither plenitude nor vacancy", (T.S. Eliot *Burnt Norton*). Photographic Image, South London, River Thames, September 2013

PART TWO:

ECOFEMINISM AND THE GEOPOLITICS OF WATER

Figure P.2: "The only wisdom we can hope to acquire, Is the wisdom of humility: humility is endless" (T.S. Eliot *East Coker*). Photographic Image, South London, River Thames 2013

Figure P.2: "The only wisdom we can hope to acquire, Is the wisdom of humility: humility is endless" (T.S. Eliot *East Coker*). Photographic Image, South London, River Thames 2013

CHAPTER THREE

WATERSHEDS IN MARXIST ECOFEMINIST ANALYSIS

Figure 3.1., "At the still point of the turning world" (T.S. Eliot *Burnt Norton*). Photographic Image, South London, River Thames, September 2013

Figure 3.2a., "Or the waterfall, or music heard so deeply" (T.S. Eliot *The Dry Salvages*). Photographic Image, South London, River Thames, April 2013

Interview with Executive Director, London River Thames Conservation Charity II, September 2013.

> ***Interviewer: If we were to carry out a case study based on these issues around the Thames Tideway Tunnel; Could you tell me how these issues relate to the Thames Tideway Tunnel?***
>
> **Interviewee:** *The Tideway Tunnel is a good example of one very positive step which society hopefully will take, because it is still going through the planning process at the moment. But, erhm, The Tideway Tunnel is designed to address one of those very big impacts on the river, which is the fact that most of the rivers in central London are still connected to a sewer system, which is a hundred and fifty years old. And regularly, weekly overflows into the river raw sewage. And in the twenty-first century that is unacceptable. That is unacceptable that there is raw sewage going into our river. So the Thames Tunnel will solve that problem. It will stop those overflows happening. And that will be a massive benefit to the quality of the river, which is something we absolutely must address. The fact that this government has approved Thames Water to actually borrow the money, invest and build it and charge customers for doing ... There has been some publicity about the cost of the Tunnel; Thames Water will make profit out*

of it; but there hasn't been a huge public negative response. So we are very hopeful that they will get the planning permission for this and work will go ahead. You know and the Tunnel will be completed.

...

Interviewer: In what ways have you supported it?

Interviewee: *There is a coalition of NGOs called Thames Tunnel Now... We are just a member of that coalition, and when we are asked by the Coalition to put our names to press statements or go along to meetings with MPs, for example, we have done things like that.*

Interviewer: As I understand, the Thames Tideway Tunnel is an engineering solution as opposed to SUDs...?

Interviewee: *Yes Sustainable Urban Drainage systems.*

Interviewer: It seems that engineering solutions have been preferred as opposed to Sustainable Urban Drainage systems?
Interviewee: *Well, the way that Sustainable Urban Drainage Systems work, for example, is that all the drains from this building, for example, would be run through a green area like wetland, erhm, or some other area which had been planted with greenery. Like a willow forest or somewhere that could absorb the nutrients from all the taps and toilets that are coming out of here, so that the water which then left that wet land or green area and went into the river was cleaner than when it came out of the tap or the toilet. And those kinds of systems can work and they do work; but they need a lot of land on which to work. You need, erhm, for a building of this size, you would need a hectare or two of wetlands, marshes, swamps, reed bed, wet forest to run your taps and toilets through. For the vegetation to absorb all the pollution before the water reaches the river. And the fact is, in a city the size of London, with its eight million people and this density of buildings, there isn't the land to put the wetlands in for the SUDs. Where would you put the green area needed for this building? ... There isn't the space to put in the greenery needed for the SUDs systems for all of the buildings in London. Which is why we agree that the only solution, or not the only solution but the major part of the solution, has to be an engineering solution like building a Tunnel. There are SUDs systems around the capital and we have been involved in promoting some of them. In the suburbs, for example, in the outer suburbs in the boroughs like Barnet and, erhm, Richmond, erh, further out where there are larger green areas, there are playing fields and green spaces along the rivers there is room in some places to put in SUDs systems. But here in central London where most of the population is concentrated and where most of the, are the actual drains that overflow into the river, you've got nowhere to put a SUD system. It's too late; we've already built the city. If you want SUD systems you have to design them into the buildings that you're*

constructing. But if you have already got, you know, a one thousand-year old city like this, we built it too densely, there isn't the space for the SUD systems.

Marxist Ecofeminist Analysis

Ecofeminists perceive the exploitation of nature to be bound-up with a patriarchal ideology of domesticity, which has and continues to exploit the reproductive labour of women. Ecofeminists invariably formulate explanations of environmental degradation through recourse to the materiality of the female body and the contours of reproductive labour. Writing in London in the nineteenth century, Marx, in *Capital, Vol. III* (1894/1981:195), observed rapid waves of migration into the city; this displacement of people increased exponentially with concerns over the materiality of the labouring body and environmental hazards. Pollution of the River Thames was incurring a "monstrous expense" (ibid.). Parliamentary Papers consistently recorded the nefarious state of the River Thames, often detailing and incriminating the inefficiencies in the operations of the Metropolitan Board of Works. Marx noted this situation as exemplifying the intractable contradictions that plague the capitalization of pollution management. Over a hundred years hence, and water quality management of London's River Thames continues to be a cause for concern. According to Boris Johnson, the present Mayor of London:

> In the middle of the 19th century, London was brought back from the brink of environmental catastrophe by Joseph Bazalgette through the construction of the London combined sewer network. 150 years later we still rely on this network to drain the city … London is now a city of nearly 8 million people and many of our Victorian-era solutions, including Bazalgette's remarkable subterranean achievement, are now struggling to deal with the demands we place of them. The growth of the city and changes in the way we live our lives are placing stresses on our water infrastructure that they were not designed to deal with. These will be further compounded by our changing climate. (Johnson, 2011: 6)

In Marx's ecological theory, the injustice of metabolic disjuncture is the injustice of paid labour, and thus excludes the unpaid reproductive labour associated with women. Within the ideology of domesticity, the power of patriarchy is an oppressive imposition on the everyday lives of women. Nevertheless, in classical Marxist ecology, unpaid reproductive labour is marginalized from the class struggle against a "progress in capitalistic agriculture", which is exploitative of labour and which is "ruining the lasting sources of that fertility" (Marx *Capital Vol. I,* 1887/2003:474-475).

Insofar as Marx's ecological theory pursues capitalist exploitation in terms of the economically productive categories of labour, the geopolitical context of activism will continue to marginalize capitalism's exploitation of reproductive labour as part of the day-to-day reality of environmental degradation. Conversely, materialist eco-feminism goes beyond the market economy of paid labour so as to appreciate the vital significance of reproduction as part of the ecological limits of capitalist accumulation (Dordoy and Mellor, 2000: 44). As Salleh (2003: 65) observes, embodied eco-feminism is "materialist" in that it aligns with Marx's critique of the capitalist forces of production, and it is "embodied in that it sets out to re-frame that discourse by giving equal weight to the organically interrelated entities – man, woman, nature".

This chapter examines embodied materialist eco-feminism, and identifies a critical dialogue that is presently underway, veraciously challenging the conceptual ideas of classical Marxist ecology. Of specific concern is the flow of time which threads through Marx's ecological theory. With the exception of Mary Mellor (1992, 1994, 1996 and 2003), it is evident that embodied materialist ecofeminists implicitly, rather than explicitly, engage with capitalist accumulation in terms of the subjectification and objectification of time, despite their prosaic recognition that:

> [T]he theory of historical materialism based on the primacy of (formal) economic relationships as defined by capitalism has produced the catastrophe of the command economies which, by mimicking the productive power of capital, continued to deny the reality of much of women's lives and produced ecological devastation. (Mellor, 1996: 257)

To this end, it is necessary to conceive and formulate a time for embodied materialist eco-feminism. This chapter engages with a recent empirical theme, a discussion of the future water supply in London, as explained by the report *Securing London's Water Future* (Johnson, 2011). The chapter identifies key reports, published by the office of the Mayor of London; and each of these publications is critically detailed as part of an analysis of water time poverty issues experienced by First World women in the metropolis. Reports critically discussed include: *The Greatest City on Earth: Ambitions for London by Boris Johnson* (Greater London Authority, 2013) and *Securing London's Water Future* (Johnson, 2011). In 2011, the Greater London Authority published a strategy for securing a "world class" (ibid:13) and environmentally resilient infra-structure for the River Thames. Central to this document is the proclamation that levels of the city's water are unsustainable in the context of both rapid population growth in London; and climatic decreases in the summer rainfall. In order to meet this demand for water, "some water companies are currently abstracting more water from the environment than it can sustain" (ibid: 13) resulting in recourse to increase the use of London's reservoirs and desalination plants. According to *Securing London's Water Future* (Johnson, 2011), one significant reason why London's water supply is unsustainable is because of poor stewardship demonstrated by households in their consumption of water. Thus, the report states:

> Londoners use more water than the national average (167 litres per person per day in 2009–10 compared to 146 litres per person per day), largely because we live in small households, which are less water-efficient. Many Londoners have little incentive to save water – only one in four homes has a water meter. We also lose enough water in leakage from our water mains network to fill more than 238 Olympic-sized swimming pools every day. (ibid: 13)

The central conclusion of the report is that London's demand for water will increase exponentially with its population, alongside depleting summer rainfall and the popularity for smaller dwellings: "This means that London's already tenuous supply-demand balance will become increasingly unsustainable – we therefore need to act to balance supply and demand" (ibid.). An integral feature of the proposed action plan is "managing water use", and instituting efficiency in the "disposal of waste water". Both of these features encourage Marxist engagement, for they echo concerns that Marx expressed in the nineteenth century about environmental degradation being caused by accelerated capitalist production. While it is necessary to factor classical Marxist reflections into modern-day analysis, the persistent coupling of ecology and household consumption makes it an imperative time for the ecofeminist Marxist analysis of water sustainability. Consequently, this chapter is organized as follows: first, a study of *Securing London's Water Future* provides the empirical basis for a classical Marxist ecological engagement with the accelerated time of capitalist exploitative production; each aspect of the classical Marxist ecological debate is then interpolated with Marxist ecofeminist critical reflections so as to highlight and valuably augment the deficiencies in classical Marxist ecology. The resulting tapestry provides a theoretical basis and conceptual formulation for a Marxist, ecofeminist geopolitics of water sustainability.

Water Ecology in the Time of Metabolic Rift

> Increasing Supply. The Environment Agency's report, *Water for the Future – Managing Water Resources in the South East of England* concluded that by 2035 demand for water in the South East of England would significantly outweigh supply unless we reduce the amount of water we use or find new resources. It notes 'we can build new resources, but we need to ask ourselves how long we can go on doing this and how resilient and flexible to climate change these options will be looking forward 100–200 years …The water company WRMPs identify seven main options to increasing supply: a) Increasing abstraction; b) Desalination; c) Increasing reservoir capacity; d) Wastewater treatment; e) Raw water transfers; f) Ground water recharge; g) International import of water. (*Securing London's Water Future*, Johnson, 2011: 42)

Acceleration in the time of circulation is a fundamental premise of Marx's analysis of industrial capitalism. According to Marx (1894/1981:135), the rate of profit accumulation "is affected by the circulation time as well as by the working time, the time of circulation restricting the surplus-value that can be realized in a certain period". Writing during the Industrial Revolution, Marx observed an epochal transformation and intensification of production. Advances in manufacturing and farming technologies had tremendously increased production rates, and thus the magnitude of surplus value that could be expropriated from labour during the working day. The technical and social structural conditions of capitalist production in the eighteenth century had both centralized and intensified production; but this was achieved through an excessively arduous prolongation of the working day. It is evident that, at the inception of industrial capitalism, surplus-value expropriation was predicated on the laborious lengthening of the working day; a method Marx defined as "absolute surplus-value" (1887/2003: 299). Advances in the technologies of industrial manufacturing and agriculture precipitated a shift away from absolute surplus-value extraction towards "the surplus-value arising from the curtailment of the necessary labour-time, and from the corresponding alteration of the respective lengths of the two components of the working day"; Marx defined this latter process as "relative surplus-value" (ibid.). The increased productive capacity manifested by regimes of relative surplus-value necessitates the cultivation of new markets capable of ingesting the increased output generated by advancements in productive forces. Marx (1857/1973: 408) states that: "production of relative surplus-value … requires that the consuming circle within circulation expands as did the productive circle previously". According to Marx, axiomatic to the productive conditions of relative surplus-value extraction is a capacious reduction in the quantity of necessary labour time required in production, and a corresponding amount of leisure time. Capitalism's progress towards the manufacture of relative surplus-value altered the dynamic of capital and labour, requiring the formation of "a new, qualitatively different branch of production … which satisfies and brings forth a new need" (ibid: 409). Capital and labour posit themselves "in a new form": accelerated productive capacities coupled with capitalism's proficiency to accumulate, increased quantities of surplus labour to the point of reshaping and thus destabilizing the metabolic conditions of capitalist production. "The tendency to create the world market" (ibid: 408) is a direct consequence of capitalism seeking equilibrium in its rates of market exchange. Returning to the previous extract from *Securing London's Water Future* (Johnson, 2011: 42), one can discern elements in the "Increasing Supply" action plan that resonate with Marx's observations about the environmental outcome of increasing the time of capitalist circulation. Of significance is Marx's prediction that the accelerated time of production impels capitalism into a relentless search for new markets. It is therefore no coincidence that the water company WRMPs has proposed the "International import of water" (ibid.) as part of the strategic solution for a sustainable water supply for London.

Marx and Engels (1848/1970: 83) observed that capitalism's need to offset a declining rate of profit obliges "a constantly expanding market for its products" and "chases the bourgeoisie over the whole surface of the globe. It must nestle everywhere, settle everywhere, establish connexions everywhere." Capitalism's relentless appetite propels forward "the exploration of all of nature in order to discover new, useful qualities in things; universal exchange of the products of all alien climates and lands; new (artificial) preparation of natural objects, by which they are given new use values" (Marx, *Grundrisse*, 1857/1973: 409). In its pursuit of its aims in all corners "of the earth in all directions", capitalism displays brutal vigour "in the exploitation of the natural and human qualities", utilising scientific discovery "as much as all the physical and mental qualities" in its audacious pursuit of "the universal appropriation of nature as well as of the social bond itself" (ibid.).

Marxist eco-feminism shares Marx's observations concerning the nature of industrial manufacturing and agricultural production. The past and present of industrial capitalism have entailed a systematic, inexorable objectification of nature. The consequences of the externalizing, decimating and neglect of nature are abundantly evident in the accelerating occurrence of environmental catastrophe. From the perspective of Marxist analysis, embodied materialist ecofeminists highlight Marx's irresolution regarding the inter-section of class, gender and nature. In Marx's analysis of capitalist ecology it is evident that both the exploiters and exploited are bolted into structures directed towards the pursuit of transcendence, i.e., freedom from the reliance on the indeterminacies of nature. Conversely Marxist eco-feminism professes: "the need for a new cosmology and a new anthropology which recognizes that life in nature (which includes human beings) is maintained by means of co-operation, and mutual care and love" (Mies and Shiva, 1993: 6). The imperative of co-operation between species is dedicated to a belief in "the special strength and integrity of every living thing" (ibid.: 13). Described as "a women-identified movement", Marxist eco-feminism directly links patriarchy with capitalism's exploitation of nature. Consequently, writers such as Mies and Shiva (ibid: 14) convincingly articulate resistance against the relentless actions of "corporate warriors" in the "imperilled times" of untrammelled global capitalist production. Contiguous with these exploitative capitalist relations is the threat of catastrophic natural and human destruction due to warfare. For embodied materialist ecofeminists, the devastations of war align with the instrumental mentality of patriarchy and the avaricious dynamics of capitalism. As Mies and Shiva (ibid: 14) put it:

> Wherever women acted against ecological destruction or/and the threat of atomic annihilation, they immediately became aware of the connection between patriarchal violence against women, other people and nature, and that in defying this patriarchy we are loyal to future generations and to life and this planet itself. We have deep and particular understanding of this both through our natures and our experiences as women.

For Mies and Shiva, women experience the exploitation of nature at the level of the body and social experience. Consequently these writers extol the virtues of an embodied eco-feminism that is directed at defying "corporate and military warriors", whose assault against the environment is, Mies and Shiva argue, "perceived almost physically as an aggression against our female body". In their advocacy for nuclear disarmament and confrontation with capitalism, Mies and Shiva (ibid: 15) seek to destroy "the illusion that atomic technology was malevolent" and prompt an appreciation among women "that their consumerist lifestyle [is] very much part of this system of war against nature" (ibid.).

Elsewhere, Mary Mellor (1997: 180), in a book entitled *Feminism and Ecology*, argues that an essential feature linking the exploitation of nature and women's subjugation by patriarchy is: "the sex/gender division of labour around human embodiment". The gendered division of paid work and domestic division of labour converges in the association of reproductive labour as largely the responsibility of women. Generations of feminists have fought vehemently to dismantle the dominant patriarchal ideas governing the domestic division of labour. Embodied material ecofeminists have nothing against the essentialist valorisations of domestic labour; nevertheless they convincingly provoke recognition of the labour of "human embodiment" and in so doing embrace "immanence", i.e., our inter-connections with nature as opposed to the pursuit of "transience" from dependency on nature (ibid.). Axiomatic with this proposition is the belief that: "The needs of human embodiment are shared by all of humanity, but are disproportionately borne in the bodies and lives of women" (ibid: 183). Mellor (1997: 184) argues that women embody societal domination and the exploitation of nature; but Mellor is keen to avoid a descent into essentialism, thus she qualifies this observation by stating that society ascribes meaning and value to "the materiality of human existence". Thus, the affiliation of women with nature is not determined by physiological difference but rather is defined by the social and historical conditions through and in which women's experiences are cultivated. This belief is the basis for Mellor's formulation of an embodied ecofeminist notion of gender, time and ecology:

> Women's disproportionate responsibility for human embodiment is partly expressed in the work that women do, but also in their availability for biological needs – what I have called biological time. Whatever social lives people construct, they are always delimited by bodily existence. Equally, social lives are delimited by the ecosystem. (ibid: 184)

Embodied materialist ecofeminists consistently detail "bodily existence" (ibid.) as differentiating the way men and women experience the natural environment. This concept is most convincing when used to highlight the disproportionate impact of globalized capitalist production on the lives of women. Thus, Mellor (2003: 14) accentuates the ecofeminist focus on agro-agriculture; for it is evident that the environmental impact of industrialized farming systems are negatively affecting the livelihood of women who have for generations played a primary role in subsistence farming. The enclosures and privatizations of agro-agricultural business disproportionately affect women, because: "As readily accessible and fertile land is lost for both subsistence farming and resource collection, women are often forced onto more marginal (and ecologically fragile) ground" (ibid: 14–15). Of pertinence are Mary Mellor's observations concerning the vital role, often assigned to women, of water collection and transportation: "It is not only women's subsistence production that is at stake: women are universally the collectors of fuel and water and, as common open access land is lost, women have further and further to walk to secure these basic necessities" (ibid.: 14). In their monumental treatise, *The Communist Manifesto*, Marx and Engels (1848/1970: 83) prophesied that: "The bourgeoisie has through its exploitation of the world market given a cosmopolitan character to production and consumption in every country". While in agreement with this observation, embodied materialist eco-feminism very convincingly insists on inter-relations between: "bodily existence" (Mellor, 1997: 184), reproductive labour and the exploitative hegemonic conditions of capitalist production. These issues encourage advances of Marx's labour theory of value, so as to additionally highlight gender and the metabolic rift precipitated by the dynamics of capitalist industrial production.

Economically Productive Labour as the Currency of Water Ecology

> Many Londoners are oblivious to how much they could save by being more water-efficient. As noted previously, even homes without a water meter are able to save money on their energy bills by using less hot water … Increasing the number of homes with a water meter. Paying for the volume of water consumed is the fairest way to pay for water … London has a lower proportion of homes with a meter than the England and Wales average (at 35 per cent) … household metering reduces water use by about 10 per cent. It is unsurprising that people tend to use less water when they pay for it by volume used rather than through a standard charge that does not reflect the amount used (and perversely encourages high consumption). The need to measure water use in order to manage it better is a strong argument in favour of water metering. As the south-east of England has been identified as an area of serious water stress, compulsory metering is another tool in balancing supply and demand. (*Securing London's Water Future*, Johnson, 2011: 61)

In *Capital*, Vol. III, Marx provides a brief illustration of the attribution to the earth's natural resources of economic value (1894/1981: 954). Marx qualifies the vignette with several clear statements concerning the basis of value: "Value is labour. So surplus-value cannot be earth"; "The land's absolute fertility does nothing but let a certain quantum of labour give a certain product, conditioned by the natural fertility of the land" (ibid.). In so far as value is achieved through the expenditure of labour: "The differences in the land's fertility have the effect that the same amounts of labour and capital, i.e., the same value is expressed in differing quantities of agricultural products; so that these products have different individual values" (ibid.). In Marx's ecological analysis, the concept of labour, which attributes value to agricultural products, refers to "the entire productive activity of man, through which his metabolic interchange with nature is mediated" (ibid.). While obviously not Marxist in its philosophy, the above cited extract, from the report *Securing London's Water Future* (Johnson, 2011: 61), highlights the significance ascribed to economically productive activity as a means for securing water sustainability. Of significance is the report's advocacy for households to be universally installed with water meters. This is presented as a calculable means for monitoring water volumes; reducing excess water usage and aligning the costs of sustainability with the volume of household water consumption. The currency of water ecology presented in the discourse and practice of water metering is clearly the currency of economic labour. The omission of reproductive labour here, and in other similar statements of value and ecology, typifies the basis of the Mayor of London's *Securing London's Water Future*. Classical Marxist ecology also privileges economic categories of labour and thus is bereft of theoretical and political strategies to challenge the financial equities due to a focus, in ecological terms, on economic labour. For the model of ideal environmental citizenship that emerges from the water metering proposed by *Securing London's Water Future* will operate to marginalize or penalize those not engaged in economically productive labour. And this latter issue highlights further impediments that accrue from the report's failure to appreciate the implications of initiatives, based on paid work, for those primarily involved in non-economic reproductive labour and/or outside of the system of paid work. Classical Marxist ecology reproduces these problems because it has as its basis a conception of ecology, and values premised upon Marx's labour theory of value. As is illustrated, by Marx, in the following account of the value of wheat:

> But how is the earth to have a value, how can it create a socially specific quantum of labour, and the particular portion of value of its own products that forms rent at that? The earth, for example, is active as an agent of production in the production of a use-value, a material product, say, wheat. But it has nothing to do with producing the value of wheat. In as much as value is expressed in wheat, the wheat is considered simply as a certain quantum of objectified social labour, this labour being quite indifferent to

> the particular material in which it is expressed or to the particular use-value of this material. (Marx, *Capital,* Vol. III, 1894/1981: 955)

In *Capital,* Vol. III, Marx demonstrates a distinct appreciation of the nature of "land"; as this latter form constitutes a paradoxically shifting feature of Marx's account of value and social production in the nineteenth century. Specifically, Marx argues that: "Capital-profit (profit of enterprises plus interest), land-ground-rent, labour-wages" constitute a "trinity form" which "holds in itself all the mysteries of the social production process" (ibid: 953). "Capital, land, labour!" are acclaimed by Marx as the essential features of "the Trinity Formula", albeit of different significance to the expropriation of surplus-value in capitalist modes of production. Indeed, Marx paradoxically states that "land" constitutes "inorganic nature as such, *rudis indigestaque moles* in its primeval wilderness" (ibid: 954). Needless to say, Marx's conception of an "inorganic nature" raises questions about value and the evaluation of nature, as is evident in the following deliberations concerning the relation of the earth to the value harvested from agricultural production:

> The productivity of agricultural labour is linked to natural conditions, and according to their productivity the same quantum of labour is expressed in more products or fewer, more or fewer use-values. The magnitude of the quantum of labour expressed in one bushel depends on the number of bushels that the same quantum of labour supplies. The quantity of product that the value represents depends here on the earth's productivity; but this value is given and is independent of this distribution. (Marx, *Capital*, Vol. III, 1894/1981: 955)

Significant conceptual ideas, central to the labour theory of value, are evident in the above extract. First, Marx's oscillation between an "inorganic nature" and "the earth [as] … active as an agent of production" is consistent with the conceptual categories of Marx's labour theory of value. This is because while material objects can be defined as embodying utility, their value derives from the labour time invested in cultivating/producing/crafting the object Thus, Marx argues that the earth "has nothing to do with producing the value of the wheat" (ibid.); and for our purposes the same applies to the value of water. Value is conceived in objects (such as water) as quantifiable amounts of objectified labour time, with this labour assuming an abstract form, and irrespective of "the particular material in which it is expressed or to the particular use-value of this material" (ibid.). Rather the form of labour prevalent in capitalist agricultural production is a time-based "magnitude of the quantum of labour" (ibid.). Axiomatic to Marx's labour theory of value is, therefore, the marginalization and or subjugation of non-economic reproductive labour. Indeed, this is a feature of the ecofeminist embodied materialist challenge to Marxist ecology. By way of illustration, consider the following extracts from *Securing London's Water Future* (Johnson, 2011: 51), for they reveal deep-seated ideas associating environmental sustainability with reproductive labour and the domestic division of labour within the household:

> Water conservation. All Londoners can save water without needing to make any investment. Simple choices, such as only running dishwashers and washing machines when full, not brushing teeth or washing up under a running tap, or occasionally showering instead of having a bath can add up to significant financial and water saving …Water efficiency. The greatest scope for improving water efficiency is in London's 3.2 million existing homes. As well as raising awareness of potential financial savings of conserving water, improving water efficiency in existing homes can be achieved through installing more efficient fittings and appliances … water use accounts for 27 per cent of carbon emissions from the home. This is made up of 18 per cent from the heating of water for baths, showers, hand washing and washing up. The other nine per cent is accounted for by washing –using appliances including dishwashers and washing machines. (ibid: 51–2)

Salleh (2010) ingeniously situates the concept of "metabolic value" as a coterminous determinant of labour value. Whereby: "The term metabolic value denotes the value sustained and enhanced" by workers ostensibly marginal to the capitalist economic relations of work, and yet integral "in supporting ecological integrity and the social metabolism" (ibid: 212). Salleh conceives this form of work as resonating with the organic productive rhythms of the metabolic exchange with nature. Of significance here, is the supposition that meta-industrial labour is technically basic, often assuming the utility of candid, intuitive, and embodied skills. Emblematic of the meta-industrial labour to which Salleh refers is the primitive foraging and domestic labour of voluntary sector river conservationists, demonstrating their civic responsibility by provisioning rivers. In as much as Marx's labour theory of value emphasizes the time of economic labour, it can be conceived to be partaking in a patriarchal logic "separating off production from reproduction and from nature" and thus colluding with patriarchal capitalism's cultivation of "a sphere of 'false' freedom that ignores biological and ecological parameters" (Mellor 1992:51.). Marx's labour theory of value, thus, neglects a second paradox of capitalist exploitation, "the way in which the male/capitalist sphere of production is materially dependent upon women and nature" (ibid.).

Securing Sustainable Water: Freedom from the Realm of Necessity?

> Rainfall intensity in London appears to be increasing. An analysis of rainfall records from a weather station in east London show that before 1960, only one day experienced rainfall exceeding 40mm, compared to ten days after this period. A day with 45 mm rainfall had a 30-year return period before 1960, and now has less than a one in six year return period of occurrence. The increase of heavy rainfall days and the increase in hard surfaces from new or re-development mean the existing surface water drains can no longer cope with the rise in runoff. In turn, this can lead to a greater risk of flooding as surface water drains are overwhelmed. Because so much of London's surface is concrete and tarmac, and therefore impermeable to rainfall, we are very reliant upon our drainage system to keep us dry …The Mayor will work with partners through the Drain London Forum to manage surface water flood risk and ensure a consistent approach across London. (*Securing London's Water Future*, Johnson, 2011: 16)

Securing London's Water Future presents an intriguing account of water sustainability in the context of managing London's rainwater. Rainwater, in its natural flows, "is either lost through evaporation, seeps into the ground to replenish groundwater levels, flows over the ground and returns to streams and rivers, or enters the drainage systems" (ibid: 16). In Outer London, rainwater flows gather from pavements, tarmac road surfaces and rooftops, pouring into exterior water drains, and eventually flowing into the provincial rivers and streams. In central London, rainwater flows into its nineteenth-century municipal sewage system, before proceeding into sewage treatment works. But decades of concreting surfaces, tarmacs on roads and the ailing nineteenth-century municipal sewage system undermine London's ability to rely on natural drainage; and even its attempts to be freed from the necessity of natural drainage is wrestling with a complicated legacy of privatization. At present, the responsibility for maintaining water drainage resides with a cacophony of agencies: "including water and sewage companies, the London boroughs, Transport for London, the Highways Agency and private landowners" (ibid.: 17). *Securing London's Water Future* (Johnson, 2011) in its determination to achieve freedom from the necessity of natural drainage recognizes that "fragmented ownership, together with the realisation that surface water flood risk is probably the greatest short-term climate risk to London, led the Mayor to convene the key stakeholders to work together to understand and manage the risk" (ibid.). But is freedom from the necessity of natural drainage the correct solution? Indeed, is freedom from necessity contiguous with sustainable environmental practice?

In classical Marxist ecology, nature, defined by Marx as "inorganic", is characterized as an inert material form; moulded by human mediation and outside of labour is an indolent property of capital accumulation. As Marx states, "Value is labour. So surplus-value cannot be earth. The land's absolute fertility does nothing but let a certain quantum of labour give a certain product, conditioned by the natural fertility of the land" (1894/1981: 954). Thus, "inorganic nature" constitutes an inert aspect of the material conditions of capitalist production; and as with all other conditions, it is on the one hand a precondition of capitalism, and on the other, integral to the dialectical process of transition from capitalist production. Indeed, an axiomatic dynamic of capital accumulation is that its civilizing vicissitudes expropriate surplus value in a manner that is less inclined towards the exhaustion of surplus labour; rather, capitalism expropriates surplus value "in a manner and in conditions that are more advantageous to social relations and to the creation of elements for a new and higher formation than was the case under the earlier forms of slavery, serfdom etc." (ibid: 958). As Marx illustrates, a curious feature of capital accumulation is on the one hand, the primitive accumulation, privation and enclosure of the commons; and "on the other hand it creates the material means and the nucleus for relations that permit this surplus labour to be combined, in a higher form of society, with a greater reduction of the overall time devoted to material labour" (ibid.). And this latter feature has direct bearing on the nature of time in Marx's writings, as is evident in the following extract:

> The realm of freedom really begins only where labour determined by necessity and external expediency ends. It lies by its very nature beyond the sphere of material production proper. Just as the savage must wrestle with nature to satisfy his needs, to maintain and reproduce his life, so must civilized man, and he must do so in all forms of society and under all possible modes of production. This realm of natural necessity expands with his development, because his needs do too; but the productive forces to satisfy these expand at the same time. Freedom, in this sphere, can consist only in this, that socialized man the associated producers, govern the human metabolism with nature in a rational way. (ibid: 958–9)

It is evident from the above extract that in Marx's dialectic, nature is eventually overshadowed by the technological advances of productive processes. This is because it is human labour power which determines the quantity of surplus value available for expropriation by the owners of the means of production. As Marx says: "It then depends on the productivity of labour how much use-value is produced in a given time, and also therefore in a given surplus labour-time" (ibid: 958). In accord with his concept of "relative surplus value", Marx argues that in capitalism, the profit gained from labour is not purely contingent on the duration of surplus labour time "but rather on its productivity and on the more or less plentiful conditions of production in which it is performed" (ibid.). Given that "The realm of freedom really begins only where labour determined by necessity and external expediency ends" (ibid.: 958–9), it is clear that Marx advocated a notion of civilizing development in which technological advances in the modes of production untether the labourer from necessity, and its concomitant entanglement with nature. This domination of man over nature is made all the more complex because: "This realm of natural necessity expands with his development, because his needs do too" (ibid.). Nevertheless disequilibrium is produced by the correlative expansion of the development of "the productive forces" necessary to satiate "natural necessity" (ibid.). Freedom in post-capitalist society, according to Marx, can only exist when:

> socialized man, the associated producers, govern the human metabolism with nature in a rational way, bringing it under their collective control instead of being dominated by it as a blind power; accomplishing it with the least expenditure of energy and in conditions most worthy and appropriate for their human nature. (ibid: 959)

Significantly, Marx discerned that even within conditions of technologically advanced production, "the human metabolism with nature", inevitably "remains a realm of necessity" (ibid.). Marx argued that: "The true realm of freedom, the development of human powers as an end in itself, begins beyond it, though it can only flourish with this realm of necessity as its basis" (ibid.). Evidently, and according to Marx, the development of technologically advanced production ideally does not replace the need to work at crafting, shaping and forming nature into materialities that address necessity, but nature never advances beyond its primitive wilderness, thus nature "always remains a realm of necessity" (ibid.). Thus true freedom of the species involves continuously working up nature so as to embattle "conditions most worthy and appropriate for their human nature" (ibid.). Time is the reward seized from man's erudite technological advances in its embattled relationship, and time capture enables: "The true realm of freedom, the development of human powers as an end in itself, begins beyond it, though it can only flourish with this realm of necessity as its basis" (ibid.). Ultimately, and according to Marx: "The reduction of the working day is the basic prerequisite" (ibid.). There are several reasons why Marx's conception of freedom from necessity frustratingly limits the scope of the Marxist ecological gaze. In their book, entitled *Ecofeminism*, Maria Mies and Vandana Shiva (1993: 6) critically discuss Marx's account of freedom from necessity. Central to their praxis is the rejection of Marx's: "notion that Man's freedom and happiness depend on an ongoing process of emancipation from nature, on independence from, and dominance over natural processes by the power of reason and rationality" (ibid.). Mies and Shiva illustrate the perilous pitfalls that arise when formulating ecological utopias based on the imperative to superintend nature as part of the development of human capacities in the relentless "march from the 'realm of necessity' (the realm of nature), to the 'realm of freedom' – the 'real' human realm – which entailed transforming nature and natural forces into what was called a 'second nature', or culture" (ibid.). Axiomatic to the utopia of freedom from necessity is the idea that harnessing technological innovation enables a freeing of time, which can then be used as a realm for the cultivation of the self. An ecofeminist riposte coincides with modernist sceptics who identify the environmental devastation ensured by the improvidently utopian celebration of scientific technological application. Mies and Shiva are particularly scathing about the translation of the utopian narrative that positions freedom from necessity in terms of domestic labour-saving technologies, which promise to "emancipate" women from the reproductive labour time of housework. Such positioning is seen as paradoxical: "Today we realize that much environmental pollution and destruction is causally linked to modern household technology. Therefore, can the concept of emancipation be compatible with a concept of preserving the earth as our life base?" (Mies and Shiva 1993:7). In Marx's account of freedom from necessity, the actual material reproduction of the conditions for freedom implicitly constructs the reproductive labour of an "Other". Moreover, the transcendence from humanity's embattled entanglement with nature implies a circumvention of necessity that is only achievable through the reproductive labour time of an "Other". Recognizing this, Mies and Shiva state:

> Within a limited planet, there can be no escape from necessity. To find freedom does not involve subjugation or transcending the 'realm of

> necessity', but rather focusing on developing a vision of freedom, happiness, the 'good life' within the limits of necessity, of nature. (ibid:8)

Mies and Shiva (ibid.) define this approach to nature as "the subsistence perspective", for it recognizes the implausibility of achieving equity among human beings while also seeking to achieve freedom from necessity. In contrast to the pursuit of transcendence, it is argued that: "nature's subsistence potential in all its dimensions and manifestations must be nurtured and conserved. Freedom within the realm of necessity can be universalized to all; freedom from necessity can be available to only a few" (ibid.). Let us consider again *Securing London's Water Future* (Johnson, 2011: 16) and its account of managing London's rainwater:

> The Mayor's vision is that we adopt a more creative approach to managing flood risk from rainfall in London, taking opportunities to slow the progress of water from 'rain to drain' and using rainwater for non-portable uses to reduce demand for treated mains water. (ibid)

It is imperative that an ecological understanding of water sustainability interrogates the market form of environmental degradation, it is equally as important to expand the notion of economy so as to encompass "the multiple contributions of women's reproductive labour to the maintenance of capital" (Salleh, 2010: 208). Nature "enfolds human being" (Mellor, 1997: 184–5), rather than presupposing an idea of transcendence from nature, Mellor (ibid: 185) convincingly proclaims the imperative of assuming humans as "immanent beings". The condition of immanence is located outside of an Archimedean timeline, and thus it is recognized that: "The dynamics of the natural world exist in their own right with or without conscious human intervention … humanity cannot have a "grand narrative" to explain its own position … Ecologically, humanity exists in a condition of radical uncertainty" (ibid.).

Waste Water and Capitalist Accumulation

> In tandem with the Thames Tideway Tunnel, ensure sustainable urban drainage schemes are developed to mitigate the likelihood of surface water flooding and reduce rain entering the sewers.
> (Greater London Authority, 2013)

> Introduce innovative new planned waste water recycling options to meet the increasing demand for fresh water.
> (ibid)

> Disposal of wastewater. Vision. The Mayor believes that wastewater should be seen as a resource and not a by-product that is best kept out of mind. Opportunities should be sought to not just reduce the greenhouse gas emissions from wastewater, but to use it as a source of low-carbon energy. In the mid 1800s, Sir Joseph Bazalgette designed and initiated the building of London's combined sewers. Still in use today, they remove wastewater and rainwater in the same pipe from properties in central and inner London.
> (*Securing London's Water Future*, Johnson, 2011: 17)

> Putting rainwater to good use. Using rainwater rather than let it all go down the drain can help to relieve pressures on the drainage system – reducing flood risk and the demand for fresh water. We want to encourage:
> ▪ green roofs – for example, plant-covered roofs that make good use of the rain that falls on them
> ▪ rainwater harvesting – collecting rainwater to water your garden, for example
> ▪ grey water recycling – for example, using wastewater from baths to flush loos
> ▪ sustainable drainage – which helps reduce the volume and speed of water flowing into sewers.
> (Mayor of London, London Assembly, 2013d)

An oft-cited name in the history of London's municipal wastewater system is that of Sir Joseph Bazalgette. He is credited with having designed, during the Victorian era, the system of overflow outlets which collects from London's sewers and efflux into the River Thames and its tributaries; the Thames Tideway. London has 57 combined sewer overflows (CSOs) which work to stream untreated "diluted storm sewage (excess sewage and rainwater) … into the Thames Tideway after intense rainstorms" (ibid: 17). According to *Securing London's Water Future* (Johnson 2011.: 17), "Discharges occur at some CSOs between 50 to 60 times each year." Evidently, increased rainfall is the cause of "over a million tonnes of untreated sewage and rainwater legally discharging directly into the rivers" (ibid.). This situation is chastised and repudiated in the report *Securing London's Water Future*, for it states that "Despite much improvement in the Thames this is clearly unacceptable in the 21st century and contravenes the Urban Waste Water Directive that requires all wastewater to be treated before it is discharged" (ibid.: 17). Significantly, Karl Marx (*Capital*, Vol. III, 1894/1981: 195), in the nineteenth century, also chastised London's municipal sewers, stating that: "there is a colossal wastage in the capitalist economy in proportion to their actual use. In London, for example they can do nothing better with the [human waste] produced by 4½ million people than pollute the Thames with it, at monstrous expense."

As capitalist industrialization expands, so too does the intractable problem of waste disposal, with much of this problem resulting from the capitalist system itself. Of primary concern is capitalism's market economy of competing agents. Marx observed how capitalist manufacturing, for example, "is marked by the waste of much material" and because companies compete against each other, the production of waste accumulates: "since nothing is done according to a social plan, but rather depends on the infinitely varied circumstances, means, etc. with which the individual capitalist acts, this gives rise to a major waste of productive forces" (*Capital*, Vol. II, 1885/1978: 252). Marx observed the paradox and contradictions that also blighted capitalism's endeavour to recycle the refuse of production, "its so-called waste products, back into new elements of production, either in the same branch of industry or in others" (Marx, 1894/1981: 172).

There is "a colossal wastage in the capitalist economy" (Marx 1894/1981: 195). Capitalization of processes, by which the waste outputs of production are reintegrated into the cycle of production, is predicated on social labour; and thus mediated by the existing structural inequalities between capital and labour. Capitalism's colossal output of waste is an intractable consequence of its expansive assemblages of production, and thus "results only when labour is carried on a large scale" (ibid: 196). Nevertheless, this imposing scale of waste output presents the avaricious enumerations of capital with additional realms for commodification; for just "as the capitalist mode of production extends, so too does the utilization of the refuse left behind by production and consumption" (Marx 1894/1981: 195). Capitalizing on waste production and transforming refuse into exchange-value continues to be a feature of municipal water management, and is an evident feature of London's water strategy. Consider the following reflections on recycling wastewater, presented in *Securing London's Water Future*:

> It is clear from what has been said previously that there is little scope for developing new sources of river water or groundwater in and around London. However, reclaiming wastewater from sewage treatment works (referred to as 'effluent reuse') is a significant potential new resource that Thames Water and other companies are currently investigating. It would be a dependable source but it may have an 'image' problem. Ensuring drinking water integrity, understanding customer attitude, and minimising

> the increased energy consumption are the key associated issues. (*Securing London's Water Future*. Johnson, 2011: 43)

It is interesting that a consumerist "image problem" is presented by *Securing London's Water Future* (ibid.) as the primary aversion from "developing new sources of river water". Nevertheless, elsewhere in the report it is evident that in the: "Disposal of wastewater ...The Mayor believes that wastewater should be seen as a resource and not a by-product that is best kept out of mind", evidently "Opportunities should be sought to not just reduce the greenhouse gas emissions from wastewater, but to use it as a source of low-carbon energy" (ibid.: 17).

Capitalism orchestrates the conversion of waste output into valuable raw material components of production; these materials function to reduce the expenditure of raw constituents, because they generate capital from their resale; thus reducing the cost of "normal wastage, i.e., the average quantity that is lost in the course of processing" (Marx 1894/ 1981: 173). To the extent that capitalism, through resale of waste, is able to reduce quantities of constant capital, its accumulation of profit is correspondingly escalated. Marx observed how the capitalization of waste output required manufacturers to be involved in the design of augmented machinery "so that materials that were previously unusable in their given form are converted into a form suitable for new production" (ibid: 196). In the realm of science, chemistry's consummate achievement of formidable discoveries in the extraction of the useful properties of waste material has progressed the broad conditions for the reutilization of waste output. Marx (ibid: 196) observed that economies of scientific application, with respect of waste management, were also achievable in small-scale horticultural agriculture, but also that "agricultural productivity is obtained in this system only at the cost of a great productivity in human labour-power withdrawn from other spheres of production". Consequently, economically recycling the waste output generated by large-scale production requires corresponding scales of processing. To this end, capitalism directs some effort towards machinery, thus recognizing that a reduction in the output of waste material "and the maximum direct use of all raw and ancillary materials" employed in the process of production, are contingent on the calibre of the machinery utilized. As Marx (ibid: 197) states: "The most important thing, however, is that it depends on the quality of the machines and tools that are used whether a greater or lesser part of the raw material is transformed into waste by the production process." Thus, advances in the machinery provided gave rise to the possibility of advances in waste reduction; but herein resides a paradox, because increased efficiencies in the ability of machines to expedite tasks heighten productive capacities and accelerate the demand for resources. Thus, improvement in the capacities of machinery to reduce waste has the paradoxical effect of depleting nature, rather than rejuvenating natural resources.

The paradoxes and contradictions which besiege waste capitalization, have featured significantly in ecofeminist critiques of neo-liberal models of environmental sustainability. In an essay entitled: "No Sustainability without Justice: A Feminist Critique of Environmental Citizenship" Sherilyn MacGregor (2006) articulates the long-established feminist criticism of the masculine bias in political practice. Of particular relevance is the generic schema of citizenship and its predilection towards masculine attributes. MacGregor (ibid: 102) convincingly argues that both classical liberalism and neo-liberalism have consistently disguised gender inequalities upon which foundational assumptions of citizenship are formulated; that is to say they disguise the masculine subject of citizenship. As MacGregor puts it:

> The most significant problem from my feminist perspective is the assumption of a generic model of citizenship that masks realities of gender (and other forms of) inequality while depending on a division of labour that frees citizens to participate in the public domain.

It is in the sphere of environmental citizenship that the masculinized attributes of the citizenship ideal ensnare and entangle the inequalities of gender, time and the domestic division of labour. This has most evidently involved:

> [the] paradoxical coupling of labour and time-intensive green lifestyle changes with increased active participation in the public sphere; silence on questions of rights and social conditions that make citizenship practice possible; and a failure to acknowledge the ways in which injunctions to make green lifestyle changes … dovetail into neoliberal efforts to download public services to the private sphere. (ibid: 102)

It is essential to be aware that, fundamental to the classical and neo-liberal conception of ideal citizenship, is the notion of the rational subject, who is independently equipped to participate in the public civic sphere; and who achieves this freedom through the reproductive labour of an *Other*. Thus, it has been observed that: "Fundamental to this citizenly identity are autonomy, self-determination, rationality, and the maintenance of a clear boundary between public and private life" (ibid: 103). In classical Greek philosophy, women were excluded from the category of ideal citizenship, because their association with reproductive labour was believed to predispose them towards emotion as opposed to reason; moreover, childbearing duties prevented women from participating in the public civic sphere. Despite generations of successful appeals to enable equal participation in politics and equal rights to be heard in public civic engagement, feminists convincingly identify enduring and stifling patriarchal legacies limiting women's involvement as complete citizens of "their political community" (ibid.). Of particular concern is the coupling of time and duty as part of the rights and responsibilities of environmental citizenship. According to MacGregor (ibid: 107), at the heart of the neo-liberal governance of environmental sustainability is a paradoxical tendency to presuppose that the rights and obligations of the environmental citizen, involve voluntarily assumed and obligatory ascribed lifestyle changes and duties. These lifestyle changes, articulated through the public sector and governmental advertising campaigns, prescribe that the good citizen assumes "both greater self-reliance and eco-friendliness and greater participation in the green public sphere" (ibid.). Consider the following extracts, prescribing due diligence in water usage, provided by the Mayor of London's office:

> As for water, we get through far more than other parts of the country – an astonishing 167 litres per person per day, as opposed to 146 litres for the average non-Londoner. No doubt some will claim that is because Londoners take more and longer showers – but it seems more likely that we are simply more wasteful; and with demand for fresh water likely to exceed our supply by 125 million litres per day, we need to rein it back. We need to meter our water consumption more effectively, and beneath our feet we need to address the crisis in the sewers. (Greater London Authority, 2013:47)

> The strategy calls for organisations involved in the city's water management to: … [s]upport and encourage Londoners to take practical action to save water, save energy and save on their utility bills (a standard package of water saving measures can save a household around 35,000 litres of water per year and £90 off their bills). (Mayor of London, London Assembly, 2013c)
>
> At the heart of the strategy is a six-point plan to reduce London's water demand. At a time of decreasing supply and increasing demand for water, it makes sense to use the water we have more wisely. The strategy promotes increasing water efficiency and reducing water wastage to balance supply and demand for water, safeguard the environment and help tackle water affordability problems. (ibid)

In media campaigns broadcast by local and national government agencies, the environmentally responsible citizen is defined as a subject that assumes their rights to a healthy environment by actively taking responsibility for: recycling refuse, utilizing stewardship in the modest use of utilities, and demonstrating a commitment to sustaining the ecology of their local community. Neo-liberal constructions of the ideal environmental citizen clearly require immense amounts of time to be dedicated to the daily routine of recycling, moderating, and self-reliantly contributing to achieving environmentally sustainable communities. Consider the following prescriptions for investment in the communal activities necessary to secure London's water future:

> The strategy also promotes actions to enhance London's environment and build community capacity to help manage flood risk. Green roofs, green spaces and naturalising rivers can help to absorb and retain water, as well as improve the environment for people and wildlife. The Mayor will support communities at potential flood risk to increase their resilience to flooding through helping them develop their own Community Flood Plans. (ibid)

Viewed through the lens of ecofeminist politics, it is clearly evident that the model of citizenship prescribed by neo-liberal environmental governance is a subject whose active citizenship is enabled by gendered inequities in the domestic division of labour. As MacGregor (2006: 107) states: "This [model] reveals a lack of consideration for the politics of the private sphere, for the likelihood that these injunctions will result in an intensification of the activities that are already divided unequally between men and women". Consequently, what is particularly disconcerting about the "green duty" agenda is its disregard for the reality of the domestic inequalities that suffuse the private sphere (ibid: 113). It is evident that the ascription of rights and responsibilities to the issues of environmental sustainability involve a domestication and relegation in the private sphere, of a wider international obligation, as is evident in the following extract from Mayor of London's *Maximising Water Efficiency in London* prescriptions:

> Less than a quarter of London households have water meters, which means over three-quarters of London households are losing out on the benefits. Water meters not only help improve water efficiency but also energy efficiency – which is doubly good for the planet and for your pocket. Research has shown that on average household metering reduces water use by between 10 and 15%. It's at the heart of improving water efficiency, which is why we are working with water companies and other partners to

> support the introduction of water metering throughout London. Heating water for washing and cleaning creates 27% of households' carbon emissions. We believe that all houses in London should have meters installed by 2015, and all blocks of flats by 2020. All new flats should have an individually metered water supply. Charges for metered water should encourage water efficiency while protecting vulnerable and low-income households. (Mayor of London, London Assembly, 2013b)

Focus on the private sphere of waste recycling and moderate household consumption of utilities, shifts attention away from national governance and the intractable environmental crises precipitated by global capitalist production. MacGregor (2006: 114) convincingly highlights convergences between neo-liberalism's aim to dismantle the welfare state and its pronouncements for good citizens to independently assume rights and responsibilities to enable environmental sustainability. These observations relate to the right and left of the neo-liberal spectrum; for environmental "citizenship has become a solution to a problem; it is regarded as a way to both enlist public participation in the management of national affairs, and relieve the duty of government to provide goods and services to the population" (ibid: 114). Synergies exist here between the advocacy of good environmental citizenship and the responsibilities espoused in the 2011 London Mayoral prescriptions for the management of waste water:

> In the areas of London with separate surface and foul drainage systems, problems occur when the two systems are interconnected, known as misconnection … Householders, or professional plumbers often inadvertently, illegally, misconnect household appliances or waste pipes, as part of small-scale domestic improvements … Fat, oil and grease (FOG) contribute significantly to blockages in sewer systems and these often result in flooding of properties and/or the pollution of watercourses. (*Securing London's Water Future*, Johnson, 2011: 98–9)

The prescriptions for the management of wastewater appeal to the good citizen's sense of duty to demonstrate stewardship in the moderate use, recycling and reduction of water usage. These duties are presented as ways for the citizen to demonstrate proficiency in their civic conduct. Ecofeminists reveal a more insidious feature of these campaigns; for the time-laden duties that they prescribe are invested with "a gendering of environmental duty" (MacGregor, 2006: 117). Even the most cursory perusal of the signifying practices, visual culture and discourse of environmental advertising media, makes transparent the gender distinction in their subject and audience appeal. As MacGregor says:

> Insofar as women are household consumers, they too become the intended audience for morally based prescriptions for greening the household … Advertisers also know to whom they ought to target their ostensibly green products (e.g., unbleached cotton diapers, nontoxic and biodegradable cleaning supplies, organic produce). Here, the focus on individual choice confuses the meanings of citizenship and consumer. Either way, when the future of their children (and of course their children's children) is used as the reason for being ecologically responsible, women are apt to feel guilty; their compulsory feminine altruism is thereby exploited for the public (and increasingly corporate) good. (ibid: 117)

MacGregor presents a convincing account of the gender inequities that beset neo-liberalism's "green duty" agenda. In its determination to inculcate a "duties discourse", the *Securing London's Water Future* strategy partakes in a domestication of water sustainability. The rallying call for Londoners to assume rights and responsibilities in the achievement of water sustainability has relevance to the manner in which neo-liberal governance mobilizes the discerning stewardship of consumer citizens. Consequently, water sustainability is reformulated into a discourse of proactive "green duty" as opposed to a critical discussion of capitalist market economies of water supply.

International Water Imports: Concurrence in Primitive Accumulation

> In the case of an extreme drought, a drastic supply measure is to import water from another country, for example as Barcelona was forced to do in 2008. Thames Water's WRMP identifies shipping water from Norway as an emergency measure. (*Securing London's Water Future,* Johnson, 2011: 44)

The vast proportion of London's water supplies derive from: "the rivers Thames and Lee, with about 70 per cent of all the water taken from the freshwater River Thames upstream of Teddington Weir" (ibid: 31). Water sourced from these rivers is then stored in the capital's numerous reservoirs. Additional sources of water supplies include reserves abstracted from the "confined chalk" which is submerged underneath the clay of the city of London's basin. It is estimated that the summer temperatures experienced by London have increased by an average rate of 0.73°C per decade over the duration of the last thirty years (ibid.: 40). *Securing London's Water Future* predicts that London's water supplies will be subject to the increased potential of drought as a consequence of the climatic impact of more extended drier summers. *Securing London's*

Water Future presents the international import of water as an imperative response in "the case of an extreme drought" (ibid: 44). In an associated discussion entitled "Global Water Footprint", the report highlights the extent to which London's water consumption has long since exceeded its national boundaries; as noted in the following extract:

> Water is involved in everything that we use and consume e.g., the manufacture of goods; their transport; and possibly their use and maintenance. When the water used to produce and supply the goods is taken into account, the true demand for water from Londoners is far greater than the 167 litres per person per day. It's over 28 times greater, at around 4,645 litres. The majority (about 60 per cent) of the water comes from outside the UK – embodied in goods and services that are imported and consumed in the UK. (ibid: 66)

The concept of "embodied water" has environmental relevance, as the report acknowledges. Thus, "embodied water" is defined as "indirect water use" or "embedded water" (ibid.). Water embedded in the manufacture of goods directs attention to the global impact of local water sustainability, for goods, and their transportation involves a global expansion of London's water dependency. *Securing London's Water Future* (2011: 67) recognizes the increasing global expanses of London's water consumption:

> If present levels of consumption continue, two-thirds of the global population will live in areas of water stress by 2025. Climate change, population change, and growing levels of income with a corresponding demand for water-intensive goods are putting pressure on global water sources. As manufacturing falls in London and the rest of the UK, we are effectively exporting our demand for water. (ibid: 66)

Marx (1887/2003: 713) observed "merciless vandalism", expropriation, dispossession and desolation as alarming features of primitive accumulation; a treacherous state in which the essential prerequisite of social labour, i.e., investment in the means of production, was seized from "the peasant of the land which he cultivates", in conjunction with their "free individuality". Capitalism's reign of control over the fructiferous forces of nature involved "the expropriation of the great mass of the people from the soil, from the means of subsistence and from the means of labour" (ibid: 714). According to Marx, primitive accumulation tears the self-earned producer away from an embodied social metabolism with nature: "the fusing together of the isolated, independent labouring-individual with the conditions of his labour", transposed "by capitalist private property, which rests on exploitation of the normally free labour of others, i.e., on wage-labour" (ibid: 714). It is evident that Marx perceived primitive accumulation, i.e., the "painful expropriation of the mass of the people" as eventuated at a specific epoch in time, i.e., it "forms the prelude to the history of capital" (ibid.). Having extirpated the conditions for independent subsistence labouring, it is consistent with Marx's ecology to have observed that capitalism set about transforming land and its natural resources into relentlessly exploited objects of capitalist accumulation. Expropriation is a process of the centralization of production capital, labour which in due course extends geographically, leading to "the entanglement of all peoples in the net of the world-market, and with this, the international character of the capitalistic regime" (ibid: 714–15). But capitalism propagates the very conditions of its own extirpation and self-defeating refurbishment.

In a chapter entitled "The Modern Theory of Colonisation", Marx (ibid: 716) establishes a basis for the consideration of primitive accumulation as a continuous feature of capitalist expropriation. But this latter proposition emerges through speculative analysis; indeed, Marx counterpoises the continual tendency for a capitalist regime to primitively accumulate. Thus, for example, Marx states that "In Western Europe, the home of Political Economy, the process of primitive accumulation is more or less accomplished" (ibid.). Writing in the nineteenth century, Marx had observed the rapid and ferocious march of capitalism in Western Europe, engulfing entire domains of national production, leaving only the lesser developed strata of society into which it sedimented the embryonic conditions for decay and decomposition. Having accomplished, through primitive accumulation, the annihilation of previous primitive forms of production, capitalism relentlessly marches forward into new and untrammelled territories.

> We have seen that the expropriation of the mass of the people from the soil forms the basis of the capitalist mode of production. The essence of a free colony, on the contrary, consists in this – that the bulk of the soil is still public property, and every settler on it therefore can turn part of it into his private property and individual means of production … Capitalistic production advances there with giant strides … However, we are not concerned here with the condition of the colonies. The only thing that interests us is the secret discovered in the new world by the Political Economy of the old world, and proclaimed on the house-tops: that the capitalist mode of production and accumulation, and therefore capitalist private property, have for their fundamental condition the annihilation of

> self-earned private property; in other words, the expropriation of the labourer. (Marx, *Capital*, Vol., 1, 1887/2003:719–24)

This observation involves Marx (ibid: 716) in a debate about colonization. Here, again, it is evident that Marx positions primitive accumulation as a prelude to capitalism; Marx refers to the capitalist as having "at his back the power of the mother-country" (ibid.). Rather than envisage a co-existent relationship between capitalist expansion and primitive accumulation, Marx comprehends a linear progression from an accomplished primitive accumulation through to a mature expanding capitalism. Consequently, Marx consistently describes colonization as a prelude, a precursor to capitalism's establishment; an instance in the forcible dispossession of indigenous people and the seizure of their resources: capitalism "tries to clear out … by force the modes of production and appropriation, based on the independent labour of the producer" (ibid.).

Postcolonial Marxist writers have, in recent times, identified primitive accumulation as an integral feature of global capitalism. In a chapter entitled "Environmental Justice Cross-Culturally", Filomina Steady (2009: 47) details the key principles of a post-colonial environmental challenge to global capitalism. "Environmental justice" espouses the principle that: "all people and communities are entitled to equal protection of their environmental, health, employment, housing, transportation, and civil rights laws" (ibid: 48). It is a disconcerting feature of global capitalism that the pursuit of profit dissuades companies from agreeing to increase their production outlay by adopting more environmentally beneficent and eco-friendly technologies. Postcolonial environmentalists associate these tendencies with an ideology of colonialism, which structures the prevailing "development paradigms upheld by the neoliberalism global economy of corporate globalization" (ibid: 49). Central to this enduring colonial ideology is the pronouncement that globalization accelerates the advance of economies in the developing world. Postcolonial environmentalists challenge this presupposition, for they invariably argue that "corporate globalization destroys the economies, environments, social fabric, and peoples of the Global South" (ibid.). Postcolonial environmentalists thus challenge the mainstream economist belief that environmental sustainability is enhanced through technological development; accordingly, it is assumed that the capitalization of production technologies in the Third World is the basis for growth and economic prosperity. However, postcolonial environmentalists highlight countless times in which developing nations have been subject to the devastations of global capitalist production. It is evident that:

> Western technologies have created much of the environmental problems in the 'Third World' through export or dumping of toxic chemicals and hazardous wastes, and dirty industries involving pesticides, asbestos, the building of dams, petroleum refining, mining, and the transfer of outdated and dirty technologies. (ibid: 50)

Returning to the strategy of internationally importing water, detailed by *Securing London's Water Future* (Johnson, 2011: 66), this report is evidently cognizant of the geopolitical perils that await shifting the encumbrance of London's water supply into international territories; for this requires encountering territories where "water governance and conservation may be lacking, it not only poses various risks associated with dependency on fragile global water resources for our supplies, but also creates numerous environmental impacts in these regions" (ibid: 66). Postcolonial environmentalists argue that the modernization theories that espouse technological development without heeding environmental concerns operate to legitimize the colonial domination of the natural resources of the developing Third World. While evidently cognizant of these issues, *Securing London's Water Future* (ibid: 66) is disconcertingly silent with regards to gender and the extent to which environmental degradation in the developing world is disproportionately experienced by "the feminization of poverty" (Steady, 2009: 56). Postcolonial ecofeminists embrace a fundamental interrogation of capitalism and nature in the context of the Global South. Global markets for the international importation of water have relevance to postcolonial ecofeminist concerns regarding "conservation as enclosure"; "debt-for-nature" biopiracy and the expropriation of the meta-industrial labour that is traditionally carried out by women.

Timely Reflections

> Of the rain that falls in the Thames catchment, two thirds is either lost to evaporation or used by growing plants (transpiration). Of the water that is then 'available', 55 per cent is abstracted for use, one of the highest amounts in the country. All the available water cannot be taken because some must be left to protect the natural environment including fish, river and riverside plants and water birds. Of all the water abstracted, 82 per cent is for public supply and out of that half is supplied to households and a quarter to non-households (the remainder being lost through leakage). (*Securing London's Water Future,* Johnson, 2011: 98–9)

In 2011, the Greater London Authority published *Securing London's Water Future,* a programme of initiatives, policies and ideas directed at establishing the city's water resources; managing rainfall; dismantling "the hydrophobic policies of the 1960s – which saw natural rivers and waterways encased in concrete"; and committing the governance of London to working "in harmony with water in our landscape to ease the consequences of heavy rainfall and beautify our city at the same time" (ibid: 6). *Securing London's Water Future* claims its dedication to environmental ecology and securing long-term strategies of water sustainability. The programme for securing London's sustainable water supply pledges to commence:

> The Mayor's vision ... that we adopt a more creative approach to managing flood risk from rainfall in London, taking opportunities to slow the progress of water from 'rain to drain' and using rainwater for non-potable uses to reduce demand for treated mains water. (ibid: 16)

> The Mayor believes that wastewater should be seen as a resource and not a by-product that is best kept out of mind. Opportunities should be sought to not just reduce the greenhouse gas emissions from wastewater, but to use it as a source of low-carbon energy. (ibid: 17)

While it is evident that these ecological principles resonate throughout the water strategy, it is also evident that the contradictions and conflicts of capitalist production encumber many of the well-intentioned policy initiatives. In September 2012, Thames Water published: *Why Does London need the Thames Tideway Tunnel?* (Thames Water 2012). In this publicity material, the then proposed Thames Tideway Tunnel sewage system was presented as:

> The proposed Thames Tideway Tunnel is the final part of the overall solution needed to significantly reduce sewage discharges into the River Thames. The independently-chaired Thames Tideway Strategic Study (TTSS) established in 2000, led to three integrated solutions, known collectively as the London Tideway Improvements, to solve the problem of London's overloaded sewage works and overloaded sewers:
>
> - upgrades to tidal sewage treatment works (now under construction)
> - the Lee Tunnel (now under construction)
> - the Thames Tideway Tunnel.
>
> Without the London Tideway Improvements, the annual CSO [combined sewer overflows] discharges would reach 70 million cubic metres in a typical year by 2020. Once all three parts of the solution are implemented, that figure is estimated to fall to 2.6 million cubic metres in a typical year. The annual number of CSO discharges would fall from over 50 times in a

> typical year to just four or less. The Environment Agency is satisfied that the predicted water quality impacts of these remaining discharges are acceptable. When both the tunnels are operational, overflows should only happen very occasionally, after sustained periods of intense rainfall and after the tunnels have captured the most damaging 'first flush' from the sewers. This will be a huge improvement on the current situation. (ibid: 6)

Researching water ecology and ecofeminist Marxism has involved many interviews with river conservation organisations, several of whom were, at that time, directly engaged in consultations on the building of the Thames Tideway Tunnel. The water ecology ideals espoused in support of this engineering solution are highly insightful, as is evident in the following transcription extracts from an interview with a Senior Programme Manager, London River Thames Conservation Charity III, September 2013.

> ***Interviewer: If we were to theme some of those issues around a recent area of engagement which is Thames Tideway Tunnel. So if we have that as the theme. So how has [London River Conservation Charity III] contributed to that; at the level of policy; and in terms of your initiatives in relation to local solutions? What sort of conversations have you been formally involved in, in terms of that?***
> **Interviewee:** *We were one of the founding members of the Tideway Tunnel Now coalition, which was a group of NGOs; some national, some local. You know, people like the RSPB [Royal Society for the Protection of Birds]: national charity. And people like us. Because when the idea of the Thames Tideway Tunnel was made public, there's quite a lot of resistance. Particularly in areas of the river where the work sites are going to be, because you're talking an area the size of two football pitches; you're talking about drill machinery going in and about sediment coming out. You know there's going to be a lot of disruption during the construction phase. Which became, I guess, as part of the consultation, sort of wrapped-up with people's concerns about the money that is going to be spent, because it is going to cost every Londoner [interviewer: "3.6 billion"]. Yes, so there's quite of lot of concern about that. There's this idea that perhaps it becomes this huge money-making asset for Thames Water and they're doing it to make money. Erhm, I think, the journey to find the right solution for this combined sewage overspill problem has been worked on just before I worked at [the charity]... and I have been working for [the charity] for ten years; it's not just, you know, come out of the sky. And it wasn't just Thames Water who were driving it. It was the EA [Environment Agency] and it's been, we will get fined by Europe under the Urban Waste Water Directive, or whatever it is, for our continued [pollution of River Thames]. We need to find the solution. But the debate has become mixed in with what has become quite, I guess, confused, it seems to me. So the*

Tideway Now coalition was various influential NGOs, who are either working directly or working on issues directly related to the Thames saying: 'You know we think the Thames Tideway Tunnel is the best solution; we know it's gonna cost a lot of money; we know it's disruption; and we are aware that there are other, not solutions, but there are other options to deal with excess urban water. But London is not suitable or, the problem facing London is so large, we, we need a very large solution to deal with it. That doesn't mean that things like SUDs Sustainable Urban Drainage are not important, but they cannot by themselves provide the solution, the Tideway Tunnel can. The volume of water is too vast.

Interviewer: Tell me what the Tideway Tunnel initiative is; so it's to be a tunnel?

***Interviewee:** It is to build a massive interception. It's not really a tunnel actually; it's a long thin holding tank. It's basically to build a huge tunnel under the river, which I think is about 7 metres in diameter. It's vast and it will intercept sewage from a lot of the overflows; most of the overflows which are currently just overflowing into the river. So when you have light to moderate rainfall. If you have moderate rainfall somewhere in London, it generally means that there is too much water. We have a combined sewage system; too much water goes into the system. The same pipes that take waste, household and domestic waste take surface water from the drains and roofs. So when it rains, there is a much larger volume of water trying to flow through these pipes. The pipes can't take the volume because the sewage system was based on Victorian design, built for far fewer people, using far less water. It's a glorified version of Joseph Bazalgette's solution. Joseph Bazalgette's solution to the problem was to build in extra sewers, so the Embankment, these lovely tree-lined embankments we have, were not built as a tree-lined embankment. They were built to house big tunnels that would take sewage east. So all the water, all the sewage that was running down, through the old drains and streams and sewage straight into the Thames, they built a tunnel along the bottom, so instead of running down into the Thames it will be picked up and taken east. The interceptor sewer, the Tideway Tunnel is almost a larger version of that. There are just now the overspills, because the large tunnel underneath the Thames and the big main treatment works at Beckton Crossness, you know, are going to be expanded, upgraded, and then they'll be able to pump the content of the tunnel out and treat it when they can. Rather than, so it is basically slowing down this huge volume of water. It's been held in the tunnel until they can treat it as Beckton Crossness. So it's a massive undertaking, huge.*

Interviewer: Why was there confrontation? There seemed to be confrontation in terms of whether this was a feasible solution? There seemed to be a lot of 'for and against', and I just wondered, what were the problems there?

Interviewee: *First of all, this is a personal opinion rather than an organizational one. My personal opinion, having been on the foreshore for ten years, and picking up sewage-derived debris, you know, you're picking up condoms, sanitary towels and nappies, and wipes and the evidence of what people flush down the toilets, they shouldn't find its way into the river right. And that's [referring to river clean-up where volunteers collect litter from the foreshore] a very clear way of showing people how much sewage is going into the river. Which is, you know, is wrong for a world-class city. So we have had those conversations on the river with our volunteers. Volunteers are picking up some, [saying] 'Why am I picking up sanitary towels every five metres sort of thing?' Well, that's because not far from here there is a combined sewage overspill and when it rains, that's just throwing sewage out into the river. And the sanitary towels are just collecting in the mud here, that's why you're finding them. People don't realize that. So we've been involved at that level. And also, I think there's, you know, in terms of the, you see fish kills in summer time; you see bubblers out. So this sewage, especially in the summer when there's not huge amounts of fresh water flowing down river to dilute it, you have a summer rain storm event, suddenly have a lot of overspill in a fairly discrete area. Erhm, the oxygen, the bacteri[um] that's breaking down the organic matter and sewage suck the oxygen out of the water so the fish suffocate. So you see that because we are out on the river lots, and we're well placed to talk to people about that.*

Erhm, and I think we felt that we wanted to make a public statement about the debates and that we believe this is the right solution, cos it is. It tackles it on a very large scale, on a scale it needs to be tackled. Now, it is difficult to separate, there's two-fold debate: there's people who are concerned about the disruption to where they live from the construction sites; from drive sites and then there's also the people who are saying this is not the right solution it's a lot of money or you know it is. And I think there's quite a lot overall between those two groups. And I think from our point of view, we just wanted to clarify what we thought about the reasons for doing it. The reasons for doing it were: we cannot continue to let sewage go into the river, and the impact of sewage going into the river is obvious both to how it looks and to the wildlife. And, erhm, we just felt, that you know we had to take a stand on that and that we believed this was the right solution. You know, the other solutions which we talked about: putting more hydrogen peroxide; low O_2 *events and SUDs. They were not feasible. We're working on a SUDs project, I think, in North London. You know we very much support more SUDs; but the volume we are talking about, with narrow London streets, especially in the main body of London, the older body. In the narrow streets; with combined sewage system; to firstly be able to separate the system, I mean digging up every single road to do so, where there is one pipe lay two, imagine the disruption and the cost of that. And to create SUDs needs space. You know, a SUDs system is, generally, what a SUDs system does is, you know, a pond, if you like, where the water flows into, sits and then drains away slowly, leaving behind often, where the water is sitting, settle out so sediment and pollutants settle out and you can put reeds or other plants which will help purify the water so the water drains away more slowly. You get cleaner water. So they are great. It's a system that slows down the movement of water or allows it to drain away more naturally. But that requires space and where you've got so much hard surfaces, so much, huge volumes, the space required for SUDs solution doesn't exist. I mean, it just doesn't.*

Interviewer: Why do you think that there has been an intensification of this issue? What do you think are the main drivers?
Interviewee: *It feels like all of a sudden, but it isn't. It feels like all of a sudden because they have finally got, you know, they've finally got the specifics of the solution. They've got drawings, they've got plans, they're at the point of implementation. And for, I started working at [London River Conservation Charity III] in 2001, I remember talking to the people who have spent five years looking at this solution and discussing it. Information has been available on this problem and the solutions to it since, well since the 1990s. Erhm, but I think that people don't pay attention, as much attention to a theoretical debate as they do to the point when someone says, 'Do you know what we are ready to say? I want to dig a massive hole here.' And that's when people go, woah, hang on, you know, this idea has been talked about for the last 10, 15 years but it is only when someone says we are going to start drilling the hole that a lot of people go, or many more people suddenly engage in debate. Thames Water, in all fairness to them, have made massive efforts to consult really widely, to explain why, they sort of understand that: (a) everyone is going to see a change to their bill; and (b) some people are going to see a lot of disruption on their doorstep. And so they have to talk to as many people about that as possible. So I think it's purely just a point in the project at which you are really close to something happening at which people naturally are going to pay more attention. It's the point at which people naturally are going to pay more attention, it is also a point that there's a lot more information to talk about. That's why it suddenly feels that this thing has kinda of landed. It wasn't quick. It's been torturous for years; them trying to sort of, you know, what to agree to be able to fund it through bills and it's been going on and on; it's been referred; I mean the Labour Government referred it back a few times to check. And then the new government came in and they referred it back again. They agreed it was the right solution, so it's been a long and slow process.*

Interviewer: Because it is an engineering solution, do you think there is less of a reliance on behavioural changes in relation to the consumer?
Interviewee: *In a way, yes. I think that's probably true, but I do think that, erhm, there's, whether it's going to provide a useful way of drawing people into the debate about water use I don't know, but there are certainly many*

water-related issues coming. Particularly for a big large urban collection like this, you know, we are likely to see: there's the Twenty One Hundred Project which is about protecting London from flooding in the future; we are likely to see changes in rainfall pattern; we are likely to see sea level rise. We are likely to see more issues about water usage based on drought. If we have climate change, water is going to become absolutely something which, as a city, we need to be aware of how we use it and what happens with it. And also, you know, as a space, water is really powerful place for people to connect to. People love being around water.

Interviewer: So do you think that now some of the issues about water quality have been addressed by the Thames Tideway Tunnel, that there is less responsibility for the individual?
Interviewee: *Yes I certainly think that will probably be true; but remember there's quite a long time before the thing appears. And I imagine when it's built, and in fact we will have a project ... it has already started in the next few years. Where we are working with volunteers to look at and monitor the quality of various aspects of the Thames water: quality: measuring; invasives [plants], a whole range of, so people can record the state of the Thames, over the next few years. And I suspect that when the Tunnel is built there is going to be an awful lot of very positive publicity about how well the Thames is looking which will help people go: 'Oh, fine, dealt with, it's not a problem.' There are still going to be lots of other things to think about. How we use water, how we treat water, in an early environment; how much water we use and what we use it for? Those things will follow quite soon on the back of this. I think I agree, but my, what I would say to that is: the problem is one that requires an engineering solution.*

The scale of the problem, the amount that we are talking about requires an engineering solution. Because the problem is not what people are flushing so much, as the problem is what we do with water that comes out of the sky when it rains. Now that we can't stop; and I think the way London is expanding and the number of people living in London; I think it is impossible to find a sustainable solution for what to do with that water. I absolutely think that any new developments, new construction and every person should be encouraged to think about sustainable water usage, you know, if there is a new development it should absolutely include SUDs as part of the design for it. There should still, people should still be encouraged to use, collect rain water to water their gardens or use their bath water in a sustainable way. I think absolutely we should be doing that; but those things would never, never alone involve the problem that the Tideway Tunnel will go towards solving.

Interviewer: Just two more things in terms of the local solutions: How are the local solutions translating into the engineering solutions in relation to the Thames Tideway Tunnel? How is there a synergy between

your local solutions, community and support for something that is a massive engineering solution? Are there crossovers and bridges?

Interviewee: *Yeah, little ones. But the part of reminding people about the story is reminding people what it means to have lots of sewage going into the river. And while this solution exists, it is quite a long way before; there is a lot of ongoing debate. It's a long time before you start building it. So, as I say, I have spent so much time on the foreshore picking up stuff, and seeing impact. I think that it is absolutely important that we continue to remind people, that from now until the thing is turned on, you know today, for example, there will be raw sewage somewhere in the river; because of the rain that's falling. You just need to keep reminding people of that. So you know that work is still to be done. Despite the fact that the tunnel is this engineering solution that is coming; and also I think, erhm, I think, as I say, we have got a project coming up which is specifically designed to encourage local people to be involved in monitoring the health of the Thames. But also we do have a big project, sort of pilot project on a tributary of the Lee looking at creating SUDs projects in parks that are close to one of the tributaries of the Lee. Looking at local people saying where they can go and understanding what they are going to be [referring to the locations of the SUDs]. And being involved in the creating; and I think that's also really important and that we should continue to do that. Erhm, that's not, that's not so much about the Tideway Tunnel specifically; but of understanding, you know, what happens to water and the impact water has on rivers. And why we want healthy rivers; and the value of healthy rivers not just for wildlife, but to people. And our contribution is more than just. You know is more than just, making sure they look green and pretty, you know, we can have an impact on water that goes into them. It's not just about having nice trees; it's about ensuring that if the rain water is running off an oily road that oil is going into the river. We need to slow it down and prevent it. People need to understand this. I absolutely think we need to be directly involved in that sort of work too...*

Interviewer: I know that part of the pedagogy is active participation, and going into extract from the river, so how would you sum up your pedagogy? Is it to provide them with the information to become proactive politically?

Interviewee: *Yes, I think, well not necessarily politically. An NGO can develop a certain level of trust with a community or with the local people, that's the local council or national policy-makers, that's often a lot: oh, it's coming from the council. So we've worked quite hard to have a track record that demonstrates that we genuinely want to see change happen and the sort of change we want is change for the better: of both the river and the local people. So that's helpful when it comes to campaigning and messaging behaviour change. So that direct delivery stuff is what, you know, it gives us the authority or it gives an authority to then talk about the issues in a way that hopefully people will have respect for what we say.*

> *You know, we've demonstrated our commitment through our direct delivery work, so when we start talking about behaviour change; when we start thinking about campaigning; when we start being involved in sort of policy; when people see our name as a kind of 'consultee' on a policy document and they understand that what we have to say is of value and we have earned the right to sort of say those things, cos of our direct delivery*
> ...
> *As I said in the beginning, we were never... we didn't start off very strategic. We didn't have a particularly strategic approach to improving rivers. We didn't look at a river and go: well, if you spend five years working on this river, we can measurably improve the catchment in a particularly way. We started working because people wanted to see a particular improvement to this bit or there's a bit of rubbish there and I think we are finding our way in this higher level of campaigning and directing. But in terms of our attitude and our action, the focus is on delivery but we want to be sure that that delivery can be channelled to a respected opinion that we've earned and we sort of represent the people who will do stuff on the ground. We represent their feelings; we represent their concerns sort of accurately at policy level. So that's the sort of direction we're travelling in and I don't think we've properly defined that yet. But that is, that's the way we want to go, so if anything that's our pedagogy if you like. That's our aim for moving in that direction.*

According to Marx, transcendence from environmental necessity involves the use and development of technology so as to achieve the most efficient evolution away from the realm of necessity, and this innovation precipitates acceleration towards 'the reduction of the working day' (*Capital*, Vol. III, 1894/1981: 959). But capitalism produces a 'colossal' amount of wastage (ibid: 195), and its attempts to capitalize waste output are beset by intractable contradictions that are inherent in capitalist production. Marx observed that reduction in the output of waste material 'and the maximum direct use of all raw and ancillary materials' employed in the process of production, are contingent on the calibre of the machinery utilized (ibid: 197). As Marx (ibid: 197) puts it: 'The most important thing, however, is that it depends on the quality of the machines and tools that are used, whether a greater or lesser part of the raw material is transformed into waste by the production process.' Marx's ecology clearly advocates technological advancements in machinery as a vehicle for achieving environmentally sustainable manufacturing. Engineering solutions, such as the Thames Tideway Tunnel, might initially appear to resonate with Marx's Enlightenment view of nature. Indeed, a tension poignantly demarcates what, elsewhere, has been described as Marx's 'social ontology' (Sheasby, 1997). That is, an 'aesthetic conception of a humanized nature that set Marx apart from the Enlightenment utilitarian tradition' (ibid: 39). In the philosophy and culture of the Enlightenment, 'ideology and myth are to be conquered by knowledge and natural necessity is to be overcome through technical innovation and scientific rationality' (Mawhinney, 2000: 52). To this end, Enlightenment thinking was governed by a crusade for the mastery of nature through its subjugation to rational control. Scientific rationality had attained an unprecedented knowledge about the determinants and stabilizing patterns of the natural world. Such optimism reinstated the nonchalant asseveration that '"Nature" now exists "for us"' (ibid: 50). In its invocation as the inheritor of the Enlightenment dream of freedom, 'modernity's drive for mastery' over nature affirmed the constitution of nature as the 'oppositional "other" that must be opposed, denigrated, and conquered, but always preserved as the outside' (ibid: 50). Marx's ecology emphasizes capitalist production as beset by contradictions and profit-driven conflicts, which disrupt capitalism's ability to provide sustainable environmental solutions. Consider, for example, the following extract on Thames Water's disinclination towards Sustainable Drainage Systems (SuDS):

> SuDs involve a variety of measures to reduce the amount of rainwater entering the sewerage system and to slow down the rainwater that does enter the sewerage system. These include green roofs (such as grassed 'living' roofs) and soakaways. They require a lot of space and are generally both costly and disruptive to retrofit. We fully support the use of SuDS as they can enhance the environment, can effectively manage surface water flooding and have a low carbon footprint, not to mention very low whole-life operating costs. However, there are limitations to sustainable drainage, particularly in the Greater London area, where the drainage systems are complex, most of the land is already developed and there is huge potential for flooding. There is not enough space in London to retrofit sufficient SuDS to control the CSO discharges and meet environment objectives within the required timescale ... Retrofitting SuDS in the densely populated urban environment of London would have a detrimental impact on virtually every household, driveway, road and open space in every borough. The maximum practical level of retrofit SuDS would take over 30 years to implement and cost several times as much as the Thames Tideway Tunnel. The cost is estimated to be at least £13bn, and would not solve the problem. (Thames Water, 2012: 20)

Given that capitalist profitability is accumulated through the expropriation of relative surplus value production, commercial imperatives will inevitability lead capitalism to accelerate production, thus producing more waste output, and subsequently venturing to capitalize output through the extirpating forces of market economy.

This chapter has applied classical Marxist analysis to recent government water quality management and water sustainability reports including: the *Securing London's Water Future* (2011) programme. This report is of particular interest because there exist synergies that link the proposals with some of the most disconcerting features of capitalist accumulation, and furthermore, the report constantly situates London's present and future municipal water system within the context of its existing Victorian water network accredited to Joseph Bazalgette's design, in the nineteenth century, of the city's combined sewer network. As Mayor Boris Johnson expressly proclaims: "These strategies will help us to stand on the shoulders of Bazalgette and future-proof London for the challenges ahead" (ibid: 6). This chapter has been concerned with highlighting how the general conditions of capitalizing sustainable water supply require the achieving of transcendence from eco-centric metabolic exchange with natural water supplies, while also exploiting the national and transnational meta-industrial labour of people who are marginalized from "economic sectors" and "are exploited not so much through low wages but by their provision, "free of charge" of service and material inputs to capital" (Isla, 2006: 60). Materialist ecofeminists counterpoise the critique of ecocentric idealism (Foster and Burkett, 2001) through emphasizing the imperative to engage "directly with the artificial humanity-nature divide" and to do so through 'a shift away from "production" toward "reproduction" (Salleh, 2001: 447). Marxist ecofeminist analysis of water sustainability mobilizes activists "to be more reflexive about how they absorb and reinforce profoundly gendered forms of alienation" (Salleh, 2003: 65). For in so doing, it becomes clear that integral to eco-Marxism should be a focus on meta-industrial and reproductive labour as human energies are expended and exploited in the capitalization of water resources.

Figure 3.1b., "At the still point of the turning world" (T.S. Eliot *Burnt Norton*). Photographic Image, South London, River Thames, September 2013

PART THREE:

SUB-POLITICAL GEOGRAPHIES OF RIVER CONSERVATION

Figure P.3: "'O sacred time! Thy moments goes not down, But I go with it! Sixty coming hours, Are with us poor expectants of more price" (Ann Yearsley *Soliloquy*, 1796), Photographic Image, South London, River Thames, September 2013

Figure P.3: "'O sacred time! Thy moments goes not down, But I go with it! Sixty coming hours, Are with us poor expectants of more price" (Ann Yearsley *Soliloquy*, 1796), Photographic Image, South London, River Thames, September 2013

CHAPTER FOUR

ORGANISATIONAL VALUES OF RIVER CONSERVATION CHARITIES

Figure 4.1., "The river is within us, the sea is all about us" (T.S. Eliot, *The Dry Salvages*). Photographic Image, South London, River Thames, September 2013

Interview with River Warden Co-ordinator, London River Thames Conservation Charity I, September 2013.

> ***Interviewer: Could you tell me about your charity?***
> ***Interviewee:*** *So we started in, we celebrated fifty years last year. It was founded in 1962. And it was founded, strangely enough, by a man who worked for Richmond Council; he was one of the early environmentalists. He was a kind of, erhm, inspector of health, health inspector. And he was worried about the pollution in the river. He contacted various people and organizations and he realized that there wasn't one organization covering all the different groups in the river. There were plenty covering rowers,*

sailors, boaters, the commercial side, and so on. And that led to a meeting and it was quite well attended, and from there they decided to form [London River Conservation Charity I]. Basically to try and protect the river, promote the river, get involved with campaigns, and it has developed ever since then really.

Interviewer: And its main ethos?
Interviewee: *The main ethos is to promote and preserve the Thames. So the aim is basically to get people interested in the river. Make sure if at all possible, which is difficult, that none of the planning issues affecting the banks of the river impinge upon the river, which is not an easy thing to do in central London. Try and prevent pollution of the river; to promote the, support the environment as much as we possibly can. So it basically covers a whole gamut of activities. What we do, we tend to support local campaigns because we cover the entire river from the source to the sea. It's impossible to get involved in everything so what we do, we kinda support local campaigns, about a particular issues like Sunbury Lock. People around there don't want hydro-electric schemes inside the lock. We look at them and decide whether we are going to support them. That's basically what we do.*

Interviewer: Could you tell me how you work with communities?
Interviewee: *We have five branches: upper Thames; Middle Thames; Teddington to Royal Windsor; upper Tideway and central Tideway estuary. Those five branches, they are still quite large areas. They work with people that probably are members of different civic organizations. So we work with the community in that way. And when we get involved in campaigns, like a couple of years ago, the Environment Agency wanted to sell off a lot of the lock-keepers' houses and that became really a kind of issue for local people, who quite like their local lock-keeper living in a house; all kinds of emotional and also safety reasons. We got very much involved with local people as campaigns. And the same with, erhm, we get involved with clean-ups, you know. If there's rubbish, pollution, in an area you try and get involved with local groups. In other words, not try and do it ourselves; get local people to do it, but with us.*

Interviewer: … I'm interested in issues around education. What's the educational role of [London River Conservation Charity I]?
Interviewee: *We have an education officer, who is a retired head teacher and he has prepared material for school children … Key stages where there is emphasis on the environment. He's produced material on the Thames. He also has, in fact, I do and other people, we are quite often invited to give talks to schools. Particularly schools on the kind of field trip, you know. They want somebody to walk along the river with them, point out things and go on the foreshore, hopefully if it is in a tidal area. And, of course, we often get invited by adult groups, ordinary groups like*

the Rotary Association, to go along to talk about the Thames and our interests really ...

Interviewer: It'll be nice to hear that talk [history of London River Conservation Charity I].
Interviewee: *So 1960s, we were formed. In those days, in the 1960s, London, the tidal river was facing the closure of the docks and that had implications for what was going to happening to all the buildings, the warehouses, infrastructure, what was going to happen to the people who lived in the docks. And that really went through, I suppose, 1970s, 1980s, it was really a burning issue, the Isle of Dogs. People who, men, mainly men, who worked in the docks forever, they were out of a job and you've got all the movement to make it a business zone and, you know, an enterprise zone. People worried about whether they are going to be squeezed out by Yuppies. The so-called Yuppies coming in, etc. Pollution in the Thames has always been an issue, and I pointed out; again, certainly before the society was formed, 1950s, the Thames was virtually dead. If you fell in the Thames and you had an infection, you probably would go to the hospital straight away, because there was so much industrial pollution and, erhm, all the ships in the Port of London used to discharge sewage and everything into the river. And so we campaigned to try and improve the whole sewage system, update the kind of Victorian system. And, for example, in the 1980s, all the sewage south and north of the river used to go down to Beckton and Crossness, and was treated and then, if you can imagine it – don't think about breakfast for a moment – then everything went, was put into boats. They were sludge boats and it was taken out to the north, to the estuary and dumped. Of course, the trouble with that is a lot of it all came back up the river with the tide. So we campaigned against that and now, all that stuff is totally treated and so it's clean. And we are now supporting at the moment the Thames Water plan to build this Tideway Tunnel under the Thames from Hammersmith right the way down to Beckton and Crossness, which will, hopefully, prevent even now sewage going into the Thames. So pollution was an issue. And there were successes in that the Environment Agency played a key role ... Then there was regeneration of the river. To the extent now, well over a hundred species of fish in the river. But even so, they are endangered by these occasional [problems]. When there's heavy rainfall, you get these sewage flows coming in, so that's the last chink in the, chink we've got to block up ... On the planning side, it's been difficult because at one time, we used to receive applications concerning the river from every single local authority. When you think along the whole river, there is something like a hundred or so, so it was impossible to cope with; you know, everything came in. So we basically decided that we would, as I said earlier on, work with local people. So where local people were worried, say, in Oxford, something was going to be built that local people objected to, then we would support that campaign – so rather than initiate it ourselves. We also represent the*

kind of consumer of the river, to put it crudely; in other words, the people who are the walkers, the cyclists, the runners, the boaters, canoeists, and so on. We represent them on such bodies as the Environment Agency. They have to have a Thames Forum. The Port of London Authority has similar forums where stakeholder groups go along; we go along to Thames Water meetings. So we try and influence the policy-makers. I sit on the Mayor of London's Waterways Commission which advises him on issues to do with the river and the canals and waterways. And we also developed, what you mentioned when you approached us, our river warden scheme. And the idea was, and it's been successful, was to try and get a network of volunteers who would walk along the Thames Path, basically from the source right the way down to the Thames Barrier and, erhm, people who probably did it anyway, you know, walking their dogs, and every couple of months, get them to report to the coordinators if they had any issues. And those issues range from trees falling across the path; foliage overgrowing the path; pollution; trees falling in the river obstructing the boats; Thames signs vandalized, turned round the wrong way: a whole range of issues. And then, we then take those complaints up with either the Natural England, who look after the Thames park outside London and inside London, the various local authorities. And that's worked; we've got 50–55 or so wardens. And that's been quite a success really ...

Interviewer: So do you think that [London River Conservation Charity I] is mainly about individuals?
Interviewee: *The members are mainly individuals but we try and work with other communities. And so, for example, we start with the Environment Agency; we started a body called the River Thames Alliance about five or six years ago and that was on the non-tidal river. And the attempt was to try and get local authorities involved. Because, although local authorities – like Windsor is a tourist Mecca and they go there to see the river – the local authority has very little input as far as the river is concerned; financially they don't give much back to the river. In a very complicated history of how the Thames was administered, there was a body called the Thames Conservancy and that involved local authorities. And if you were a citizen of Oxford, a penny off your rates went to Thames Conservancy. So there was a kind of commitment; that doesn't exist any longer. So the River Thames Alliance was to try and involve councils and say to them, look, you get a lot from the river in terms of tourism, people cruising and walking, and so forth, along the river, how about putting something back? Or even, when you issue, when you give planning permission, to be sensitive to the river, which often isn't the case. And as I said before, if the Barnes Society came to us and said, 'We are really worried about, they are going to paint the Barnes Railway Bridge yellow and red.' We'd say, we agree with you, it'll spoil the look and then we'll kind of write a letter to whoever is responsible and try and stop it. We have limited resources, but we do our best.*

Interviewer: … Charities encouraging individuals to have a sense of responsibility?
Interviewee: *Yes, I think that's true, yes. I mean, even as something as mundane as rubbish. You know if you go along the Thames at low tide, the tidal, you'll see the amount of rubbish that is just dumped. Now, that's come from people. Plastic bags, you know. And so we try, whenever we can, whenever we get involved in these clean-ups, we try and help put leaflets through people's doors, and so on and say: 'Do you realize the river's looking pretty ghastly where you live? Come and help us clean up and, by the way, you do yourself look after it, don't you?' It's that kind of thing really. Yes, of course, individuals are, they are the ones who write to their MPs, the Mayor, and so on. Yes, wherever we can, we encourage them to have an interest in the river.*

Interviewer: And what about the responsibility of the commercial world, of manufacturers?
Interviewee: *Yes, certainly, they have. I mean as far as the, sorry, I keep on dividing the Thames into the tidal, and non-tidal, because it is in many ways just different. One of the big issues, which is a commercial issue, is the amount of water abstracted from the river by the water authorities; Thames Water particularly, but there are others, when you get up into Oxfordshire and Gloucestershire and they have an arrangement with the Environment Agency. They can take literally millions of gallons each year. Now when you have a period of drought, taking all that water out, which we all need, admittedly, it does have an effect on the river. The depth of the river is reduced, [meaning] the boats quite often find it difficult to navigate. It does have an effect on the vegetation, and so on. So we're keen to try and ensure that bodies such as Thames Water, who are private companies, they spend money on stopping leaks, looking at building reservoirs where they can, and to try and balance, or if you like, reduce the amount of water coming from the river. And also for them to be vigilant about some of their sewage works. Last month there was a terrible incident in Marlow, a lovely little leafy town. And raw sewage just came into the river, and that was because the waterworks there, the treatment works, are sort of nineteenth-century machinery; something was broken. So we are looking at those commercial interests to be vigilant. On the tidal river there isn't much industry left, until you get down to beyond the Isle of Dogs. Where the ships, you start seeing ships at Silver Town, Tate and Lyle, and so on. Most of the work now is down at Tilbury and there's a new port opening in a place called Thames Haven. So there we're interested in trying to make sure that the Port of London Authority is vigilant about the oil discharge, about tankers cleaning their tanks out. And we are also, whenever we try and persuade local authorities to keep an eye on fly tipping, which is a problem. There are certain parts of the Thames where it is quite easy for a lorry to back down onto a slope and dump the rubbish. So you just try and persuade the local authorities, perhaps put a camera*

there to monitor it, and so on. Wherever we can, we try to get the commercial interests to accept their responsibility. But things have improved, in the old days, everything, everything was dumped in the river. They just dumped it ...

Interviewer: ... What about the domestication of the river?
***Interviewee**: I wouldn't say domesticate it. There's only a limited amount individuals can do in terms of cleaning up the river. We don't want them to put rubbish in the river in the first place. There individuals can best be encouraged to take part in the various clean-ups that are organized. We are not trying to domesticate it, if you mean make it just pretty. It's a natural river; to some extent it isn't really natural any longer. It's man-made. But to try and make sure to keep the balance between the environment that's developed over years on the river banks and in the river and in the foreshore, make sure that's preserved, and cultivated and can grow.*

Water is not a commercial product like any other but, rather, a 'heritage which must be protected, defended and treated as such' (European Parliament 2000: 2). In Europe, a 'first wave' of water quality management entered into Community governance in 1975 through the establishment of water legislation prescribing standards for rivers and lakes that were used for drinking water abstraction (European Commission 2013). By 1980, this early prescription was enshrined into binding agreements for the framing of drinking water quality objectives (ibid.). Additional quality targets, derived from the 1980 legislation, aimed to regulate the emission of pollutants into aquatic ecosystems: the Dangerous Substances Directive (ibid.). In 1988, European Union (EU) water quality management policy began to engage with the regulation of pollution derived from urban waste water and from agricultural production. An outcome of the Community Water Policy Ministerial Seminar in Frankfurt in 1988 was the Council Directive 91/271/EEC Concerning Urban Waste-Water Treatment, which was adopted in May 1991. When it appeared in The Hague in 1991, the Ministerial Seminar for the deliberation of the issue of groundwater emphasized the need to intervene 'to avoid long-term deterioration of freshwater quality and quantity, and called for a programme of actions to be implemented by the year 2000 aiming at sustainable management and protection of freshwater resources' (European Parliament 2000: 2). Central to this direction in the Community's water policy is a focus on cultivating citizenship:

Whilst EU actions of the past such as the Drinking Water Directive and the Urban Waste Water Directive can duly be considered milestones, European

> Water Policy has to address the increasing awareness of citizens and other involved parties for their water. At the same time water policy and water management are to address problems in a coherent way. This is why the new European Water Policy was developed in an open consultation process involving all interested parties. (European Commission 2013)

In May 1996, the European Commission hosted a Water Conference, attended by delegates from Member States, water providers, industry and agriculture, alongside environmentalists and consumers (ibid.). The aftermath of the conference was the agreement to merge the existing water policy initiatives into a centralized framework of legislation. Evidently, in response to this assertion, the Commission constructed a proposal for a Water Framework Directive, which would aim at "getting the citizen involved more closely" (ibid.). The European Union's Water Framework Directive is an outcome of these initiatives; and, in its formal guise, is known as "Directive 2000/60/EC of the European Parliament and of the Council of 23 October 2000 Establishing a Framework for Community Action in the Field of Water Policy". London's river conservation charities, interviewed as part of this research, consistently referred to the Water Framework Directive, for it "establishes a legal framework to protect and restore clean water across Europe and ensure its long-term sustainable use" (European Commission 2008: 1). Primarily informed by semi-structured narrative interviews with river conservation charities, this chapter considers the centrality of the Water Framework Directive to the formulation and articulation of organizational value. I was specifically interested in the institutional and organizational water policy framework that informed the local community activities in which river conservation charities were increasingly involved. My interviews with organizational members provided detailed insights into the processes of translation through which water policies dictated by the European Commission were ascribed, valued and enrolled into the river charities' existing programmes of local community water ecology. Of particular interest were the pedagogic strategies employed by river conservation charities, as part of their cultivation of active environmental citizens.

Interview with Senior Programme Manager, London River Thames Conservation Charity III, September 2013.

> ***Interviewee:*** *We're an environmental charity ... we have a long history that goes back to '94, I think. And essentially we were created because there is no statutory duty on aquatic litter. So the story goes that someone fairly senior from, I think, the Environment Agency and the Port of London*

Authority were on a boat trip. They were getting off the boat, looked down into the river, saw a load of rubbish all collected by the pontoon where they were getting off, and looked at one another [and said]: 'Should you be dealing with that? Shouldn't you be dealing with that?' And of course it is neither of them; because no one has to deal with aquatic rubbish. Erhm, as a consequence, a project was tried out to apply Keep Britain Tidy capital standards methodology, of assessing how badly littered things were, to the river. So we started off as part of the National Litter Charity: Keep Britain Tidy. Which is why if you look at our logo, you can see the Tidy Man. The little man dropping things in the bin you see on all the crisp packets. So it started off as a litter management organization. The first job was to assess how badly littered the foreshore of the Thames was, and the second job was to see if local people would help do something about it. And it started off as one volunteer in the London Keep Britain Tidy Office and has expanded from that point. Erhm, and so rubbish still forms a fairly sizeable core of our work, because there is, since the 90s, still no statutory duty on aquatic litter. Erhm, we became an independent charity, independent from ENCAMS as they were then, but Keep Britain Tidy as they are now. So we became independent from ENCAMS in about 2004. An independent charity and up until that point we had been part of the organization. And so our focus over those years and up to now has been diversifying, every year, so you know litter is still something that concerns local people. But [we] as an organization [are] concerned with local people and water. So our main aim is to connect local people to their local waterway. Now, how that connection is formed now has many different strands to it: so it can be still collecting rubbish; it can be doing river maintenance on the banks, you know, cutting back nettles and brambles or helping trim trees. It can be river improvement works, so building river restoration: you know, improving the banks or putting in river features. Erhm, it can be education, working with schools just to do pond dipping to learn about their local river. Or we've done water quality testing with local people in some areas. So there's a range of projects. But essentially for the same aim: local people engaged with their local piece of river so they are involved in taking care of it; they are the people who value it; they are the people who have the clear ideas of what they want to see happen to it; and they're the people we help facilitate [making] those changes. So that's, you know, that's what we have been attempting to do for all these years. Often, certainly not in the early days, not particularly strategically: we were working with people, where people wanted our help. We were a relatively small organization and because we work on the tidal Thames and all the tributaries, that's a massive amount of waterway. So, erhm, you know, in terms of our capacity to deliver, it's quite labour-intensive, particularly as our method of delivery is to turn up with all the equipment that is necessary, and have staff there to supervise, so that volunteers can just turn up in old clothes and we will have organized the activity – often with them, in discussion with them – but [we] organized the activity, organized the equipment. Volunteers are covered by our health and safety, our public liability insurance, so that people can volunteer very easily. So a lot of our work is in that sort of the donkey work of organizing and then people can get involved very easily. So, as a consequence, as I say, we weren't particularly strategic. We weren't aiming it specifically at improving rivers in a certain way. We were aiming at engaging people. Now I would say in the last few years, we've been attempting to become slightly more strategic: so looking at the overall outcomes for the river, though still keeping this focus on the communities. But at the same time in the last few years, there's been quite a, sort of a, more intense focus on the Water Framework Directive, which I don't know if you have heard about in your research? So the Water Framework Directive is a huge piece of European legislation, which we have signed up to – the UK, has signed up to – which, we have been, or the UK has been busily implementing for, I don't know, ten or so years. But we are at a point in that implementation where Europe wants to see us make changes. And we're at a point in that implementation where the Environment Agency, the government's statutory control body that is responsible for implementing the Water Framework Directive, have categorized all the current waterways. So the point of the Water Framework Directive is to bring the water quality of the UK's inland waters up to 'Good' status. So for the last few years the EA [Environment Agency] have been deciding what, you know, categorizing the waterways and trying to decide what 'Good' means, and so, you know, what is the quality ecological potential. So there's a lot. And then also looking at the waterways and seeing those that aren't 'Good', why aren't they good? So there's been lots of measuring and categories and cataloguing. But they are now at that point where they are looking to create plans which should see improvements. So we've got much more involved in this movement towards, you know, a strategic movement towards improving waterways, particularly water quality, because of the Water Framework Directive. And one of the ways that government, DEFRA particularly, and therefore the Environment Agency, see that we implement this improvement as part of the Water Framework Directive is what's called the catchment-based approach. The catchment-based approach is a way of looking at rivers as a whole catchment, and engaging all the organizations who are in and around that catchment. So instead of it just being the Agency wandering in and saying, 'Here's the problem, this is how we're gonna solve it', [instead they are] talking to all the local people; talking to local councils; talking to local NGOs [non-governmental organizations]; user groups and interest groups. And creating a plan that's based on local knowledge; based on local problems and then creating local solutions. So this is now policy: this, what's called the catchment-based approach ... Because of the way we have been working for the last however many years, because we've been a kind of community engagement focus around waterways, we are ideally so placed to be involved in this catchment-based approach. You know, which is creating a more strategic side to our future working and giving us more of a focus on water quality. Whereas perhaps before it was just a focus on how local people felt, though that's still important. But so, that gives you a very quick understanding of the perspective of how [the charity] has been operating.

Figure 4.2., "We cannot think of a time that is oceanless, Or of an ocean not littered with wastage, Or of a future that is not liable, Like the past, to have no destination" (T.S. Eliot, *The Dry Salvages*). Photographic Image, South London, River Thames, September 2013.

The Water Framework Directive and Public Participation

Established on 23 October 2000, the Water Framework Directive affirms a legal framework to safeguard, enhance and restore clean "inland surface waters, transitional waters, coastal waters and groundwater" across Europe and secure long-term sustainable use (European Parliament 2000: 8). The Water Framework Directive institutes an enterprising system of coordinated administrative arrangements for water management sub-structured on river basins. This is evident in Article 3 of the Water Framework Directive, which stipulates that Member States identify and catalogue the "river basins lying within their national territory and, for the purposes of this Directive, shall assign them to individual river basin districts" (ibid.: 11). The adoption of river basins as the focus of EU water policy is guided by the belief that: "The best model for a single system of water management is management by river basin – the natural geographical and hydrological unit – instead of according to administrative or political boundaries" (European Commission 2013).

In England, the Water Framework Directive was instigated in December 2000 and inaugurated into United Kingdom (UK) law in 2003, according to the following regulation "The Water Environment (Water Framework Directive) (England and Wales) Regulations 2003 (Statutory Instrument 2003 No. 3242) for England and Wales; the Water Environment and Water Services (Scotland) Act 2003 (WEWS Act) and The Water Environment (Water Framework Directive) Regulations (Northern Ireland) 2003 (Statutory Rule 2003 No. 544) for Northern Ireland" (JNCC 2014). The lead authority for the application, in England, of the European Commission's Water Framework Directive, is the Environment Agency (EA). The EA is assigned the task of applying the rules of the Directive; mostly by making operational "the programmes of measures specified in the river basin management plans", which are formulated as part of the River Basin Management, of the river basin districts that reside within the EA's territory (European Parliament 2000: 12). Specifically, Article 13 of the Water Framework Directive states that: "Member States shall ensure that a river basin management plan is produced for each river basin lying entirely within their territory … River basin management plans shall be published at the latest nine years after the date of entry into force of this Directive. River basin management plans shall be reviewed and updated at the latest 15 years after the date of entry into force of this Directive and every six years thereafter" (ibid.: 22).

The Water Framework Directive interposes a fixed series of six-year cycles for the delivery of river basin management plans. The first cycle forecloses in 2015; it will be followed by further consultation, during which the river management plan will be revised and reissued (Environment Agency 2013b). The Water Framework Directive is particularly prescriptive regarding the content of river basin management plans. It stipulates that these plans must integrate environmental objectives; raise social concerns; and explore economic factors for all bodies of water that exist within the river basin district (European Commission 2008a). It is a feature of river basin management plans, that they incorporate procedures to report "the bodies of water not meeting the Directive's environmental targets and describe the reasons for this, such as what is causing problems and risks" (ibid.: 2). Furthermore, these plans are required to incorporate a programme of measures detailing the actions and procedures to be accomplished in order to address threats to the river basin district's aquatic ecosystem and to enhance water quality (ibid.). Pivotal to the formulation of the river basin management plans are the river basin district liaison panels, which are constituted by "representatives of businesses, planning authorities, environmental organizations, consumers, navigation, fishing and recreation bodies, and central, regional and local government, all with key roles to play in implementing the plans" (Environment Agency (2013j). River conservation charities are part of this remit in their capacity as environmental organizations. According to the Environment Agency, river basin district panels are integral to helping it "work with local stakeholders to identify the actions needed to address the main pressures on the water environment" (ibid.). River conservation charities interviewed as part of my research often referred to their participation in river basin district panels as adding value to their existing programmes of community engagement. As the Senior Programme Manager previously interviewed says:

> ***Interviewee:*** *So the Water Framework Directive is an EU Directive and the implementation of that is the responsibility of the Environment Agency, who are the government quango responsible for pollution control and pollution management and environmental improvements, and they are part of the DEFRA [Department for Environment Food and Rural Affairs] family. So the Department oversees the operation and the funding of ... EA's funding comes from DEFRA, so both DEFRA and EA have been working at a national policy level for how to implement, or how to implement Water Framework Directive outcomes. And one of the methods they're settling on for doing it is called the catchment-based approach. The catchment-based approach is this collaborative involving local stakeholders' approach to looking at river catchments. And so instead of saying: 'This stretch of river, because it happens to be in this council's borough, or this county's borough ...' you look at the catchment – because rivers don't care where human settlement is, or where the boundaries have been drawn: rivers go where they go – so, looking at the whole catchment, involving everyone in the catchment, in the solutions to the problems.*
>
> ***Interviewer: ... The catchment-based approach involves local stakeholders in terms of 'local solutions' ... And that's what I am interested in, because what I'm interested in is: how do you translate these local solutions into sort of initiatives in relation to working with communities?***
> ***Interviewee:*** *There are two areas that we particularly are involved in. One is in the area which is consultation, which is informing, because water quality issues particularly in urban rivers can be quite complicated. You*

can have lots of pollutants that are non-points also. You can have road runoff and misconnections and people dumping stuff. And, so there's lots of routes that can lead to a river being degraded. It is often not as straightforward as going, 'This river is in problematic status because there happens to be this thing, this bit of pollution'. It's not like, 'Oh, there's an industry on this river and what's coming out of that pipe is what's causing the problem to the river.' In a lot of these urban rivers, there's little bits from here, there and everywhere. So that becomes, that is quite a complex thing to explain to people, particularly if you want local people to be involved in some of the solutions. Erhm, so, the consultation, the explanation phase, that's something we can be involved in; because we already have a lot, we already work with a lot of people: local people. All those people that come to us and give up their time as volunteers are clearly interested in the river. Now, they may be interested in the river because it's in their local park and they come down there with their kids and they want it to look nice. But they are there for a much more receptive audience: if you are trying to have conversations about misconnections, for example. Misconnections is where your home and personal plumbing isn't, hasn't been done properly; and so your waste pipes are connected into surface drains. So that toilet waste or erhm, you know, dishwater sink waste is going in the surface drain which is then going directly into a river, by-passing treatment. Instead of plumbing, instead of your plumbing being plumbed into the sewer system it is plumbed into the drainage system. And people may not realize that and having that conversation is quite difficult. And also trying to explain to people that the river they like may not always, you know, function in the way they want and some of the things they like about it; for example, there may be a nice little weir in a park where the water trickles over merrily, but that actually might be a barrier to the movement of fish, because it is a concrete weir that has been created artificially and actually it would be better to remove that weir. Again if you are making significant changes to something that is a local feature, people need to understand why it is happening, because otherwise change generally creates friction ... like, why you doing this? What is this happening for? You need to explain that. So we are well placed to communicate some of these complex messages, and the other thing we also – because we are about practical engagement, we are about delivery, we are well based to get people involved in creating some of those solutions. So we get to a point whereby we have teams of volunteers who are, for example, building, you know, building a new river feature. So in a lot of cases urban rivers have been widened and straightened for flood risk purposes. You widen and straighten them to get the water through fast. So because an awful lot of water, when the rain falls, hits the hard surfaces of roofs and roads, goes down a drain and gets into a river very quickly and the engineering solution to prevent local flooding was to make them wide and straight so the water can go charging down as quick as you like. Now under normal conditions, it is not raining hard that means you have a very thin strip of water, which has very few features and doesn't actually function as a biological system, very well. It's not a great river. So you can start putting features in: you put in the low level features that narrow the flow channel so, instead of having this big wide shape, you might put in a little island here. So your water is now flowing in this direction; it is flowing faster and deeper and more natural. This little island you put in vegetates; creates a more natural feature; and then you get your high flow; the river just comes up and flows out over the top of it. So those sorts of adding features are something that volunteers can get directly involved in: they can help [in] building them. They are not high technique. They are fairly low technique. If you're local, it's your local park. You can start, you know, once a week you have little volunteering sessions, ten to twelve people come down. They'll bang in stakes; they'll weave willow; they'll start building these features; that's great. There's a buy-in; there's a real sense of achievement; there's a real sense of ownership. These features get looked after by local people and they understand why you are adding something to the river. We have done water quality testing there too, so that people can have an understanding of some of the problems. Sometimes that is about making sure the information is out there. So, you know, on the River Lee, for example, where there's a lot of over-spill from big sewer systems, so we have large sewage pipes, which, when it rains, the water goes into these large sewage pipes. Sometimes the amount of rain means the sewage pipe is full, it can't take any more volume, so to prevent flooding you have over-spills where they just, so the sewage system just overflows directly into the river, so you have raw sewage that goes directly into the river.

Article 14 of the Water Framework Directive stipulates the requirement that Member States assiduously engage "Public Information and Consultation" as part of methods to encourage the active participation of stakeholders in the implementation of the Water Framework Directive and the associated river basin management plans (European Parliament 2000: 12). Public access to environmental information is governed by: "The convention on access to information, public participation in decision making and access to justice in environmental matters" (European Commission 2008a: 2). Consultative participation is designated by the Water Framework Directive to be manifest through mechanisms "that government bodies use to consult people and interested organizations (stakeholders) to gain from their knowledge and experience and to jointly develop solutions to problems" (ibid.). Consultation processes may assume several formats, which include written correspondence from interested parties and other stakeholders; and oral dialogue between authorities, stakeholders and members of the public. So as to facilitate the contribution of the public and stakeholders in the revision and development of river basin management plans, the Water Framework Directive encourages the provision of "proper information to the public of planned measures before final decisions on the measures are adopted" (ibid: 2). Furthermore, those directly involved in the construction of river basin management plans are required to make available documentation relevant to the development of their respective plan. Moreover, "once a plan is in place, authorities need to report on the progress of its implementation to the public and stakeholders" (ibid.). During my interviews with river conservation charities, a form of didactic informationalism evident in the Water Framework Directive's consultation model was the focus of concern, and issues were raised about the effectiveness of the didactic pedagogic. An indication of this is evident in the following interview extract with an Executive Director, London River Thames Conservation Charity II, September 2013.

Interviewer: Could you tell me about your objectives, in relation to conservation?
Interviewee: *For the past six years our objective has been to help the implementation of the Water Framework Directive. Because when I started working for the [River Conservation Charity II], I, basically I said to the trustees, you have a very broad remit. You have a massive job here. Society has a massive job still to do on the River Thames. Even though it has been cleaned up quite a bit over the last 50 years, it's still a long way from meeting the requirements of the Water Framework Directive, both in terms of water quality and water quantity and also the ecology of the river itself. There's a long way to go and you cannot do it alone as one charity. Even the government's Environment Agency with all its resources is going to struggle to meet the goal of the Water Framework Directive, which is getting the River Thames to 'good' ecological status by 2027. So I suggested to them that what we should do, as a trust, is focus on working with the other organizations to help implement the Water Framework Directive, because that directive is the one overarching piece of legislation that will drive getting the river where we all want it to be. And so they agreed that and we have spent the past six years on a whole range of activities to help implement the Water Framework Directive, which are on our website, like demonstration projects and other activities.*

Interviewer: Could you tell me about some of the activities?
Interviewee: *So our activities have been erhm, of two kinds mainly: one policy and planning work. And the main activity there has been representing the NGOs on the Environment Agency's liaison panel and helping to develop that river basin plan. Because it is the river basin plan which will drive – or is driving, should be driving – everything that everyone is doing to improve the river: whether they are a government*

agency, or a charity or a local authority. So that plan should be driving what everyone's doing. We've put a lot of time and investment into helping produce a plan which is practical and useful and comprehensible to all of the people involved, all of the players involved.

Our second area of activity has been to help produce some demonstration projects of how to improve or restore rivers to the standard required by the Water Framework Directive, which isn't clear in the Directive itself; I don't know if you have ever read the Water Framework Directive? But the reason why it's called a framework directive is because it sets out broadly the goals that should be achieved and the time scale for achieving them. But it does not tell EU Member States exactly how they should achieve these goals. Erhm, so exactly what you do to your river, to improve it, is actually down to each individual river basin management plan. Partly because every river is different and every river has a different set of problems, so every river's plan will be different. Though the European Commission, they were quite wise in not trying to define exactly what you should do, because it will differ in Scotland to Sicily hugely. But the UK government has made very heavy weather of implementing the directive, and in my view and the view of many NGOs, have spent far too long, and been far too cautious about what it should do, with the result that our progress has been very slow. Erhm, anyway, it isn't easy to change government policy or get government to hurry up. It isn't easy to get government to invest in the environment, while we've just gone through the biggest financial recession that the country's ever had. So it's been an uphill struggle over the last six years. Erhm, anyway, that first river basin plan comes to an end in 2015 and, as I have said, we have already started the process with the Environment Agency and others of developing the next six-year plan, which hopefully will be a lot better than the current plan.

Interviewer: In what ways will it be better?

Interviewee: *Well, the current plan isn't really good enough. Erhm, for a lot of reasons; it's a long story. There's a very long story here. But basically when the Water Framework Directive came into force in 1997, the environmental NGOs encouraged the government to take a bottom-up approach to this Directive. We said we know that more than 75 per cent of our rivers do not meet this Directive; we know they're in bad condition. So what you should do, oh, government, is use the Environment Agency to first of all go out and measure the ecological quality of all the rivers and then sit down with us and local people and form an agreed view on what is causing the problems on each river, and what should be done about it, and then we can draw up a plan of action to improve it. And to cut a long story short, the government spent more than ten years saying: 'No, no, no, no, we don't need to do that. We know what's wrong with our rivers; we have all the information held centrally. We will produce the river basin plan from the existing data and then we'll come and talk to you and local people about what to do about it.'*

Erhm, the net result was that in 2009 when they produced the first river basin plans, they were widely criticized by everybody outside of government; including business and industry and the water companies, even their own staff privately were criticizing the plans.

Interviewer: What were the main criticisms?

Interviewee: *Well, it was too little, too late. They did not have the data held centrally about the ecological condition of our rivers. They hadn't been, ever been measuring the right things. Erhm, they did not fully understand what a change in emphasis the Water Framework Directive was. Previously the whole focus of water management in the UK had been around water quality: how clean is the water? Is it clean enough to drink? And so all their monitoring systems and all their monitoring stations were set up in places to measure purely the quality of the water: was it polluted or not with a various range of substances? The Water Framework Directive isn't just about water quality. It's about quantity and ecological quality. Which includes things like: are the banks natural or not? And they have no systems to measure these things. They had no idea, along a tributary of the Thames such as the River Wandle, for example, how much of the banks were natural or not; whether if they were natural, that they met 'good' ecological status under the Water Framework Directive. They actually knew nothing of that, erhm, but they didn't want to admit it, erhm, until it was too late and in the year 2009, when they had to produce the plan for the Thames and all the other rivers around the country, they actually came to the liaison panels. In fact, the first, they came to the DEFRA national Water Framework Directive committee where again I was representing all the river trusts, and they presented their plans for the plan. 'How to do river basin plans around the country', and we all sat there and said, 'That's rubbish; that's just a load of rubbish. You've spent ten years working on this and every time somebody like me asks: 'What are you doing? How are you doing it? Can we see the plan for the plan? What are your criteria gonna be? How are you gonna involve stakeholders?' For ten years the answer [was]: 'Don't worry, it'll be all fine; you'll see it in the draft plan; yes, we've taken account of what you say, it'll be in the draft plan, in the draft plan.' When we finally saw the draft plans they were just a load of rubbish. So even DEFRA, the Department for Environment, Fisheries and Agriculture, who chair the national Water Framework committee, the DEFRA chairman said to the Environment Agency: 'I'm sorry, this is too little, too late. This is basically, this is rubbish ...' A new team [was] brought in with less than a year to produce a plan for all the rivers around the country. So they produced something which was better than what we had seen; but still not good enough ...*

Interviewer: As part of the Directive, to what extent is there a focus on the local supervision and maintenance of rivers?

Interviewee: *None, basically. The Water Framework Directive is the requirement on Member State, and Member States have to implement it. Erhm, there is one part of the Water Framework Directive, which is Article 14, which says that Member States have to involve the public in the preparation of the river basin plans. But how you involve the public is a matter for each Member State to decide. So this is one of the areas where we NGOs have been battling with the government. Right back since 1997, we've been saying that it is absolutely fundamental that you involve the public and representatives like us in the preparation of the river basin plans. For many good reasons, but from the point of view of the government agencies who are charged with implementing the Directive, you know if you do not get local people to agree what you plan to do with the river at the end of their garden, then, come the day when you turn up with your bulldozers and diggers and you start doing something, you're gonna find that you're gonna get a load of complaints and opposition and work will have to stop. You know, to us, it is fundamental that first of all you get people on your side. You sit down with people and say, 'Look, unfortunately your river doesn't make this European Directive. We've got to do something about it, this is what we think we could do, so it can look nicer and smell nicer and there'll be wildlife in there, and kids can paddle and catch fish. Would you like that?' And, then, when you get the majority of people signed up to it, saying yes, [you then say], 'Oh, by the way, we're going to have to drive a bulldozer across your garden for two weeks. But we'll put it all back together again'. You get people to agree up front what is done at the end of their garden. Erhm, but no, unfortunately, government didn't take that view.*

Interviewer: So part of the Water Framework Directive Article 14 insists that the government has to work with local NGOs?
Interviewee: *All it says is that they have to involve the public in the preparation of the river basin plans. So the Environment Agency, for example, back in 2009, how it proposed to implement Article 14 was simply putting a mass of information on its website. They said: 'It's all there on the website. If anybody wants to know what they'll be doing at the end of their garden, they can go and look on the website and find out.' Actually, when you go and look on the website, you can't find out what might be happening at the end of your garden. There's a whole mass of technical jargon; you couldn't connect one bit of the site to another; there was just lists and lists of tables; it was terrible, it was absolutely terrible. Under pressure from the NGOs, they started to improve it because they realized they had to. Because they started getting complaints from people: 'I tried to look at this; I cannot find out what's happening in my local river; I cannot understand what's happening at the end of my garden; this is not good enough.' So then they came running to us, saying: 'Local people don't like this, we're getting complaints about this; can you help us*

sort this out?' And we said: 'Yes, you need to do what we have been telling you to do for the past ten years.'

Interviewer: Which is?
Interviewee: *Well, work with people like us to develop leaflets; comprehensible websites; public meetings; walks for the public along their local river; and talks about what is wrong with it; what needs doing to it. You have to find ways of communicating with people face to face. You know, real-life people. Not just someone through a website. You have to find ways of getting your staff or other organizations working on the river, and your staff to go and talk to the people on the river, people that are interested, about what they want done with it. And I know that takes time and energy and money, but if you don't do that, you won't succeed in what we know is going to be a very long-term goal. We're talking 30 years or so, to improve our rivers. You've got to invest time and money up front in getting the public on board with what you want to do. Otherwise you cannot plan a 30-year programme of activities because at any point, someone might turn up and say: 'No, I don't want that done at the end of my garden ...' So it took them a long time to realize what we call the bottom-up approach of working with local people of getting them involved.*

Interviewer: ... Article 14 suggests that governments have to involve the public in the preparation of the river basin plans.
Interviewee: *How they do that has been left to individual governments ... And most Member States have tried to do this on the cheap. Especially because we have had six years of recession, they wanted to do this Directive on the cheap. So they've said: 'We don't need to go and talk to anyone, we'll just put this information on the website. And then one day we'll turn up and start digging their river away and that will be alright, won't it?' But governments all across Europe have found people don't like having their rivers messed around; people feel very attached to their local rivers and passionate about it. And they do want to see them improve but they want to be involved in the process. Because, after all, they are the ones that are going to have to live there after it's done. They're going to have to live with the result. So people do care very much about what is done to their local environment.*

Interviewer: And that's what interests me. So the governments were left in terms of having to realize those processes of communication, with the local people. And some of the governments just chose a system of providing information?
Interviewee: *Well, most European Member States started off developing their plans purely as desk-top exercises by government officials in their offices; not talking to anybody outside and then just putting information on their website ... It has taken them since 2000, and we are now in 2013, to realize, no, there is no short cut here. You've got to go out and talk to*

people on every river that you want to work on and get the local people on your side, and if you do:(a) they will, you know, help you do the work; but (b) they will literally help you by volunteering for NGOs like mine or giving money to NGOs like mine, which means we can do some of the work so that government doesn't have to do all the work or pay for all the work.

Figure 4.3., "Into different lives, or into any future" (T.S. Eliot, *The Dry Salvages*). Photographic Image, South London, River Thames, September 2013

Figure 4.4a., "We had the experience but missed the meaning, And approach to the meaning restores the experience" (T.S. Eliot, *The Dry Salvages*). Photographic Image, South London, River Thames, September 2013.

Figure 4.4b., "We had the experience but missed the meaning, And approach to the meaning restores the experience" (T.S. Eliot, *The Dry Salvages*). Photographic Image, South London, River Thames, September 2013.

Informationalism and the Limits of Public Participation

The Water Framework Directive's Article 14 stipulates: "Member States shall encourage the active involvement of all interested parties" in constructing, implementing and updating river basin management plans (European Parliament 2000: 23). The Directive advises that: "government authorities need to establish adequate consultation mechanisms at river basin district level" (European Commission 2008a: 4). My interviews with river conservation charities highlighted problematic features manifest in the informationalized mode of public consultation, initially adopted by EU Member States. NGOs and interested participants resisted a method of consultation, which sociologically can be defined in terms of Paulo Freire's "banking concept". A leading light in Marxist philosophy, Freire identified capitalist ideology as sustained through acceptance of a pedagogic field of instrumental informational didactic relations. According to Freire, a noticeable feature of the teacher–pupil dynamic is its narrational character. Characteristic of informational pedagogy is the existence of a narrating "subject" and a listening "object". It is Freire's observation that in this educational environment: "The contents, whether values or empirical dimensions of reality, tend in the process of being narrated to become lifeless and petrified" (1996: 52). Referring to the above transcription's account of the informationalized presentation of environmental material, it is evident that pedagogically, this online mode of communication designates a dynamic in which the literature purports to advance objective actualities "about reality as if it were motionless, static, compartmentalized, and predictable" (ibid.). In so doing, the informationalized event effectively transforms the topic into an object "completely alien to the existential experience of the students" (ibid.). It is unsurprising that members of the public perceived the informationalized environmental literature, hosted on websites as: "*a whole mass of technical jargon; you couldn't connect one bit of the site to another; there was just lists and lists of tables; it was terrible, it was absolutely terrible*". According to Freire, informationalized didactic knowledge transfer merely attempts to "fill" the audience with informational content that is detached from their own experiential reality: "Words are emptied of their concreteness and become a hollow, alienated, and alienating verbosity" (ibid.). Informationalized consultation furnishes a pedagogy of knowledge transfer in which the literature, communiqués, abstract deposits, in so doing project "an absolute ignorance onto others" (ibid: 53). Freire contends that depositing, or "banking" of deposits, is "a characteristic of the ideology of oppression", for it "negates education and knowledge as processes of inquiry" (ibid.).

Freire contrasts informationalized didactic modes of education with the *raison d'être* of libertarian education. In contrast with the banking concept's propensity to render apathetic critical consciousness, "the humanist revolutionary educator" endeavours "to engage the student in critical thinking" in the "quest for mutual humanization" (ibid: 56). This interpretation breaks radically from individuating didactic communications, because education as the exercise of critical dialogue "does not take place in ivory tower isolation, but only in communication" (ibid.: 58). For it is an integral philosophy of transformative pedagogy that: knowledge "has meaning only when generated by action upon the world" (ibid.). Such sentimentality was variously articulated by the river conservation charities interviewed as part of my research. By way of illustration, consider the following extract from my interview with the Executive Director of River Conservation Charity II, September 2013.

> ***Interviewee:*** *... One reason why we are so keen on organizing river clean-up activities is: (a) because it's something we can do and local people can do. You can go and take a lump of rubbish out of your local river. And once you've done that, and you've seen everyone else do that, you see that, actually, OK, there's no more shopping trolleys in the river, but, hey, it still doesn't look good and it still smells bad, and there are no fish or wildlife in there. So why did I bother to take the trolley out? You know, someone else is still doing some damaging activity to that river. Who is doing that damage? Where does that pollution come from? Why don't they stop? You know, we get local people thinking about these issues and then get them involved in the bigger campaigns like forcing Thames Water to build a super-sewer because actually it is them, or rather the 150-year-old sewers that they inherited, which are putting the worse stuff in your local river. So we have to force them to do something about it. So in itself, removing the rubbish from the river is a minimal; a minimal activity, a minimal improvement. But it's the engagement of the people and the awareness-raising that it caused, and then the involvement of those people in the bigger issues, that's why we are doing it; that's why we are doing it. It's for the bigger picture. As well as, OK, it looks nicer without the rubbish on the edge. But, hey, you know, the rubbish on the edge is just the seed, just the start, just the acorn of a huge oak tree that needs to grow to improve that river.*
>
> ***Interviewer: ... You're saying that it is only the acorn, only the initial step to a whole sequence of policy initiatives?***
> ***Interviewee:*** *And eventually beyond that to people understanding that what we're asking is politicians to respond to, in fact, something that we are all doing every day. It's our washing and bathing and pulling the toilet chain which cause all that waste water to be produced in the first place.*

Thames Water as a company only exists to deal with our waste, but that waste comes from us. As individuals we are responsible for the condition of that river. And one of the great things about my job, for me, is seeing people who initially had no connection with the environment, or activity in the environment, or local work at all, but saw us working on the local river and came along and asked why we were doing it and joined in and then they, eventually, they go through that whole journey of understanding why the river's polluted; why it's in concrete banks; why has their house got sewage flooding through the basement. And understanding that there are government agencies that manage these things and politicians need to change the system; Thames Water needs to build a tunnel. But at the root of it all, why all those things exist, because we, they, that person, uses a lot of water and produces a lot of waste water like everyone in this city. And most people in this city never give a thought to where does that water come from or where does it go when I pull the chain in the toilet? It's when they understand that water cycle and when they understand that they are part of the problem. And then they go: 'Oh, right, now I have to be part of the solution ...' [River clean-up], that's just one means, one mechanism, to get people in their river to do something. You've then got them as a captive audience to try and explain this whole bigger picture and how they are an important part.

Interviewer: Do you think that some people just find meaning and value in just that very small picture, of just going down to the river? Do you think they all get the bigger picture?
Interviewee: *I think you'll be surprised about how many people do get the bigger picture. And this has been my criticism of the government's Environment Agency. Because one of the things that they said, back in 2000, when the Water Framework Directive came in: 'Oh, we don't need to go and talk to everyone about this because ordinary people, they won't understand, it's too technical; it's too scientific; it's too big; they won't understand it; so let's not waste time with them; let's just go and do it.' And our experience was the opposite; it was, you will be amazed about how many people standing near that bus stop outside – if you went and engaged them in a discussion about river ecology and water management and pollution – you'll be amazed about how many of them (a) care about it; and (b) do get it when you explain it to them; and then (c) say, 'Yeah, actually, I want to do something to help. I want to be involved in this ' Our experience is it's a lot of people, so wake up government, pay attention to those people and involve them.*

The interviewee here provides insight into the philosophical and social context of the outreach community activity undertaken by river conservation charities. One gains an appreciation of the holistic ambitions of environmental pedagogy and their modes of consciousness-raising.

Through vignettes and allegory, the interviewee furnishes a vision of the environment as the focus of a dialogical process of political literacy. Community outreach exercises alongside engagements with local stakeholders establish knowing relationships, as opposed to didactic informational broadcasts. This involves challenging and clarifying "the opaque dimensions" of governmental environmental and development practice (Freire, 1985: 100). To actively engage in praxis necessitates dispensing "with the naïve idea of 'consuming' what we read" (ibid.). Consciousness is conceived as "an 'intention' toward the world" (ibid.). Interestingly the interviewee describes outreach river conservation events (e.g., river clean-up) as charting a pathway that intersects with the administrative guidelines for environmental education. In particular, river conservation knowledge is perceived in terms of humanistic civic activism; an evolution of knowledge that Freire describes as involving "a constant unity between action and reflection upon reality" (ibid.). Spatial practices are a feature of the environmental pedagogy articulated by river conservation charities. Indeed, it is my observation that the engagement with the space of place defines and distinguishes river conservation charities. For, as Freire puts it, "Like our presence in the world, our consciousness transforms knowledge, acting on and thinking about what enables us to reach the stage of reflection" (ibid: 100–1).

The Executive Director of River Conservation Charity II describes a similar situation, in which critical praxis emerges from critical reflection on our biophysical presence in the river environment. Education provides some bridges between learning and actively doing, thus ensuring that we "take our presence in the world as the focus of our critical analysis" (ibid: 101). In this respect, educational activities, described by the Executive Director, appear to be less concerned with resolving local environmental issues, focusing more on the cultivation of civic citizens articulate in the securing of environmental agenda. Epistemology in contrast to ontology appears to be the emphasis here: "we grasp the knowledge of those experiences" (ibid.). The value of community engagement is thus directed toward uncovering "reasons to explain why we are as we are", in order to "grasp the reason behind our reality" (ibid.). Environmental education, in this model, contributes to the facilitation of political literacy skills. Orientated toward dialogical exchange of experiential knowledge, educational activities evolve into coordinated actions of political representation. As the Executive Director says, "*So, in itself, removing the rubbish from the river is a minimal; a minimal activity, a minimal improvement. But it's the engagement of the people and the awareness-*

Figure 4.5., "Perfection, be it trifling as the mote, Which reveals in the Sun-beam, cannot own, Its essence self-originating" (Ann Yearsley, *The Materialist*, 1787). Photographic Image, West London, River Brent Tributary of the River Thames, February 2014

Figure 4.6., "Hail! Steady Friendship, Stubborn in thy plea! Most justly so, when Virtue is thy guide: Beneath your mingled ray my soul is free, And native Genius soars with conscious Pride" (Ann Yearsley, *Written on a Visit*, 1787). Photographic Image, West London, River Brent Tributary of the River Thames, February 2014

raising that it caused and then the involvement of those people in the bigger issues, that's why we are doing it; that's why we are doing it. It's for the bigger picture". The mutual dependence of river conservation charities and local stakeholders is particularly evident in this dialogical inter-play of experiential learning processes.

Figure 4.6., "Hail! Steady Friendship, Stubborn in thy plea! Most justly so, when Virtue is thy guide: Beneath your mingled ray my soul is free, And native Genius soars with conscious Pride" (Ann Yearsley, *Written on a Visit*, 1787). Photographic Image, West London, River Brent Tributary of the River Thames, February 2014

Figure 4.7a., "How oft, with thee, when life's a keen tempest howl'd, Around our heads, did I contented sit, Drinking the wiser accents of thy tongue" (Ann Yearsley, *On the Remembrance of a Mother*, 1787). Photographic Image, West London, River Brent Tributary of the River Thames, February 2014

Figure 4.7b., "How oft, with thee, when life's a keen tempest howl'd, Around our heads, did I contented sit, Drinking the wiser accents of thy tongue" (Ann Yearsley, *On the Remembrance of a Mother*, 1787). Photographic Image, West London, River Brent Tributary of the River Thames, February 2014

The pedagogic and political challenges posed by the attempt to cultivate political literacy among local stakeholders differentially guided the promotion and delivery of the river conservation community interventions. While epistemological issues of knowledge also guided the curricular programmes of the other river conservation charities, the emphasis on social-spatial formations translated into interventions aimed at biophysically resolving an environmental problem: civic activism was immediate and corporeal, whereby volunteers physically went to the foreshore in order to resolve problems caused by environmental pollutants and invasive botanical species. An indication of the pedagogical aspects of this form of biophysical engagement is evident in the following interview transcript with the Tidal Thames Coordinator, London River Thames Conservation Charity III, August 2013.

Interviewer: One of the main things I am trying to do is understand [the work of London River Thames Conservation Charity III].
Interviewee: *We are an environmental charity and we work with volunteers across London to improve London's waterways. So the charity originally started by focusing on the problem of litter in the river. So focusing initially on the litter in the Thames, cause there's no, so there's no statutory requirement for anyone to take rubbish out of the waterways. So once rubbish goes into the watercourse, it becomes no-one's responsibility to take it out. So it started originally as a project called Thames Clean, which was, sort of, a short-term. I think it was mainly a year or two-year project, to sort of try and see whether or not there was the appetite within London for volunteers to work to, sort of, to take rubbish out of, out of the Thames and then expanding out into the tributaries. So that's how it originally started. And I think the initial project was very successful. And then from that [London River Conservation Charity III] kinda of grew and became established. We still do litter clean-up work and we work on the Thames, we work on all the tributaries that run into the Thames. Erhm, we have, historically, we've worked on the canals as well. But with the Canal and Rivers Trust, British Waterways becoming the Canal and Rivers Trust, they're now taking on a more active role in terms of engaging volunteers on the canal. So we are doing less work on the canal network than we have done previously. And so we still do work with litter but we kind of expanded the work we do. It is more about the general health of the waterways, so we do kind of tackle invasive species on the rivers; we do kind of habit enhancement work; we do water quality monitoring and partner with other organizations to sort of monitor so we have a little link with the Zoological Society for London to monitor eels and their migration up the river. So it's all about engaging communities: everything we do is working with volunteers, erhm, and it is all about the health of London's waterways; so kind of engaging people in all of the issues that affect the waterways. So we work all over Greater*

London. We also work on the non-tidal sections of the River Thames, so it's tidal up as far as Teddington and then we have a project that works from Teddington upstream. So it doesn't work right to the source of the Thames, but kind of works on the non-tidal section. All of the work that we do in all the different areas is along a similar vein in terms of the variety of activities, just in kind of different geographical areas.

Interviewer: As I understand it, part of your initiative has a very significant focus on voluntary work. Could you tell me about that? Who actually volunteers?

Interviewee: *Yes, there's a huge variety of people; different people in different areas. Sometimes we work with community groups, so, like, residents associations or on the tributaries there's quite often, like, friends of different park groups. So, sometimes it's established groups; a lot of the time it's just individual volunteers. So we promote our activities. People can just come along and join in. For all of our events, we provide the equipment that anyone would need to take part; so we try and make them as accessible as possible so people could just come along and turn up and we'll provide them with the equipment that they will need to join in. So there are established groups; there's individuals. We work with companies as well, sort of doing team activities along the rivers. We do education work with schools; but also kind of work with youth groups and young people kind of initiatives. So there is something called the Challenge Network, which is all about getting 16 – I think it is mainly 16-year-olds – to kind of focus on volunteering and kind of engage with community and, sort of, social issues and things like that. So, we've done work with that, so they kind of bring the volunteers to us; but then we provide the activities, work with the young people to do different activities. So, yeah, a whole suite of people: we work with over 9,000 volunteers each year. So there's a huge appetite in London for volunteering and giving kinds of a small amount of time in order to make a difference to sections of the waterways. So it's something that we are really fortunate really; we couldn't do the work without volunteers to do it.*

Interviewer: What do you think actually motivates a volunteer to participate?

Interviewee: *It's really varied. Some people are, kinda of, for some people I think it's the environmental benefits, they kind of see waterways and they want to see environmental improvements. For other people, it's kind of health benefits of being outside. We've linked with different mental health charities in the past, 'cause there's proven research that sort of being outdoors and doing physical activity and working in a team can kinda help people who are kind of recovering from health, mental health illnesses. So we've worked with groups like that. So I think that sometimes it's the, kind of, health benefits, sometimes it's a social thing. Quite a few of our projects run regular events at the same location; it's often the same* *volunteers that come along each week. So there's the, kind of, social benefit of meeting people and, kind of, working together and doing things collectively, so it's pretty varied.*

Interviewer: In terms of what you identify with, what ethos and values do you identify with working here?
Interviewee: *... It was really that combination of sort of environmental improvement, but working with the community to sort of deliver them. So that's what really motivates me, in the, kind of, work that we do. It's, kind of, working with individuals to help them understand the issues of their local river; and enabling them to, kind of, make a change. What we try to do is make them sustainable by working with local communities on their local stretch of river, so that those people, kind of, develop a greater appreciation for what they see; and hopefully in the longer term they will have more of a vested interest in trying to make sure that the river is protected and looked after. So it is, it's all about, it's that combination of kinda working with people and communities, in a, kind of, environmental sort of charity to sort of deliver the improvements. And it's very varied. I mean, London is a hugely diverse place. All the rivers and the canals and waterways are different, as well as the different communities. My role especially, I don't have a specific geographical area to focus on, which means that I go to different areas of London, so I work with different communities. And that's kind of rewarding and challenging, and no one event is ever the same as the next one 'cause different people [come] to work with as well as the kind of different natural environments.*

Interviewer: Could you tell me about one of your initiatives?
Interviewee: *So there's a new project which I am working on which is called Thames River Watch; so, that is about trying to engage people in monitoring the health of the Thames. It's just starting. We are going to be looking at engaging volunteers in sort of doing water quality monitoring and looking at invasive species along the river and looking at litter and trying to monitor and gather more data so that we can better understand the kind of health of the river and then we can sort of talk about some issues that are affecting it and what we can potentially do to help make improvements. So that's the sort of project that I am working on at the moment.*

Interviewer: So just three things in relation to that, because I realize that you haven't much time: can you say something in terms of monitoring and also in terms of being about to organize communities, and then also who are you reporting to?
Interviewee: *In terms of the monitoring, so we are sort of building up protocol at the moment of exactly what we are going to monitor and the processes we are going to use to do that. The aim will be that we will have a system for collecting the data from volunteers; storing that data and then*

presenting it back in a sort of visual format through an online map, so that people can see where the survey has been done and they can look at the results of the survey. And that system will be accessible both to the people who have contributed, but also to members of the public who are just interested in finding out more about the health of different stretches.

Interviewer: What constitutes the survey?
Interviewee: *So we will be doing water quality monitoring. The fine details of exactly the parameters of what we are monitoring are being established at the moment. Because it is a new project, we are sort of defining what's valuable to monitor; what's the reliable processes for monitoring. As well as, kinda, what will give us kind of interesting data to analyse.*

Interviewer: How do you determine what's valuable to monitor?
Interviewee: *So it's really looking at the techniques that are available for doing the monitoring. And how reliable that data is that we can collect using volunteers. Cause obviously the more reliable it is, the better the results it produces and then we can talk about it more in terms of what it is showing us.*

Interviewer: In terms of one of your more successful project, could you tell me about that?
Interviewee: *A different project to Thames River Watch? It's so varied what we have. So, for example, one of my colleagues, it's not something that I've directly been involved with, but a project on the Ravensbourne catchment – so that runs through south-east London – and [London River Conservation Charity III] was part of an initiative called the Three Rivers Clean-up, which aimed to sort of tackle, initially, Himalayan balsam, which is an invasive plant species. And the approach that was taken was by bringing together different organizations and lots of volunteers to try and remove that plant species from the catchment. So I think it's run for four years now. In the first year, there was a lot of balsam growing along different stretches of the river. Year two there was slightly less; and year three there was hardly any of the plant that could be found.*

Interviewer: How was that organized?
Interviewee: *Different organizations, sort of, promoted events; organized and promoted events and gathered the volunteers in. All of the organizations met, in order to try and establish, who would being doing what, and in what areas people were working in. Then the programme ran over a sort of two- to three-week period in the summer. There are lots of events happening over that period of time to kind of focus volunteer efforts on removing the Himalayan balsam.*

> ***Interviewer: … What events were happening?***
> ***Interviewee:*** *… The task itself is, sort of, pulling balsam from the ground. So it's quite shallow rooted, so you can, sort of, get hold of the plant stem and pull it out. So that was the main activity of the river. So it would be either walking along the river banks or wading through the river in order to pull out the balsam.*
>
> ***Interviewer: Why was the balsam in the river?***
> ***Interviewee:*** *It is an invasive plant species. It thrives in damp environments. It loves growing on the river banks, erhm. It's also very effective at spreading its seeds. So once it's within a catchment, it then gets carried along by the river flow and settles onto the bank. The problem with it is that it grows very densely in small areas. Well, not actually always in small areas. It grows very densely where it grows. But it's an annual plant, so seeds, it grows up in the springtime, dies off in the summer and then it just leaves the bank bare. So it then means that they are open to erosion, because there's no vegetation growing, they outcompete the native plants that would be there. So that's the real problem with it. So by removing it, we're kind of creating space for the native plants to grow, and we also as part of the events do seeding. So we will put down native plant seeds after we've done the pulling. Erhm, we've removed the balsam and then, kind of, seeded more native plants, which will grow perennially so they'll grow, kind of, year on year, so you don't get that problem with the soil being left there.*

One can discern, from the above transcription, an appreciation of biophysical embodiment as integral to the social praxis of cultivating responsible citizenry. According to Freire, for education to facilitate consciousness-raising, the acquisition of environmental knowledge should invite "learners to recognize and unveil reality critically" (1985: 102). Civic activism in embodied social praxis unsettles the depiction of teacher/key worker as liberator: "in an education for freedom there are no subjects who liberate or objects who are liberated, since there is no dichotomy between subject and object" (ibid.). Such a transposition highlights the transmission of ideological and critical inquiry through the embodied inhabiting of spatial practices. Learners "are not marginal beings who need to be "restored to health" or "saved", learners are viewed as members of the large family" of environmentalists, engaged in social activism and community activities (ibid.). Thus direct involvement and engagement in the provisioning of the river foreshore advocates an environmental education in which learners experientially confront broader issues concerning water sustainability, locality and climate change. During my interviews with organizational members of river conservation charities, the issue of community featured consistently. Of significant relevance is the relation of the individual to the community, for it became evident that these charities are distinct in their ability to mobilize autonomous individuals to enter into loose collectivities that exist for the duration of the environmental event. Some insight into these sprawling event-based community affiliations is evident in the following transcription extract from an interview with a Volunteer Projects Manager, London River Conservation Charity II, September 2013.

> ***Interviewer: Do the website and the material that you provide, does it try to educate autonomous environmentally conscious forms of action? So not necessarily guided by yourself, but for people autonomously?***
> **Interviewee:** *Yes, absolutely. And there are many people out there who don't want to join a group. I think one of the mistakes that we make as charitable organizations is that we think, we say to people: if you want to do something for the river, join [London River Conservation Charity III] or join [London River Conservation Charity II] or become a member of [London River Conservation Charity I]. But a lot of individuals don't want to join in a group, they just want to be autonomous and so this is the sort of information [referring to literature produced by London River Conservation Charity II] where perhaps they can find out the information they need and they can go and do their little bit on their own or as a family without having to be part of a bigger organization. And I think it is important that going forward into the future we cater for those people. 'Cause, historically, we've only been looking at group activity and, yes, the activity needs some structure and system to be following through. But that's where the river wardens, whom I mentioned to you, are really pretty autonomous with [London River Conservation Charity II]. They [members of the public] know what they are there for and they'll go and do it. If they need any support, they will get it from [London River Conservation Charity II], but they don't have to be regularly doing that. [London River Conservation Charity III] is a very good example, because there you can join a group, you can go out with 50 people and you can spend the whole day picking up litter or maybe building a weir or doing something sort of engineering. But a lot of people don't want that group activity. They might just want to go themselves and take photographs or go and weed a section of the riverbed or, you know what I mean? Those things are important too, because they all play a part. They all have a role to play.*
>
> ***Interviewer: Might it be that the material, the resource materials, the book that you've produced and the website; does that enable individuals? Do you envisage your audience to be individuals or communities or people who will, through observing your websites or reading your material, your book, actually then collectively mobilize? Does it actually engender collective action or does it speak to the individual?***

Interviewee: *I think it gets the individual in autonomously. It reaches out to people who are interested and gets them interested. And then having got them interested, I believe our material channels them through to the group activity, where they can do things collectively ... Because here [referring to the charity's resource material] you can say, 'I'm an individual and I want to do something for the river; well, maybe I'll look it up, I'll ring up one of these organizations and I'll go and do something with somebody.' So I suspect, you know, always there's more result in working together; always, in every walk of life. So I think it is about collective action, it's not just about autonomy. It's about autonomous connection and collective action. I think that's what it's about.*

According to Freire (1985: 102–3), in critical political literacy, "Answers for their situation do not reside in their learning to read alienating stories, but in their making history that will actualize their lives." This in turn points to the realization that environmental education is politically effective when it emphasizes "humanity in its relationships with the world" (ibid: 103). While it is self-evident that no aspect of social reality is exterior to the natural world, a key feature manifest in the social praxis of provisioning the river foreshore is the impossibility of escaping from nature. It is difficult to lose oneself in "abstract visions of the world" (ibid: 103) when the visceral corporeality of the body is immersed in the interplay between institutionalized environmental policy, organizational praxis and the mud-drenched wilderness of the river foreshore. This suggests the impossibility of political literacy "without practice, the impossibility of thinking without a transforming action in the world" (ibid: 104). As Freire further states: "to know and transform reality are reciprocal prerequisites" (ibid.). Mobilizing volunteers in social-spatial formations geared towards resolving local environmental issues accentuates an appreciation of "people as a presence in the world, as beings of praxis, of action and thinking about the world" (ibid.).

Timely Reflections

Interview with Marketing and Communications Officer, London River Conservation Charity III, September 2013.

Interviewer: Could you tell me about your charity?
Interviewee: *... To begin with, its focus was just on the River Thames. It began with just clean-ups on one section of the River Thames, the Isle of Dogs. Do you know the bend in the river in East London, where historically is collected a lot of litter? So it used to be really just a carpet of plastic bags, this bend in the river, and there is still waste there, but that*

was the kind of motivation behind the project. And it kind of grew and grew and they realized that people, the few people that set it up, that in every part of London and maybe everywhere, people are actually interested in their waterways. I think that, something that I have kind of thought personally and since being here, is that people are drawn to water and I think that's regardless of whether it is a, like, a manufactured square space, like a Canary Wharf, for example, so it's not what you think of as a natural environment, that's still appealing to us. People like to sit near that space. So I think that people care about their rivers and they also want to get involved in community activities. So there's, kind of, I think, the charity operates with a joint focus on the environment and on communities. And I think it works best and is most sustainable when communities, it's done with the community. So they choose, they decide that they want to be part of it and how they want their river, or their area, to be. Erhm, and that's what we found the most success. So now [London River Conservation Charity III] has at any time around eleven community projects and we work right across London, so on canals, and rivers and ponds, on lakes and basically anywhere where there is a stretch of waterway. Some of them are more, they're urban rivers so there's a variety of problems. Some of them are healthier than others. But in any of those spaces, a whole range of ways to engage people and get people passionate about their area, and I think there's a multitude of benefits to that; there's the environmental impact of working to improve biodiversity, the removal of litter, raise awareness of those issue. But also connecting people to each other and helping make an area nice, you know, a nice place to live. And sort of foster a greater appreciation for the natural environment. So the idea then is that people take ownership for that space, so we work together to improve it and through that process they care about it, and they'll continue working and caring for that river. Because one of the downsides of the charity sector and of our work is that we run, we rely on grants and, you know, finite funded projects, so it is difficult when you go to an area and say, 'You have a year, or three years or sometimes just a short series of events and then we can't keep working there if the funding runs out.' But what we are trying to do through river training programmes and forging strong links with community groups is allow them, the people in that community, to take over and we'll support them to do that, so it's an independent kind of autonomous group or action. So that you don't have that issue of, it just, it's all just done when we run out of funding. So I say that's one of our key values and it is definitely something that we need to keep working on. But I think that it is important to us that it isn't, that everything is kind of as sustainable as it can be.

Interviewer: How do you, and this is once again about education, how do you teach an individual that their activities, that their proportion of the communal responsibility will be effective? ... How do you actually teach them to understand their place in, or understand, their contribution?

How do you teach them to understand their contribution in relation to the whole and the whole is the environment?
Interviewee: *I think that's difficult, actually, I think that's a really difficult, a really difficult issue. Yeah, some people use the expression: think global, act local. And I guess that's kind of what we would aim for. And because we are only London-focused and in some time-specific projects, we are only community-focused; maybe even one tributary, in one section of a river. So it is hard to see how that makes a difference overall. And I think with environmental issues the news can be so damning; it can be so, it can be so hard to bear. I think people often feel completely disempowered and, you know, don't feel the energy or motivation; they don't feel that they could do something. So I do think that's an issue actually. One of the ways that we would, we would demonstrate that success is to try and congratulate our volunteers; let them know in measurable terms how much litter was removed, what effect we had on one event and then at the end of the year, for example, we will send out newsletters and things and say, 'Together we removed this much litter and we've made this much difference.' So I think you need to, you need to thank people and demonstrate the effect they've had. And we also link in with national and international organizations, say, for example, coming up we've got the [river clean-up event] and we are linking that to the International Coastal Clean-up, which is US-based but I think is in over two hundred countries over the world; they're doing clean-ups on the same day on rivers, lakes, and oceans and sea. And so that's a way of linking, I think, the individual, and the country to the global issue. And I do think that's important to, kind of, point out, particularly with waterways that they are actually all connected so it's the coastal clean-up. We are not coastal here, but the River Thames is a tidal river and it does go into the sea. So when people remove litter from there, they are in effect preventing litter going out to sea.*

Interviewer: … And just staying with how the individual sees, understands, their contribution in relation to the immense scope of the environmental issues, because I can imagine people feeling quite overwhelmed. How do you harness, how do you inspire them? How do you encourage them to have a spirit despite how big the project is?
Interviewee: *So I guess it's about, firstly, educating people about the issues and then through training. So we have a training programme which is kind of more formal sessional-based training, but then we also offer training, really, throughout the event so the project officer there will support the volunteers. And I guess the idea is that by contributing, erhm, and learning about their issues, they're feeling empowered, that they do feel empowered. And that they have a sense of achievement in actually seeing what they've done and being part of a group. So I guess it is working to make the people feel valued and to engender a sense of group work, I guess, of team work. I think that can be really powerful, especially*

if these people don't – and often they do know each other – but often they don't know each other, but they do form a bond throughout the activity. And I think that's quite a powerful thing actually, that people with a common interest but from different walks of life, coming together for a shared cause, and people come for different reasons. I think people have different motivations to volunteer. And in a way that doesn't matter; there can still be a real sense of achievement at the end. I think the other important thing, maybe, is that [London River Conservation Charity III] doesn't claim to be a kind of answer to all the environmental problems and I think that no one organization can be, because actually if we do think about all the environmental problems, it is really overwhelming; it's difficult. And so, I mean, I think you've got to be realistic about what you can achieve and we're, I think we are setting out what we are trying to achieve, so I think that's important as well; we are not, we don't keep missing the bar. So I guess to an extent our, you know, you could say removing litter from one small section of river, that's quite a small part of the world, the global environmental problems. So I guess we wouldn't put it in that context. It's about helping that local community and helping the local environment and people connect into their local environment which, erhm, in London is often, it's often obvious where the environment is if you live in a really built-up area where you don't have much green space, then, erhm, people might not know much about nature or feel any sense of affinity with, it, so, erhm, so yeah, so I think that's a really important step in as much as, erhm, even if that doesn't solve an environmental problem, it connected someone to their environment. So it's kind of small steps, I guess.

Interviewer: I was wondering, is there a project that you are particularly proud of that you can share?
Interviewee: *I represent [London River Conservation Charity III] on partnerships; such as to put a charge on plastic bags, for example, or to call for the Thames Tunnel. But when I feel most is when I sort of join in with or local community groups and meet the volunteers and see the work first hand ...*

Interviewer: And in the marketing and communication campaigns, how are you marketing and communicating, in maybe a recent campaign? The plastic bag one sounds interesting.
Interviewee: *So that's a coalition campaign called Break the Bag Habit and we are working with the Marine Conservation Society, the Campaign to Protect Rural England, Surface Against Sewage. So it's a broad alliance to call for a charge on plastic, on, sorry, all carrier bags, which has happened in Scotland and has happened in Wales now, and so our kind of line is: we are the only home nation with Ireland; we are the only home nation not to have implemented this, so it's really needed. We've seen reduction in bag use numbers and we know how many bags; because we*

particularly – we [London River Conservation Charity III] and the Marine Conservation Society – because we do a lot of clean-ups, we know how many plastic bags end up in rivers and seas. So yesterday, we had a rally in Westminster actually and we worked with an organization called EcoTales which brought in a small group of pupils who read a poem about plastic pollution. They had a huge albatross, and the albatross is significant because a huge number of them die through ingesting fishing lines and plastic bags and things like that. So they made an albatross out of plastic bags, and they delivered a letter to David Cameron and got a really positive response; he's recognized that this charge needs to happen and he's said that twice in two years now but still hasn't brought in the legislation, so it's looking positive but they are not actually committing to it. I think it's the Treasury that doesn't want to commit to it. So I guess that feels quite powerful, like a good step.

Interviewer: The onus of responsibility in terms of charging for plastic bags: retailers or the consumer? Who should bear the responsibility?
Interviewee: *Well, I think that, the idea would be that, there would be a 5p charge on plastic bags. But not everyone can afford that and that unfairly penalizes people on low incomes. But the idea would be that people reuse bags. So I guess it's a shared responsibility between the retailer and the consumer; and I guess the other point we would make is that the retailer meets that cost, doesn't have to bear that cost completely alone and it's something they can incorporate, I guess.*

Interviewer: And, yet, in terms of that shared responsibility, is there now a notion of a consumer citizen alongside your environmental citizen, you're teaching for or you're helping to cultivate? … The person who agrees with the plastic bag charge, are they a consumer citizen?
Interviewee: *OK, I guess my understanding of it would be – and I'm thinking about how I make purchasing decisions – someone who considers the social and environmental impact of the things that they buy, as well and maybe alongside what they want to buy; the need for it and the cost of it. So, I mean, I think that it is difficult for people on low incomes to make ethical decisions about what they purchase all of the time. But I think that we can all definitely do more to purchase ethically. So I guess that's what I think of when you say consumer citizen. It's like thinking about the sustainability really of the things that you are purchasing, so not just looking at the product; realizing where it's come from. Where's it actually been made or produced or sourced from? How's it got there? And what is it packaged in? And I guess you might also think: where are the profits going? And sometimes if you think of all those things, all of the time, it leaves you feeling that you can't buy anything? Which, I mean, some people do take that approach, to not buy. I've got a friend who will buy nothing new … and she doesn't like any packaging … I definitely think there are things like that, that each of us can do and that makes us feel like*

we are doing something. But ultimately, it has to be on a much bigger level than that, because there are so many of us and, erhm, and even if we were all to reduce the plastic bags that we buy here, and the packaging that we buy here. And other parts of the world, they are growing and just getting more packaging and coming in line with the way we do things and so, yeah, we need a kind of more joined-up approach and those, kind of, when individuals make those decisions, that's quite a personal thing. I guess, it's probably more to do with how they, helping it make themselves feel like they are doing something I suppose. I do feel that's important, but I don't know if it can be a solution to all the problems.

Interviewer: … It is sometimes suggested when we are focusing on misfitted pipes, or when we are focused on packaging, sometimes these can be very domestic issues. Sometimes they can be very domesticating of the environment. So when we see, and I empathize, the amount of packaging, and I'm really into visual culture, advertising etc., etc., in relation to politics, but when we are focusing on those issues and we are taking those on board as the way that we help, do you think that we are overly domesticating the environment and making it one in which it's domestic, private, in terms of the private sphere, it's about the homestead? Do you think those sort of issues domesticate the environment?
Interviewee: *Well, I guess it does. But that's the reality of how we live. I think that it is very important to connect people to their environment and to talk about the wider context of the natural world. But we do live in houses and we do, we are kind of citizens of this urban environment. Generally, more and more people are living in cities, so I think that's the reality of how we live and probably we'll continue to live in this way more and more. So I think that, and the other, the other thing is that those issues are having a direct impact: so we know that misconnections are having a real impact on the health of people's waterways, so I guess that does bring it back to individuals and the domestic situation and the very local issue, but that's really the source of the problem in that instance. With misconnections it is very, it is very specific. I mean, they are able to find, where, down to what house a misconnection is coming from or what street a misconnection is coming from so I guess in that instance, ultimately, it is domestic.*

Based on ethnographic narrative interviews, this chapter has explored the organizational values and culture of three London river conservation charities. Each one of these charities is dedicated to the River Thames water ecology and sustainable environmental conservation. In the numerous interviews conducted with organizational members, it became evident that the issue of environmental education is a consistent feature of the responsible citizenship that is advocated. Furthermore, outreach community engagements constitute a primary vehicle for transforming action through education. The essential point to highlight is the role played by the European Commission and the Water Framework Directive in structuring the form and function of community engagement practised by river conservation charities. A prominent feature of interviews featuring the Directive was the issue of demystifying information so as to facilitate political literacy. Paulo Freire's observations on "education for political literacy" provide an apposite theoretical framework: "If we don't transcend the idea of education as pure transference of a knowledge that merely describes reality, we will prevent critical consciousness from emerging and thus reinforce political illiteracy" (1985: 104). The chapter considered pedagogic practices that characterize outreach community engagement. In the writings of Freire, critical pedagogy relates to a process of "conscientization (based on the Brazilian *conscientização*), the process by which human beings participate critically in a transforming act" (ibid.: 106). Axiomatic with the philosophical belief in conscientization is a provocative "recognition of the world, not as a 'given' world, but as a world dynamically 'in the making'" (ibid.).

Framed within the philosophy of education espoused by Freire, this chapter has described and engaged with River Thames waterways charities in conceptual terms: using informational democratic actors; utilizing channels of political education; ascribing value to traditional ecological knowledge; uniting the local and global through environmental education; engaging in local environmental problems; uniting government, commerce and local communities in the endeavour to cultivate sustainable urban waterways; engaging with environmental issues within the context of neoliberal educational reform; promoting an appreciation of nature in the context of urban environment; and being innovative and inspirational in cultivating education as social praxis.

Figure 4.8., 'Would the sage Tempt Nature's mineral depths, or trace the stars Thro' their nocturnal course" (Ann Yearsley, *Familiar Epistle to a Friend*, 1787). Photographic Image, West London, River Brent Tributary of the River Thames, February 2014

Figure 4.8a., 'Would the sage Tempt Nature's mineral depths, or trace the stars Thro' their nocturnal course" (Ann Yearsley, *Familiar Epistle to a Friend*, 1787). Photographic Image, West London, River Brent Tributary of the River Thames, February 2014

CHAPTER FIVE

VOLUNTEERING OUR TIME IN THE SUB-POLITICAL GEOGRAPHIES OF RIVER CONSERVATION

Figure 5.1., "The brief sun flames the ice, on pond and ditches, In windless cold that is the heart's heat, Reflecting in a watery mirror" (T.S. Eliot *Little Gidding*). Photographic Image, South London, River Thames, September 2013

Interview with female volunteer, South London, River Thames foreshore, September 2013.

> ***Interviewer: I'm just conducting some research and I'm just trying to understand why people erhm, are so committed to cleaning the river, to maintaining the river. And I'm just trying to understand the values***

behind it. And I wonder if you can tell me. How did you find out about this event?
Interviewee: *I've been talking to a staff member two weeks ago, because I live exactly next to the river in that house there. And every weekend I've been walking along the river. Every weekend I just find it so relaxing, and it's beautiful and yea there's a certain attraction to water, I think. So I saw them and I was asking them for a river cleanup because I've wanted to do it for the last probably a year; but I can't do it on my own. Because I have a lot of rubbish right in front of my house, because I'm next to the river and ever single day I am basically like looking at it and it's like that's disgusting, I wanna do a cleanup. But you can't do it on your own. I mean I probably will in the future, because they can't organise it as frequently as I wish them to. And I did work in the environment for the last seven years myself so erhm it's quite a matter to my heart anyway; so I want to do that all the time. So living next to the river I think it just happened to be the river, apart from my Monday to Friday job where I'm working for the environment anyway. So it is something which matters to me, it's protecting the environment.*

Interviewer: And do you think a community is developing around these issues?
Interviewee: *I think it is. It is growing more and more, because I think it has become more public in the headlines and newspapers. Everything, you know, like from contamination and climate, to water, to all the pollution levels. People get, I don't know, this little wake-up call thing by now. And they see more and more and they see catastrophes around the world happening and so I think there is more and more people readily growing into it. It's always hard for me to say because I grew up in the context, I was studying environmental engineering. So I always was surrounded by people who were generally interested in maintaining the environment how it is and keeping it healthy. I'm not sure if that is representative for me, because I am surrounded by people who do care. I think it's more a question of how you can get to other communities of people, who haven't developed their sense yet; because I think it is mostly their rubbish we are picking up here. It's certainly not mine, because I know I flushed no, not anything down the toilet. Like panty liners, finding little Q-tips or cleaning equipment or blister packages. Like that's not me anyway. Somehow people throw it away and that's a lack of education more than anything. I'm sure they don't do it because, ah, yeah, I wanna harm the environment. It's just a lack of education.*

Interviewer: Do you think this is educative then?
Interviewee: *Yes, because people... I think it has two benefits. People come along and see us. They know exactly what it is without even; we don't have to explain our selves that we are cleaning up. Everyone can see that clearly. And they are thinking: why do they do that? And they maybe*

> *realise as well how full the crates are [used to collect the extracted material], like, 'Oh my God that's really bad!' And it just stays. And that's like a thought developing in your head. So I think that's already an education although you don't even talk to people. This way, people walk by here they see you and they're like, 'Ah look at that, that's really good'. And I think that's really nice.*
>
> *I think it should be paired up with, I don't know, not sure how you do that, you can't have classroom services. Who would come along? Let's invite all the people to city hall and have a talk. Maybe some people show up. But the people you wanna reach for sure they wouldn't show up. So you need to find a way of communicating with people who normally you would not reach, because they are not interested in these kinds of events. But erhm, I think that's a real challenge. So I mean, it's a challenge for me as well in the real world when I am talking to clients and making them think about the things they do which are harmful to the environment. So and I think ... that and I am definitely very happy that they do it. I tried to organise the next one myself, because I am really pushy, pushy. I wanna have that part cleaned up there... (Female, South London, River Thames Foreshore, September 2013).*

"Human influence has been detected in warming of the atmosphere and the ocean, in changes in the global water cycle, in reductions in snow and ice, in global mean sea level rise, and in changes in some climate extremes" (IPCC 2013a:12). Scientific evidence consistently states: "It is likely that anthropogenic influences have affected the global water cycle since 1960" (ibid:13). In measuring the global scale variations in precipitation rates above ground, scientific evidence states that: "Anthropogenic influences have contributed to observed increases in atmospheric moisture content" (ibid); and "Extreme precipitation events over most of the mid-latitude land masses, and over wet tropical regions, will very likely become more intense and more frequent by the end of this century" (ibid:16). One might enquire as to how these drivers of climate change relate to the above transcription; for the interviewee is merely describing a local incident of accumulated debris polluting the river foreshore. The respondent's intention to be proactive rather than rely upon insurantial interventions is highly enlightening. In previous research I have written critical challenges to the "women's market" in financial services. The "women's market" was integral to the profligate financialisation of the domestic sphere initiated in the 1990s post-deregulatory market of financial services. My previous research formulated a post-structural Foucauldian account of gendered time as part of a critical intervention into the marketing knowledges and advertising technologies through which linear representations of insurantial risk are gendered. Given the insurance industry's past and present drive towards financialising the domestic sphere, the respondent's proactive endeavour to cultivate a community-based non-financialised environmental intervention is an intriguing new direction in environmental activism. Moreover, scientific measures indicate an increasing acidification of ocean surface water (ibid:8) and it is statistically evident "that more than half the observed increase in global average surface temperature from 1951 to 2010 was caused by the anthropogenic increase in greenhouse gas concentrations" (ibid:12). These IPCC (2013a) statistics provide an indication of the wider ecological issues central to the interviewee's concern, but it is difficult to explain why the climate change statistics have made nature the focus of our individual lives. According to Ulrich Beck (2007), the global scale of climate change is the counterpart of industrial society. In late modernity, the social and political risks of global climate change increasingly exceeded the capacities of institutional response; thus, "the dangers of industrial society begin to dominate public, political and private debates and conflict" (2007:5). Based on interview data with volunteers involved in river cleanup activities this chapter explores the concept of environmentalism in terms of the sub-political geographies of river foreshore conservation.

The Sub-political Geography of the River Foreshore in Global Context

> In the RCP projections, thermal expansion accounts for 30 to 55% of 21st century global mean sea level rise, and glaciers for 15 to 35%. The increase in surface melting of the Greenland ice sheet will exceed the increase in snowfall, leading to a positive contribution from changes in surface mass balance to future sea level (high confidence). While surface melting will remain small, an increase in snowfall on the Antarctic ice sheet is expected (medium confidence) resulting in a negative contribution to future sea level from changes in surface mass balance. Changes in outflow from both ice sheets combined will likely make a contribution in the range of 0.03 to 0.20 m by 2081-2100 (medium confidence). (IPCC 2013a:18)

"Warming of the climate system is unequivocal, and since the 1950s, many of the observed changes are unprecedented over decades to millennia" (IPCCb:1). Indications of climate change effects on water ecology include thermal temperature increase of the oceans; reductions in snow and ice accumulation in Antarctica, rising sea levels and escalating concentrations of greenhouse gases in the atmosphere (ibid). The devaluation and neglect of environmental sustainable development confronts the assertions and contours of capitalist production and ruptures open the transition to "the other side of the obsolescence of the industrial society" i.e., "the emergence of the risk society" (Beck 2007:5). At its most extreme, the concept of the "risk society" claims that the repercussions of an entire modernist epoch are persistently exceeding the superintendent capacities of institutions ordinarily designed to safeguard civil society and facilitate health and well-being. Industrial capitalism championed the technological advances gained through the combustion of fossil fuels; now the aftermaths of this environmental misdemeanor relentlessly reverberate: "The atmospheric concentrations of carbon dioxide (CO_2), methane, nitrous oxide have increased to levels unprecedented in at least the last 800,000 years. CO_2 concentrations have increased by 40% since pre-industrial times, primarily from fossil fuel emissions and secondarily from net land use change emissions" (IPCC 2013b:1). Approximately 30 per cent of anthropogenic carbon dioxide emission is absorbed by the ocean, a primary cause of ocean acidification (ibid). Confronted by the immediacy of environmental degeneracy, the paucity of corporate society's centralist democracy reveals stifling limitations, which "begin to dominate public, political and private debates and conflicts" (Beck 2007:5). "Over the period 1901-2010, global mean sea level rose by 0.19 [0.17 to 0.21] m" (IPCCb:1) and yet the responsibilities for insuring against the risk of flooding have become an obtuse actuarial wasteland. Indeed, actuarial knowledges, and insurance "institutions of industrial society [have] become the producers and legitimators of threats they cannot control" (Beck 2007:5). Previously distant risks become realizable threats to everyday security, strengthening cognizant connectedness to the natural environment. Antarctica is no longer a distant otherworld; it is difficult to ignore scientific evidence that: "Over the last two decades, the Greenland and Antarctic ice sheets have been losing mass, glaciers have continued to shrink almost worldwide, and Arctic sea ice and Northern Hemisphere spring snow cover have continued to decrease in extent" (IPCC 2013b:1). Increasing proportions of the world's people inhabit territories that are vulnerable to unprecedented denaturing in "the global water cycle" (ibid). Differences "in precipitation between wet and dry regions and between wet and dry seasons will increase" (ibid), uniting territorial divides in their shared encounter with the extraordinary processes of climate change. Proponents of this perspective maintain that proliferating environmental hazards reveal the contradictions and limitations of industrial capitalism. Acutely embedded within industrial capitalism's economic structures, are "certain features of industrial society become socially and politically problematic" (Beck 2007:5). Needless to say, state institutions continue to govern, but their operational activities "are clouded over by debates and conflicts that stem from the dynamism of risk society". According to the IPCC (2013b:2): "Cumulative emissions of CO_2 largely determine global mean surface warming by the late 21st century and beyond. Most aspects of climate change will persist for many centuries even if emissions of CO_2 are stopped. This represents a substantial multi-century change commitment created by past, present and future emissions of CO_2".

Figure 5.2a., "I do not know much about gods; but I think that the river, Is a strong brown god – sullen, untamed and intractable". (T.S. Eliot *The Dry Salvages*). Photographic Image, South London, River Thames, September 2013

Figure 5.2b., "I do not know much about gods; but I think that the river, Is a strong brown god – sullen, untamed and intractable". (T.S. Eliot *The Dry Salvages*). Photographic Image, South London, River Thames, September 2013

Figure 5.3. "By worshippers of the machine, but waiting, watching and waiting" (T.S. Eliot *The Dry Salvages*). Photographic Image, South London, River Thames, September 2013

Figure 5.4., "The more delicate algae and the sea anemone. It tosses up our losses, the torn seine" (T.S. Eliot *The Dry Salvages*). Photographic Image, South London, River Thames, September 2013

Figure 5.5., "We must be still and still moving" (T.S. Eliot East *Coker*), Photographic Image, South London, River Thames, September 2013

Figure 5.6., "The sea has many voices, Many gods and many voices" (T.S. Eliot *The Dry Salvages*). Photographic Image, South London, River Thames, September 2013

Figure 5.7., "The menace and caress of wave that breaks on water" (T.S. Eliot *The Dry Salvages*). Photographic Image, South London, River Thames, September 2013

Figure 5.8., "And the sea yelp, are different voices, Often together heard: the whine in the rigging" (T.S. Eliot *The Dry Salvages*). Photographic Image, South London, River Thames, September 2013

Figure 5.9., "older Than time counted by anxious worried women" (T.S. Eliot *The Dry Salvages*). Photographic Image, South London, River Thames, September 2013

Figure 5.10., "Lying awake, calculating the future, Trying to unweave, unwind, unravel, And piece together the past and the future" (T.S. Eliot *The Dry Salvages*). Photographic Image, South London, River Thames, September 2013

Figure 5.11., "When time stops and time is never ending; And the ground swell, that is and was from the beginning, Clangs, The bell" (T.S. Eliot *The Dry Salvages*). Photographic Image, South London, River Thames, September 2013

Figure 5.12.a, "Mild contemplation from the barren rock" (Ann Yearsley *Soliloquy*, 1796), Photographic Image, South London, River Thames, September 2013

Figure 5.12.b, "Mild contemplation from the barren rock" (Ann Yearsley *Soliloquy*, 1796), Photographic Image, South London, River Thames, September 2013

Figure 5.13., "Of statues form'd by man for future worlds" (Ann Yearsley *Soliloquy*, 1796). Photographic Image, South London, River Thames, September 2013

Figure 5.14., “Mycias, behold this bird! See how she tires, Breaks her soft plumes, and spring against the wires!” (Ann Yearsley *The Captive Linnet*, 1796). Photographic Image, South London, River Thames, September 2013

Sub-politics in the Age of Proliferating Environmental Risk

> The global ocean will continue to warm during the 21st century. Heat will penetrate from the surface to the deep ocean and affect ocean circulation. (IPCC 2013b:2)

The denaturing of environmental sustainability is observably accelerating; and continuously bringing into question the capability of capitalism to oversee and control the proliferating risks of global climate change. Industrial modernization involves exploitative subjugation, and the capitalization of the natural environment is an integral feature of this capitalist dynamic. It is evident that the preceding modes of industrial capitalism modernized while disregarding issues of sustainability in the expenditure of renewable resources. The "risk society" is the unintended outcome or reverse-side consequence of industrial capitalist opulence. The "risk society" refers to a second order of modernity in which the risks precipitated in a previous order of modernity exceed traditional mechanisms of control; it "designates a developmental phase of modern society in which the social, political economic and individual risks increasingly tend to escape the institutions for monitoring and protection in industrial society" (Beck 2007:5). According to Beck (ibid), the conditions of risk society precipitate changes at the level of individual apprehension. Beck (ibid: 2) conceptualizes this as the "sliding" into a new order of "reflexive modernization". Beck is keen to clarify his use of "reflexivity", for it relates to "self-confrontation" rather than a notion of "reflection". The expansion of environmental hazard in the "risk society", involves a transition in capitalist modernization. Insurance risk calculation is part of the transition in the way risk is apprehended; this is because "insurance can be defined as a technology of risk" (Ewald 1991:198). Insurance transforms calculative possibilities into "values or capitals possessed or represented by a collectivity of individuals, that is to say, a population" (ibid: 199). Thus, as a technology of risk, insurance is primarily a mode of disintegrating social reality and rearranging parts so as to link individual elements to entire populations. Rational, probability calculus frames the insurantial knowledges and practices established in the context of modernity. In the modern age, insurance operates as a practice of rationality, rather than "a practice of compensation or reparation" (ibid). Thus, one only insures oneself against risk; and what constitutes risk is limited by the calculus of probabilities. Consequently, insurantial calculation is an active part of risk apprehension, for insurers are not merely "passively registering the existence of risks" (ibid: 200). The risk industry is productive; it "produces risks…makes risks appear" (ibid). It is an idiosyncratic feature of insurance that it engages in the objectification of designated events into risks; and there is no limit to the aspects of the material world that can be subject to insurance. Thus it has been suggested: "Everything can be a risk, insofar as the type of event it falls under can be treated according to the principles of insurance technology" (ibid). Global climate change presents risks that are markedly unusual in form, for they unpredictably proliferate and extend into the minutia of our everyday lives. Beck (2007:5) states: "The transition from the industrial to the risk period of modernity occurs undesired, unseen and compulsively in the wake of the autonomized dynamism of modernization, following the pattern of latent side effects". Consequently, in contrast to previous formulations of risk, Beck (ibid) contends: "The constellations of risk society are produced because the certitudes of industrial society… dominate the thought and action of people and institutions in industrial society". In the risk society, hazard intrudes into everyday life; confronting risk is not a possibility that responsible citizens can choose to avoid; "cumulatively and latently", industrial modernization produces risks which threaten eventually to confound the substratum of industrial society (ibid:6). Thus "reflexive modernization" refers to "self-confrontation with the effects of risk society that cannot be dealt with and assimilated in the system of industrial society" (ibid: 5).

In the context of environmental crisis, reflexive modernization refers to a condition in which ecological hazards exceed designation as the predicaments of the natural world, for they become "a profound institutional crisis of industrial society itself" (ibid: 8). Rationalist models of the economy appear irrational when confronted with the inter-generational glacial time-frames of global warming, for example the IPCC (2013b:2) reports that: "Global mean sea level will continue to rise during the 21st century. Under all RCP scenarios the rate of sea level rise will very likely exceed that observed during 1971-2010 due to increased ocean warming and increased loss of mass from glaciers and ice sheets". Global climate change unsettles the legitimate authority of industrial approaches to the environment. The risk society is a "self-critical society", in which experts perpetually contradict each other as part of the erosion of "social forms (role patterns) that produce binding orders and security fictions which are relevant to action" (Beck 2007:12). In the conjuncture of industrialisation and modernization is arising a crisis in the authority of "security fictions", and this situation invites unprecedented new opportunities for civic activism. The innovative disposition of contemporary civic activism is accentuated by exigency and immediacy in the demand for action. Beck (ibid) describes the urgency in the "compulsions for action"; in the wake of disintegrating "security fictions" the appetence for immediate solutions becomes ever more palpable (ibid).

Locally Mobilizing Global Responsibility

Axiomatic with this social and political transition is individualization at the level of individual biographies. It is evident that traditional forms of social solidarity are in decline and being substituted by an evacuated liminality in which sociality is detached from traditional prescriptions. Identities are cultivated through the internecine conjunctures of a post-traditional order. Erupting in these spaces is "the compulsion to find and invent new certainties for oneself and others" (ibid: 14). Individualisation is accentuated by the deteriorating proficiency of the state, and the corporatist entrepreneurial superstructure. Integral to this premise is the notion of a descent into a post-traditional political order, involving the eruption into politics of previously marginalized interests. As Beck (ibid: 18) expresses it: "the political constellation of industrial society is becoming un-political, while what was un-political in industrialism is becoming political". Suffice to say that the location of politics has permutated; politics erupts into disjunctures and internecine spaces of the risk society. It is incongruous to forage for politics inside the archaic institutions of industrial society; for "Politics [now] determines politics, opening it up and empowering it" (ibid). Nowhere is this more apparent than in environmental activism. Environmental sustainability has become an all-inclusive perturbation; locally uniting activists into global campaigns for environmental sustainability. The mobilization of local communities around global environmental issues is evidence of the extent that environmental deterioration capillaries into political consciousness (ibid: 19). Care for the environment continues to be articulated through political institutions; but "they also withdraw from them, with at least part of their existence, their identity, their commitment and their courage" (ibid: 20). The immediacy of environmental degradation in everyday lives enfolds the space of place into humans' interconnection with nature. Action on environmental degeneration continues to involve conventional politics, but the exigency to respond to the immediacy of hazards means that activism is also migrating towards the "new niches" and materialities of political consciousnesses (ibid). Consequently, the process of individualization from the stewardship of institutional environmental justice campaigns involves continuity, change; and manifests "contradictory multiple engagements … which mixes and combines the classical poles of politics … Everyone is … an activist in partial aspects of his or her self" (ibid: 21). Consider, for example the following transcription of a female volunteer on the foreshore, September 2013, conducted as part of my ethnographic research into river conservation along the southern regions of the River Thames:

> ***Interviewer: And the reasons why you do this? [Respondent is cleaning up litter from the river foreshore]***
> **Interviewee:** *Part of the reason is I live locally and this is where I bird watch and I see birds sort of playing with the rubbish on the foreshore. And erhm, so all these images of birds with loads of you know plastic stuff in their stomach and dying of it. And I thought, you know something needs to be done about it. So that's one of the reasons why. Plastic's good fun, [but] so erhm.*
>
> ***Interviewer: And it's a great way of meeting people?***
> **Interviewee:** *It is also a great way of meeting people. Yes I've made some friends this way over the years.*
>
> ***Interviewer: And do you think that we are making a difference?***
> **Interviewee:** *It is making a difference, yes.*
>
> ***Interviewer: In what way?***
> **Interviewee:** *It is actually a lot cleaner. You need to come back and think that it is a lot cleaner. And also it raises awareness and people stop; start changing their behaviour a little bit.*
>
> ***Interviewer: And do you think it helps politically as well?***
> **Interviewee:** *That I don't know.*
>
> ***Interviewer: But it helps people's awareness, in what way?***
> **Interviewee:** *Erhm they come here and erhm some times it's part of a building exercise [corporate events] so it's not necessarily their choice to come and do this; and then they notice the amount of rubbish that people just threw down the toilet and maybe they think, 'Oh that's where it ends up, maybe I'll stop doing it'. And erhm next time we'll have a few less tampon applicators and you know. Actually I haven't had too many this time but erhm sanitary towels are one of the things you keep on finding all the time.*
>
> ***Interviewer: So you think that's an issue?***
> **Interviewee:** *It is a big issue. Yes because it's plastic. The cotton part of it disappears very quickly. But the plastic part of it will be there for*

> *centuries; and does end up, wherever, in the water, being ingested by the fish, by the turtles, by everything. Even plankton is eating the plastic, when it starts getting tiny, tiny little bit[s] (Female, South London, River Conservation Foreshore, September 2013).*

The interviewee illustrates how situated subjects enfold global meaning into their locally determined civic activism. During my research, river conservation activists frequently emphasized the imperative to mobilize citizens in order to address the immediate issues of environmental deterioration. They were "consumer[s] of the river", actively investing rights and responsibilities in the civic activism of river conservation. No pre-formulated scripture appears to guide civic confidence in the capacity of activism to make a difference. At the same time, traditional indicators of political proficiency appear alien, obtuse and irrelevant to a politics streaming through the exigency of the moment. Such political engagements are in themselves creative deconstructions of orthodox convention; and in the process of river conservation, subjects are personally and communally united by the rhythm of the river's ecology. A noticeable feature of connection, aligning the river conservation events attended, was the differences in the backgrounds of participants. Academics and civic activists readily discuss transfigurations in the constituency of political communities. The reinvention of the political subject is axiomatic with sub-politics. Beck (2007:22) illuminates distinctly new agents: "outside the political and corporatist system" professionals, scholars, NGOs, experts and members of the intelligentsia are amongst the non-conventional "allowed to appear on the stage of social design". It is precisely because the location of politics is increasingly taking place in the internecine spaces of modernist disjuncture that ordinary citizens readily jostle for position alongside "social and collective agents" (ibid). Some illustration of this hustling for political representation, in the new geographies of sub-political engagement, is provided by the following interview of a male volunteer, on the south London foreshore, September 2013:

> ***Interviewee:*** *I think it's urgent we give back to, we are very careful about the planet. I've been working on those issues of environment and I think you have to start with doing this take your gloves off, gloves on and erhm and work hard. What I like also is the organisation of [London River Conservation Charity III]. I expect to, I'm just interested in the way they organise, it's amazing. It's nine thousand people that volunteer there. I just wonder how other European countries are organised as well and maybe I would like to go, to being much more involved professionally on these issues maybe in Continental Europe or doing more research on...*

Interviewer: So do you think it's more about community? Is it about developing a community around these issues?
Interviewee: *Yea it's about community. But it's really, I think we have big issue[s], it's urgent. Because we are globally we are, destabilizing a lot of ecosystems. And plastic has a dreadful, I am so upset with all the packaging we are using in supermarkets and companies continue to do it. So there's, there's going to be something to do really seriously. So some very smart guys in Switzerland are doing a lot of work on small ecosystems which are really damaged at the moment. And they are very pessimistic about the next year's, erhm the second half of the century...So we just have to be serious about it. And it's not a question of marketing it's really helping the next generation to move on.*

Interviewer: And by doing these small tasks do you think that we are actually making a bigger difference?
Interviewee: *Yes because first it is going to help me be involved at a higher level; because I'm going to [be] able to explain that erhm: you see that I'm not high in the sky; it's useless to be in the pyramidal organisation. You have to be close to the reality.*

Interviewer: So it helps you to be able to speak and represent?
Interviewee: *Yea, speak and represent; take action, erhm generate high strategic ideas. There's a lot to do.*

Interviewer: And so it helps to actually get politically engaged?
Interviewee: *Yea,*

Interviewer: How; in what ways?
Interviewee: *Well, think at the level of companies. There are very simple things to do and erhm people, erhm especially you see: we are collecting so much plastic here today, OK. So usually when you go, for example in[to] a shop, you use your plastic bag for twenty minutes only and then it go[es] in the bin. So we are-... and we are using petro chemical products, fossil energy to help you carry your bread and your vegetable for twenty minutes only. This is the average, it doesn't make sense. Ok? All the chain is damaged so we need to go out of our comfort zone to accept responsibility.*

Interviewer: And this is a beginning?
Interviewee: *This is the beginning, it is the beginning. It has to be, much, much more organised and I think everybody needs to be responsible. It is not a game. You have even on the investment management of the, you have some companies who claim they are socially responsible and so it's fine they are a kind of global; but sometimes [it's] a kind of marketing thing. So there's a lot to do about that. (Male, South London, River Conservation Foreshore, September 2013).*

One sees here illustration, of a sub-political activism that is reconfiguring the politics and culture of environmental sustainability. Beck (2007:23) observes how the sub-political flow of influence "from below" undermines enforcement power and accelerates the "shrinkage and minimization of politics". The interviewee, above appears to be dexterous in their endeavour to cultivate social capital, via the formation of networks within London River Conservation Charity III. The interviewee's intention to outlay social capital so as to network across environmental organisations is illustrative of how the dynamically changing political landscape provides new opportunities for self-transformation and expression. The concept of social capital has clear relevance to the values and procedures that individuals utilize in their approach in volunteering time to river restoration conservation charities. Pierre Bourdieu (1984/2010) refers to social capital as social connections, capacities and inter-personal competencies which are accumulated throughout the individuals' habitus and experiential encounters. In this sense, social capital is "a capital of social connections, honorability and respectability that is often essential in winning and keeping the confidence of high society, and with it a clientele, and may be drawn on, for example, in making a political career" (ibid: 116).

For, Bourdieu, society can be envisaged as delineated into co-existent fields, each constituted by forms of material and immaterial capital, competitively valued in the social structuring of the field. Bourdieu's primary contention proposes that the central differences that distinguish "the major classes of conditions of existence, derive from the overall volume of capital, understood as the set of actually usable resources and powers – economic capital, cultural capital and also social capital" (ibid: 108). Within the context of voluntary sector organisations, one may assume, from Bourdieu's perspective, that: "The distribution of the different classes … thus runs from those who are best provided with both economic and cultural capital to those who are more deprived in both respect" (ibid). Referring back to the previous interviewee's line of justification for networking with river conservation charities, as a means of achieving a more advanced vocation in international environmental institutions, it is evident that the river conservation voluntary sector operates as a "field of forces, but it is also a field of struggles tending to transform or conserve this field of force" (Bourdieu 1995:30). In the above interview, the volunteer recognises that entry into the field of international environmental institutions will be contingent on their ability to network the "objective relations between positions" and their capacity to orient "the strategies which the occupants of the different positions implement in their struggles to defend or improve their positions (i.e., their position-takings)" (ibid). In so doing the volunteer has to make calculative choices in determining which voluntary sector environmental charity to network into and volunteer their time. Thus the individual's choice of voluntary sector organisation involves an exchange relationship in which material and immaterial capital is invested into the accumulation of social capital. This is because:

> Every position-taking is defined in relation to the space of *possibles* which is objectively realized as a *problematic* in the form of the actual or potential position-takings corresponding to the different positions; and it receives its distinctive *value* from its negative relationship with the coexistent position-takings to which it is objectively related and which determine it by delimiting it. (ibid)

Consequently, the respondent's decision to join any one of the many river conservation charities has an element of risk associated with it. This risk would be realised if the symbolic currency of social capital, proffered by the voluntary sector organisation, becomes less competitively valuable, i.e. "the position remains identical" (ibid). But the opportunity to appreciate social capital is predicated on position-taking, and thus the process of choosing has risks and rewards, which are only realizable through position-taking, as Bourdieu indicates: "the meaning of a work … changes automatically with each change in the field within which it is situated" (ibid:30-31). Thus the process of position-taking within river conservation charities is not a stable system of selection between the "state of the system" and the "repertoire of possibilities which it offers" (ibid: 34). It is a key feature of my findings that for a significant proportion of actors, volunteering time in river restoration involves decisions that engage the actor in the cultivation of symbolic capital. In accordance with Bourdieu's conception of immaterial capital: "the principle of position-takings lies in the structures and functioning of the field of positions, this is not done so as to return to any form of economism" (ibid:35).

Participation in river restoration, involves a cultural economy of choice, for it is not merely a decision of which activity to invest in, treading, weeding, collecting litter, all amount to the same thing. Rather, deciding to volunteer involves a cultural economy of time, in which the actor decides "the right moment to invest or deinvest, to move into other fields, when the gains in distinction become too uncertain" (Bourdieu 1984/2010:85). Thus, individuals do not progress through the social field of river restoration charities "in a random way" and this is "partly because they are subjected to the forces which structure this space (e.g., through the objective mechanisms of elimination and channelling)" (ibid:104). Bourdieu is particularly insightful in his conception of "symbolic violence" as a mechanism through which, that which is arbitrary can be defined as legitimate. The operation of symbolic violence in legitimating the cultural arbitrary, highlights the extent to which the subject's progression through the field is facilitated or hindered by the arbitrary evaluative imposition of the dominant culture. In contrast with Marxist frameworks of analysis, Bourdieu does not automatically define this situation as economically determined; nevertheless, Bourdieu does refer to the operation of markets, for he argues that:

> Because they are acquired in social fields which are also markets in which they receive their price, cultural competencies are dependent on these markets, and all struggles over culture are aimed at creating the market most favourable to the products which are marked, in their manners, by a particular class of conditions of acquisition, i.e., a particular market. (ibid:89)

Following Bourdieu, it is my observation that a further condition that is responsible for structuring the progression of the subject within the field is the decision to invest in the immaterial capital derived from the volunteering of time. It is evidently simple to conceive that a time pressured volunteer might resist attempts made by a voluntary organisation to acquire more quantitative time; but I am referring here to the longue durée of time formed through the inter-subjective networks of social capital. When volunteering time is conceived in terms of social capital, it is plausible to envisage a situation in which a volunteer resists the attempt made by an organisation to ingratiate them into the organisation's social networks of capital; and this latter action can be defined in terms of a process of translation, in which the social capital derived from volunteering gains currency in terms of position-taking within a cultural economy of time. As Bourdieu observes:

> If the foregoing argument suggests an 'analysis of essence' … that is because most groups have sought to lay down absolute, final differences by means of the irreversibility of time, which gives inflexible rigour to every form of social order based on the order of successions. The holders and claimants to succession–father and son, owner and heir, master and disciple, predecessor and successor–are separated by nothing, except time; but there is every sort of social mechanism to make this gap unbridgeable. Thus, in the struggle between the different 'manners', i.e., the different

> manners of acquiring, the dominant groups are always on the side of the most insensible and invisible mode of acquisition, that is, the oldest and most precious one. (ibid:65)

"In the wake of sub-politicization", new geopolitical gateways provide platforms for expression, beckoning forward individuals "hitherto uninvolved in the substantive technification and industrialization process" (Beck 2007:23). Networking, from the foreshore of the river, into the centre stage of institutional decision making, was not an unusual precedent. Indeed, several stories of dexterous cross-institutional networking were conveyed during my ethnographic studies. For example, the following interview extract describes volunteers who mobilized their networks with London River Conservation Charity III, so as to gain political representation within the local authority:

> ***Interviewee:*** *We've now got projects right across as far as Reading and Hillingdon and different project officers. Apparently there's only 22 members of staff; so we've got at least 12,000 volunteers. And they're working on events. There's more than, at least one maybe two events every day of the year. So wherever you look, you've got events taking place: either on the river; on the canalized rivers; waterways and other areas. So it means that we're capable of actually looking at litter clearance, water quality testing, having areas where we tackle the invasive and wild-invasive species like Himalayan balsam. We have balsam bashing [events] and remove that from areas where it is taking over. Because this species, being non-native, will take over an area and actually it will improve bio-diversity if some of these species are removed. Or areas like Japanese knotweed, again which is another well-known [interruption: interviewer: "Invasive species"] another invasive species. We have been out on the ... where we have had floating pennywort, which is another species, which, if it grows, it spreads like a carpet on top of the canal and it prevents oxygen reaching the fish in that area.*
>
> *We also have projects like [name of colleague]. He's actually an event leader for a project ... event support for those events at Beckton Lake, for example, which is, kind of, more a residential area. You go down to your local park and there's a lake; but again you look at the water quality and there's algae in the water. In hot weather the pump wasn't working so you could see fish that were being starved of oxygen, maybe over-fished potentially. They plan to set up some kind of angling club society. They've already got one but to try and regulate the fishing and make sure the council ... They did get involved, they came down one day, this is Newham Council, to actually aerate the water. And it's, it's improving. But there's also problems with fly-tipping on the borders of the park, near the warehouse where I've been leading a group, small group, when I've been working on [name of colleague] event. Working in the woodland area,*

warehouse, where there's just been fly-tipping in the woods; just chucking things over from this warehouse next door. So there's hidden problems which the council can't always tackle; so, we have to work together with the council and supplement what they're doing. And also work on areas where the council won't take responsibility, like on the foreshores along the Thames and the banks of rivers and near streams and waterways, lakes, and so forth.

Interviewer: No, that's interesting. Having that angle as well in terms of supplementing areas that are somehow marginal to local authority responsibility … And this idea of working with local communities, local residents. And there seems to be this area that seems to be not sufficiently the council's responsibility?

Interviewee: *Yes, but I think local residents can form pressure groups, which is an ideal way to get involved. If they approach [London River Conservation Charity III] for training. If they, for example, live by a river or canal and they see it as a problem, they can always approach [London River Conservation Charity III] for training and help lead their own events to help clean up that area on their doorstep and it's about empowering local residents to take pride in their area and taking responsibility.*

Interviewer: Can you give me an example of when local residents have approached [River Conservation Charity III] for training?

Interviewee: *Well there's this example at Leighton Jubilee grounds where I've been working with another organization, where I recommended [London River Conservation Charity III] because this park is actually bordering onto Dagenham Brook. There's a big motorway, on the other side there's the Middlesex filter bit and the River Lee, the other side, very close by in East London. And I recommended [London River Conservation Charity III] because you know, in terms of the training I know that these local residents wanted to do something about the litter problem bordering on that area. Formerly it was Leighton Marshes Areas, it's been renamed.*

So they approached [name of colleague], the training officer at [London River Conservation Charity III]. And they are now looking at getting training and actually forming a group to help lead an event along the Dagenham Brook. And their local residents, they're coming in, in the London borough of Newham, to that area, to that local stream. So following that principle, you know, you can work with other little groups as well. And to me, it's all holistic, you've got to, say, for example, with this event, you've got the tow-path which is bordering, you've got the foreshore, and you've got to work with the tide. So it's going to be very organized by a river. With a canal, you haven't got the same perimeters of time, because you don't have the tidal river. But you do need to have

> *sufficient resources and outreach. You can do community engagement with events, go door-to-door knocking; get other local communities, other residents; contact other litter groups who are working in areas near waterways. And get more people involved as well as promoting through the website ... So it's all a matter of good PR and communication strategy as well. (Female, Organizational Member of London River Conservation Charity III)*

It is apparent that in the milieu of sub-politicization, these forms of political engagement are not isolated, but defined in relationship to "a decrease of the central rule approach" (Beck 2007:23). The respondent describes complex features of retrenchment and ambiguity in the ascription of responsibility for the litter management of London's waterways. The issues raised about fly-tipping are emblematic of "a decrease of the central rule approach". Financial pruning of local authority expenditure coupled with a neoliberal re-engineering of its duties is resulting in situations in which "processes which had heretofore always run friction-free, fizzle out in the resistance of contradictory objectives" (ibid). River conservation charities provide valuable vehicles for local residents to mobilize in the absence of the means supplied by the constitutional state. It is also evident that the current context of voluntary sector operation is formidably vulnerable to the prescriptive managerialism of Big Society market reform. For example, consider the following extract derived from an interview with a volunteer projects manager of London River Conservation Charity II:

> ***Interviewer: I am researching into water ecology and organizational values…Case studying organisations***
> ***Interviewee:*** *Case studying organisations, involved in conservation, sustainability?*
>
> ***Interviewer: Yes***
> ***Interviewee:*** *Because I think one of the problems for all of us River Thames charities is that there are really unclear boundaries: who does what; and there are so many organisations out there doing different things. There's a lot of duplication; and there's a lot of misunderstanding; and there's a lot of political dysfunctionality if you like. Because people are being competitive for funding sources…there is a huge sense of ownership and localization.*
>
> *You talk about organizational values; I think local knowledge is a huge factor in organizational values. When you've got something like a river that people are sustaining or conserving there is an enormous and I think quite negative sometimes, sense of ownership. It can be negative and*

positive. Positive in the sense it makes people belong to a section of the river and so they do more to look after it. And negative because they then compete against rather than work with other people in the same area. You see what I mean?

Interviewer: OK yes. And erhm could you tell me more about that community sense of value so in terms of the organisations that you have worked for? So that notion of ownership and then, and belonging, how that actually translates?
Interviewee: *The [London River Conservation Charity I] for example have fifty five River Wardens and they're all volunteers typically retirees and they look after a stretch of the river and they probably walk it two or three times a week. And they talk to people, they pick up litter and they keep an eye on what's going on in that bit of river. And so they consider themselves the sort of keepers of that river. Now [London River Conservation Charity III] who are operating in London…are doing very much the same thing and they have people called River Keepers and so already there is some sense of conflicts here because yes they should be working together. Because they're both applying to the same funding sources for money to keep these sorts of activities going they're actually competing against each other, not working together. And I think, in my mind, in terms of the geopolitical situation that's probably the biggest problem it's funding and how the funding is distributed.*

Because there is an incredible amount of people out there doing some really good conservation work. But a couple of things: they are not funded; they are very modest and they don't talk about what they are doing; so they are very quiet; and a lot of them go out completely unnoticed. And maybe they are happy like that, maybe that's fine. But there are people like [London River Conservation Charity III] that are very good in attracting bigger dollars, bigger pounds, buckets of money to them. They've got more of a commercial background. And then there are other organisations that are struggling because they haven't got that commercial ability and that's a big issue (Volunteer Projects Manager, London River Conservation Charity II, September 2013).

Beck (2007:23) prophetically describes similar sub-political activist developments, in which "various groups and levels of decision-making and participation mobilize the means of the constitutional state against one another". The interviewee provides examples of waterways litter management as a space of intersecting and conflictual responsibilities. It is evident, as Beck (ibid) observes, that the milieu of sub-politicization is intersected by competing and contradictory political agendas: "not just in the confrontation of institutions and citizens' groups, but also in the conflicts of national and local politics, between a Green-motivated administration and the old industrial management, and so on". Particularly pertinent are the confrontations and conflicts emanating from the individuating neoliberal marketisation of environmental risk and conservation. Interviewees often spoke of frustration with regards to their local experiences of national policies; especially in relation to individuating policies that seemed to work against community-based action. Consider the following interviewee extracts:

> ***Interviewer: Your reasons [for cleaning litter from the Foreshore] are?***
> ***Interviewee:*** *I think my reasons are: I find it ridiculous that it's no one's responsibility to look after it, once the rubbish is in the water. So to find a charity that like[s] doing it, I'd rather help to do it you know, it makes the river cleaner and it's historic as well and it's part of London.*
>
> ***Interviewer: And what does it mean to you then? What do you gain from it?***
> ***Interviewee:*** *I gain from it to look in, like I said we done this for about five years now. And when we first come here when the tide was out, all you could see was plastic bags. They've got photos of it, it was really bad and now look at it you've got, all the bags were buried, weren't they. You know, so we're clearing all the top stuff, it's definitely going down. So it's achievement as well yea. And also hopefully there's people here that haven't been along here before and they've been shocked by seeing the rubbish that's there so, it gets their awareness up. Hopefully that will stop people throwing stuff in or saying to their friends: oh we don't drop that in the water.*
>
> ***Interviewer: And what sort of values do you think we learn from it?***
> ***Interviewee:*** *Erhm, well value wise I think well, this is a danger you could do some damage to the environment if you carry on, you know. So you need more people like the people here to care about the environment and stop that happening. Cos not only is it damaging wildlife, it is also making everything look unsightly. I mean I wouldn't like to be a tourist, come over to this country visit Greenwich, look over there and suddenly you see all the bags of rubbish on the foreshore it's not good is it? I don't think it's good at all you know. That's what I think it makes it look nice as well, better environment, you know. (Male, South London, River Conservation Foreshore, September 2013).*
>
> ***Interviewer: Do you think it helps you think about this politically?***
> ***Interviewee:*** *It could do. Politicians seem to go, plastic bag and make you pay five pence for. What they should do is get it like a deposit so you can get it back, when you bring it back. Or like the people that bring the bags back rather than throw them away. If they pay five pence to throw them away, then they'll throw them away, won't they? There's got to be more*

imaginative ways of doing things, [rather than] make people pay money for. It's the ubiquitous easy answer to things. Make, put the cost of living up or tax you or, in one way or another.

Interviewer: Do you think there's a sense of community that people gain from being here?
Interviewee: *I don't know, there's community with people you know. I don't know about the people who live locally. They consider that it's just some event that's taking place, I think. Or maybe they're too busy to help, working.*

Interviewer: So if it's not a local community, what type of community do you think is evolving? Because people seem to be getting to know each other
Interviewee: *Some places have got more, sort of community anyway than others. You get all these new luxury flats, second home owners near and behind their locked gates and you don't get no community from it .Whereas they will moan and oppose everything like boats in Lime House Dock.*

Interviewer: And what do you feel you've gained from being here today?
Interviewee: *...done something today.*

Interviewer: So what have we done today?
Interviewee: *Cleared up the Thames*
(Male, South London, River Conservation Foreshore, September 2013).

Volunteering Our Time in the Context of Flooding and the Individuation of Risk

Severe winds and storms affected the United Kingdom throughout last winter. The extreme weather, in combination with high spring tides, led to widespread flooding and devastated communities. The sudden and sustained nature of the winter floods underlined the serious need for coherent policies and sufficient funding to protect homes, businesses and farmland.

Investment in flood prevention is preferable to spending on cleanup, both from an economic and a social perspective…Funding for maintenance is at a bare minimum. Furthermore, when budgets are tight, maintenance – in particular, watercourse conveyance and dredging – is the first thing to be cut. In the face of limited budgets, low-priority areas such as farmland are sacrificed in favour of urban, highly populated areas. The evidence we received demonstrates that dredging can be beneficial in certain circumstances, and as part of a portfolio of measures, but it should not be seen as an all-purpose solution. Too often maintenance is neglected until a need is created for costly one-off capital investment…We urge Defra to revisit its policy for funding allocation to recognise the economic and social value of agricultural land.

Each catchment area has different flood risk management requirements and, consequently, different funding needs. To be effective, priorities must reflect local knowledge and local circumstances. We agree with the Secretary of State, Rt Hon Owen Paterson MP, that the distinction between capital and revenue funding "is a bit of a grey area in practical terms" and we urge the Government to assess the possibility of a future transition to a total expenditure classification for flood and coastal risk management. This would allow greater flexibility to target funding according to local priorities.
(House of Commons Environment, Food and Rural Affairs Committee 2014)

Figure 5.15., "Her sweet note warbled from the blowing thorn, That hangs o'er yon cool wave; responses clear" (Ann Yearsley *The Captive Linnet*, 1796). Photographic Image, South London, River Thames, September 2013

In the winter months of 2013 and early 2014, England experienced a torrential storm of winter weather systems. The accompanying flooding caused mass devastation and exacerbated existing inequalities already distinguishing the riverscape communities of urban and rural England. In addition to the individuating environmental agendas of rural and urban riverscapes, one needs to include the proliferating environmental risk registers of insurers and the ad hoc market-based funding structures that currently define the voluntary sector's response to river flooding. With regards to the former, it is an operative feature of risk calculation, that any environmental possibility can be objectified as a risk and subject to the probability calculus of insurantial technology. Unsurprisingly, it is evident that environmental risk insurance has proliferated at one and the same time that the state is retrenching from securing reparation and indemnification for damages ensued by environmental hazard. Consider recent government policy in the area of flood insurance. Until recently, the United Kingdom's flood insurance security provision relied on voluntary agreements from members of the Association of British Insurers (ABI). In 2013, these flood protection insurance agreements, were called the Statement of Principles, and are very soon scheduled to expire (DEFRA 2013). The government has consistently referred to how vital the Statement of Principles are to individual citizens: "Without them, there is concern that many hard-pressed households in flood risk areas will no longer be able to access the level of insurance cover they need" (ibid: 1). In 2013 the insurance industry presented a scheme, called Flood Re: designed to institute, as an industry standard, a premium ceiling on the rate high-risk households have to pay for flood insurance/home insurance. According to DEFRA (ibid): "Flood Re would be an industry-run, not-for-profit scheme that would support up to ~ 500,000 of the highest-risk households in the UK" (ibid:2). To finance the scheme an industry supported levy will be introduced, which is currently estimated at £180million annually for the initial five years. The cost of this industry-backed levy will be passed onto the respective customers of the insurance provider. This method of distributing risk across the policyholders is a widely established practice within the industry and apparently "lower risk households are already subsidizing those at higher risk of flooding", at approximately the same proposed rate. (ibid:2). The government is intent on obligating the industry to assume flood insurance responsibilities and in 2013, its stated intention was to "seek powers in the Water Bill to enable the Government to stand ready to regulate…These powers would require insurers to take on their share of high flood risk households or face penalties" (ibid: 2-3). Central to the operational infra-structure of the scheme will be an alliance between the government, the Environment Agency and its correlatives in the devolved administrations working together to construct a register of United Kingdom properties designated as at high risk of flooding. It is estimated that this list will be constituted by at least 500,000 homes (ibid). It is intended that individual insurers will "be allocated a quota of high-risk properties to cover from this register (its 'obligation'), based on its market share" (ibid:3).

The government has consistently guaranteed that the ethos of the scheme will be based on market competition. Thus, for example, the DEFRA (ibid) report *Securing the Future of Flood Insurance*, states that: "The Flood Insurance Obligation would force insurers to compete with each other for the business of high-risk households in order to meet their targets". In accordance with the rhetoric and discourse of neoliberal consumer citizenship, market competition between the scheme's insurance providers "should help customers to shop around to get a good deal" (ibid). While it might be argued that such statements appeal to the necessary fiction of the rational decision-making consumer, the self-evident intention of the Government's proposals relates to reducing state responsibility for the flood protection of its citizens. It is a feature of insurance that "risk is collective" (Ewald 1991:202). Whereas the occurrence of misfortune is experienced by the individual "striking at one and not another, a risk of accident affects a population" (ibid: 203). In actuarial terms individual risk, does not actually exist; for risk only becomes calculable when distributed over a population. The role of the insurer is "to constitute that population by selecting and dividing risks" (ibid: 203). Thus the government's proposal to oblige insurance companies to provide flood insurance across risk households will have the effect of "socialising risks" (ibid). It will constitute each individual as part of the whole insurance population. The risk of flooding is thus socialized to encompass the entire population: "No one can claim to evade it, to differ from the others like someone who escapes an accident" (ibid). In the logic of insurantial probability calculus, each individual in the population "is a factor of risk, each person is exposed to risk" (ibid). Consequently the government's flood insurance proposal will have the effect of individualizing the risk (and premium payments); for "it defines each person as a risk" and "it is an individuality relative to that of other members of the insured population, an average sociological individuality" (ibid). But the "sociological individuality" created by private insurance is only in actuarial principles collectively shared amongst the elements of the population. And this produces a stifling paradox; because the duality of environmental hazard as lived by the victim (in terms of premium fees and insurance claims) and statistically shared across a population, fails to bring about equivalent relations of corporation. But the scope of environmental risk does not narrowly define the victim, thus the concept of proportionality in risk calculation is the antithesis of community engagement. The combination of these contrasting dimensions complicates the administration of community based environmental activism. Limitations are further multiplied by the ad hoc structure of voluntary sector funding.

It is evident that recent government programmes directed at the voluntary sector have been keen to engineer market dynamics, infused by an entrepreneurial spirit of self-reliance and market efficiency. Continuities are evident in the Coalition's advocacy of active citizen engagements with voluntary sector organisations; while at the same time encouraging these organisations to competitively tender for increasingly ad hoc charity funding. The spirit of entrepreneurialism infuses the Coalition's voluntary sector policies. As Milbourne (2013: 44) expresses it "In parallel, the rhetorical emphasis on entrepreneurialism, which displaces social values and community expertise, refers to multiple opportunities that can be 'seized' for those willing to adopt an entrepreneurial spirit, new skills and a more business-like approach". Thus the communitarian ethos of the Big Society combines elements of localism and community-based initiatives with market tendering for ad hoc short-term grants that subject communitarian organisations to the antithetical purposes of performance measures. In this context "seemingly contradictory ideologies and strategies – neoliberal market ideology, local co-production and citizen empowerment" provide indication and insights as to the stifling pressure that voluntary sector organisations are presently experiencing (ibid: 45).

According to Beck (2007:23) sub-political engagements are hindered by ambivalence, and this is one of the side-effects of the risk society. Similarly, recurrent themes in my research were limitations, due to funding; economies of size; individuating environmental policy and the prescriptive constraints of market-based performance indicators. Elsewhere, it was clear that multi-level confrontation within state institutions, along with citizens' groups, has had the effect of inducing paralysis, whereby "no side gets its way, neither the opponents of power nor power itself, where these concepts become as relative in actuality as they ought to be in thought". A stifling intransigence occurs, which according to Beck (ibid) is the inevitable anterior of sub-political civic activism.

But such despondency appears to be on a distant shore when observing conservation along the river foreshore. River conservation is clearly the antithesis of the paralyzed political quagmire, involving active intervention, and real-time responses to immediate disjunctures in environmental sustainability. Beck (ibid) states that the enfeebling of "the implementation process of industrialization" precipitates losses "on all levels", and that this inevitably "can slow the process". Multi-level institutional and extra-institutional conflict markedly reconfigures the formulation of sub-political civic activism, and frames its "instrument of power", to the extent that the materiality of power in sub-politics can be "congestion (in the direct and the figurative sense)" (ibid). Beck recites the tale of the "Munich motorists" to illustrate the frustrated immobility of agents hemmed into traffic jams of intransigent state officialdom. Congestion is certainly a condition of ecological posturing. The differences between agents in the drama of environmental degeneration and climate change could once be distinguished as they were separate actors; the irresponsible, unrelenting environmental assault of powerful commercial interests versus environmentally conscious citizens actively protecting a benign and helpless natural world. Big business is becoming entrepreneurially adept at capitalising both the processes and outcomes of environmental degeneration; and thus the distinct roles of villain and salvationist blur into each other.

As Beck (ibid:50) observes: "The resistance of the one half of the economy and society encounters a grand coalition of the alarmed public, the eco-profiteers and the eco-careerists in industry, the administration, science and politics". Convergence towards ecological consciousness "produces and cultivates a cultural Red Cross consciousness" (ibid). Every agent is an ecologically conscious actor. According to Beck (ibid) this has the unforeseen and extraordinary effect of proliferating our everyday realities of environmental risk: "It transforms everyday, trivial, unimportant things into tests of courage in which heroism can be exhibited". These converging conditions of environmental intervention agitate political consciousness with respect of environmental threats, creating "a substantive semantic horizon of avoidance prevention and helping" (ibid). The culture of risk morality expands with the escalating spectacle of environmental hazards, and this precipitates "a turning away from mere criticism and the transition to the siege of the status quo by alternatives" (ibid: 51). An evocative proposition of Beck's analysis is the contention that the paralysis of "the siege…by alternatives" and the involuntary strike of sub-political power "can be a precursor of an unregulated anarchic self-limitation and self-control. Perhaps 'anything goes' means 'rien ne va plus' nothing goes any more?" (ibid:23). Alongside this picture of despondent impasse, Beck (ibid) contends that:

erupting through the paralysis of intransigent sub-political indecision is the increased propensity for reinvented political subjects, thus suggesting the significance of exploring meaning, identity and subjectivity as determinants of volunteering time to the conservation of the river foreshore.

Figure 5.16., "Lo! Dreary Winter, howling o'er the waste" (Ann Yearsley *Sonnet to*, 1796). Photographic Image, South London, River Thames, September 2013

Figure 5.17., "To lure the shivering pilgrim o'er the green, He hastens on, nor heeds the pelting blast: Thy spirit softly breathes – 'The worst is past'…Thus breaks thy friendship on my sinking mind, And lures me on, while sorrow dies behind" (Ann Yearsley *Sonnet to*, 1796). Photographic Image, South London, River Thames, September 2013

Timely Reflections

> ***Interviewee:*** *I am the [London River Conservation Charity III] project officer for the River Cray, which is a tributary of the Thames that runs through south east London: So it starts from [the] London Borough of Bromley by Orpington: and then runs down towards the Thames, through Bexley it goes in past Crayford and then joins the River Darent and then goes into the Thames Estuary there...Cause the plastic that gets down here is also in the other rivers. People tend to forget, they throw rubbish down in the street and think someone is going to come along and clean it up. Yes there are street cleaners, but between that, it rains it goes into the drainage system, into the outflows, the surface water outflows designed to go into the rivers. Because a lot of the urban rivers have been taken over with flood defenses...but that rubbish will then go out and get caught up in the river. When we talk about our river it is quite heavily modified, it is an urban river. And it's very historical. So it started off with people settling by it. Then we had people doing mills, plough mills, powder mills...It became an industrial site and then with the coming of the railway, where I am, because we are in the suburbs, people moved in en masse, all the housing development came on. And then the rivers became flood defenses.*
>
> *So what's happened there is it tends to steer away from the natural effect of the river and it's now payback time to look at those rivers. Because they are, they are linear paths; ...many parts of them are open spaces, meadows, waterways...So if we can, one of the things that [London River Conservation Charity III] is doing they are getting involved in looking after those rivers and actually improving them. Turning them wherever possible to a more natural habitat: so the fish have got more chance to shelter in there; the birds and all the vegetation. So that's the additional work I could be doing in the future. [Project Officer, London River Conservation Charity III, September 2013]*

"It is likely that anthropogenic influences have affected the global water cycle since 1960" (IPCC 2013a:13). Scientific evidence proposes that human activities have constituted part of the observed escalation in atmospheric moisture and the corresponding "global-scale changes in precipitation patterns over land…and changes in surface and sub-surface ocean salinity" (ibid). Scientific evidence further suggests that "Arctic sea ice cover will continue to shrink and thin and the Northern Hemisphere spring snow cover will decrease during the 21st century as global mean surface temperature rises" (ibid:17). Directly related to these increases in surface temperature has been a decrease in the volume of global glaciers (ibid). It is evident that since the 1970s, a decrease in global glacier volume coupled with "ocean thermal expansion explain[s] about 75% of the observed global mean sea level rise" (ibid:7). The IPCC (ibid:21), proposes "geoengineering" methods, designed to intentionally alter the climate system and counter-balance global climate change. However, the IPCC concedes that: "There is insufficient knowledge to quantify how much CO_2 emissions could be partially offset by CDR [Carbon Dioxide Removal] on a century timescale" (ibid). Similarly, models of Solar Radiation Management (SRM) indicate their potential to counterbalance a global increase in temperature "but they would also modify the global water cycle and would not reduce ocean acidification" (ibid). CDR and SRM methods of geoengineering the climate system "carry side effects and long-term consequences on a global scale" (ibid). Indecisively concluding climate change reports is perturbing, though not unusual in the current science of climate change; the coalition of scientific evidence and environmental risk assessment is uncoupling, and this opens the "possibilities of a politics of politics, a (re)invention of the political" (Beck 2007:18). This chapter has proposed the sub-political geographies of the river foreshore.

Figure 5.18a., "When Melody, enchanting pow'r! Lur'd all my cares to rest: My spirit drank her melting lay, In sacred rapture dy'd away, And trembled to be blest" (Ann Yearsley *Extempore*, 1796). Photographic Image, South London, River Thames, September 2013

Figure 5.18b., "When Melody, enchanting pow'r! Lur'd all my cares to rest: My spirit drank her melting lay, In sacred rapture dy'd away, And trembled to be blest" (Ann Yearsley *Extempore*, 1796). Photographic Image, South London, River Thames, September 2013

CHAPTER SIX

GENDER AND THE GEOPOLITICS OF CARE IN RIVER CONSERVATION

Interview with two female volunteers, south London, River Thames foreshore, September 2013.

> ***Interviewer: What motivates you to actually come to help clean up the river?***
>
> **Interviewee 1:** *I'm here because my work does this volunteer thing, you know. Where, in theory, you can go and do the volunteering and they give you the time off to do it. However, this is my off day ... But no, I think it's important to keep, to have the river clean. I mean, it's not just this river, they've done wonderful stuff with the River Wandle. Erhm, it's really important.*
>
> ***Interviewer: Why is it important?***
>
> **Interviewee 1:** *Erh, wildlife bio-diversity. It's good to have a clean river. I mean, I grew up near the river.*
>
> **Interviewee 2:** *To have a clean river is to have a clean ocean. Isn't it? And if you actually see how the sea, the fish, the albatross, everything is dying with eating plastic.*
>
> **Interviewee 1:** *Yea, the plastic bags, I think is really important.*
>
> **Interviewee 2:** *How do you disintegrate [plastic]?*
>
> **Interviewee 1:** *Yea, and the damage that they do, yea.*
>
> **Interviewee 2:** *Everything washes to the sea and we live off the sea, after all, that's where life comes from. So we have to look after it, quick!*
>
> **Interviewee 1:** *Yea, I completely agree with that.*

Figure 6.1., "How patiently toils on this little watch! My veins beat to its motion" (Ann Yearsley *Soliloquy*, 1796). Photographic Image, South London, River Thames, September 2013

Interviewee 2: *Yes, it's passed through all the life cycle, hasn't it? The food chain cycle, whatever they call it. The plastic is disintegrating and it's being eaten like plankton so and that's our, yea, it's our food chain. That's really bad.*

Interviewer: And do you gain a sense of community from being here?

Interviewee 1: *I think so, yes. I mean we've only met each other today and we've worked together, it's been really good to work together, wonderful, yea.*

Interviewer: What type of community and friendship do you gain?

Interviewee 1: *Satisfaction with a job well done; doing something physical outside. Yea?*

Interviewee 2: *Yea, you know you are doing something for your earth. You're meeting people.*

…

Interviewer: You said you lived where?

Interviewee 2: *In Barcelona, but I have been living here for six months now. I wanted to talk about this and take pictures [for her blog].*

Interviewer: So tell me about your blog and what's brought you here.

Interviewee 2: *Well, the blog talks a little bit about everything. But I have been doing sustainable work for quite a long time. So I talk about, erhm, I do yoga as well. I talk about sustainability; about eco-design about how factories, the industry or the products can actually be sustainable. And how we as users or as customers have got the power, I think there is no democracy but your buying power. So if you stop buying or you push industry to produce things the way you want them, we can actually change and that's in each one of us.*

Interviewer: And how does that bring you to here?

Interviewee 2: *How does that bring me to here? To London, you mean?*

Interviewer: To the River Thames here.

Interviewee 2: *Well, it's necessary.*
Interviewer: Have you been here before?

Interviewee 2: *No, I think it's wonderful.*

Interviewer: How did you find out about it?

Interviewee 2: *By blogging. Blogs and surfing on the web. Yea, trying to look for conservation volunteers, volunteering groups. Yea, basically.*

Interviewer: And what sort of difference will it make? What difference will it make? How will things change? It looks beautiful; that big plastic you picked up was amazing.

Interviewee 2: *It doesn't look that dirty, does it? When you come here, you don't think there is so much plastic?*

Interviewee 1: *No, as soon as you dig, a little bit.*

Interviewee 2: *We are on top of plastics, aren't we? Yea, it's amazing.*

Interviewer: And how do you feel about yourself? How does it make you feel about yourself?

Interviewee 2: *Great. You feel more satisfied. It's the best therapy (she laughs). If you ever needed a therapy: go and work [she laughs]. They say in Zen, before, before illumination [enlightenment]; I'm translating from Spanish. Before your illumination [enlightenment], chop wood and get water from the well; and after illumination [enlightenment] chop wood and get water from the well [she laughs]. That's it. It's good, isn't? [Interviewee 1: 'yes']. You need to, you need to be here.*

In 2010, hundreds of Costa Rican volunteers rallied together to extract plastic debris that had accumulated along the foreshore, where Rio Tárcoles intersects with Rio Virilla in Central America. The river clean-up event was orchestrated as a result of a joint initiative between local government agencies, private enterprise, and the Costa Rican tourist industry. It is estimated that roughly 145 large garden bags of river extracted debris were collected; much of which consisted of non-biodegradable plastic. In 2013, as in previous years, hundreds of volunteers participated in litter extraction based river conservation, in the form of community-outreach projects, across the UK capital of London. Neither of these events is unusual, as river clean-up activities have interchangeable features. Nevertheless, it is precisely because synergistic forces are evidently mobilising voluntary environmental activism, (set against the maelstrom of modernization), that river conservation provides compelling illustrations of how local cultural values and experiences both intersect and challenge transnational ecological conditions. In Chapter Four I explored these themes with regards to the enrolment of European

Directives within existing voluntary sector community outreach river conservation educational programmes. Chapter Five examined the sub-political motives through which voluntary sector organisations provide actors with a vehicle to mobilize and galvanise social networks. Specifically, it focused on the milieu of privatised environmental issues; mapping the sub-political geographies of ecological responses to local experiences of national and international crises in institutional proficiency. This chapter further extends the questions of domestication; for it deconstructs the ideal feminine care ethic of river conservation, and subsequently evaluates the ecological activism that is taking place on the river foreshore. The chapter questions whether river restoration can be progressively defined in terms of a gendered ethics of care.

Symbolic Mediation and the Domestication of River Conservation

Interviewee: *I worked in Vienna…I try to do something for the community. So when I have a bit of time, I want to take in some fresh air and enjoy some social life.*

Interviewer: So what does this mean to you, being here?
Interviewee: *You're part of a, you know that other people cares about the environment and that. So you feel that you are part of a group and they have some interest in our future and our wellbeing. You feel then there is something in common to share. And also I like to enjoy the mud and the water so [she laughs]. When it's raining it's a bit less pleasant, but for some few hours a week or even you do two three times a month, that's enough. But you do a bit of difference.*

Interviewer: What type of difference does it make?
Interviewee: *If you consider then everybody is doing like you just a bit. At the end of the month there is some change. You can't do [it] every day or so it would be even too much. But if you do a bit, it's like everything. If you do a bit you have some, at the end you have some satisfaction in the long term, but you have something*
Interviewer: And is it a community?
Interviewee: *People are, like this event is like they are coming from all over London because it's a big clean-up. But in general I work like in East London around the Lee. And so each of us there is erhm an area and they are working regularly. This one is like they do twice, twice a year, this kind of event. But the one in East London it can be a regular meeting. (Female Volunteer, South London, River Thames Foreshore, September 2013).*

Interviewer: Why do you spend time here?

Interviewee: *...just volunteering. We saw an ad[vertisement] here and we are just volunteering. It's environmental and yea.*

Interviewer: And how did you find out about it?
Interviewee: *Through an email that we get through 'Londonist'.*

Interviewer: And what does it mean to you to be here?
Interviewee: *Oh well you kind of do a little bit. Every little bit helps doesn't it? Just a little bit of cleaning up. I mean plastic bags, you dig one it gets worse more and more aren't there?*

Interviewer: And does it make you feel closer to nature?
Interviewee: *Not so much no. You want to get close you dive in {laughter}. Don't want to get that close. I think it's everyone's duty to do a bit. Isn't it? You feel good doing it yea*

Interviewer: And do you feel that we're making a difference?
Interviewee: *Ah every little bit helps (Female Volunteer, South London, River Thames Foreshore, September 2013).*

Interviewer: Things we would associate normally with cleaning but we are applying it to this environment. Are there any issues we should, are there any reason why? That's a positive thing, isn't it?
Interviewee: *Yea absolutely*

Interviewer: So we are teaching how to be able to use our skills, in terms of domestic skills to actually come here, sort of domestic skills to actually come here?
Interviewee: *You can relate it to domestic because it is so many plastic bags. You're picking up the plastic bags here so maybe you wouldn't use as many or whatever...But you know the least plastic bags you use. The place is full of packaging and things isn't? But plastic bags that are easy to pull out and break are obviously a lot better than the ones that don't breakup at all.*

Interviewer: Do you think the problem is plastic bags or is there a bigger wider problem?
Interviewee: *Ah that's probably it, yes, yea. I mean that's the big one. That's the obvious one. It's erhm, it's people's erhm; they throw things over the river. It's teaching the, teaching the life skills of being tidy really isn't it? (Female Volunteer, South London, River Thames Foreshore, September 2013)*

Reflecting on the previous interview transcriptions, it is a curious feature of nature that at certain moments the qualities and characteristics ascribed to it are defined by the narrators' various relationships to nature. Indeed, a recurrent theme emerging from my interviewees with volunteers is that there is not a distinct domain or certain essential phenomenon that is nature. The meaning of nature tended to shift as and between interviewees, or even as part of the narrative form of a respondent's activities. Fluidity in the meaning of nature has been observed elsewhere; for the nature of environmentalism is formulated through the language and practice of environmental activism (Beck 1995: 37). Respondents referred frequently to the river foreshore as a terrain at risk; as under invasion; and river clean-up as a counter offensive defence, lest the foreign species take hold and denaturalise the environment. Non-biodegradable plastic bags featured prominently as an intrusive alien invader that needed to be extracted from the real natural environment. But even in these linguistic repertoires it was evident that utopian appeals to a naturalistic terrain were in actuality, self-referentially appealing to a nature that has long since been subject to human reconfiguration. According to Beck (ibid), it is a feature of late modernity that: "The process of interaction with nature has consumed it, abolished it, and transformed it into a civilizing meta-reality that can no longer rid itself of the attributes of human (co-)creation". River conservation easily lends itself to the proposition that ecological intervention is cultivating the very object it seeks to conserve. Beck (ibid) describes the emergence of touristic industries geared towards achieving the authentic and unmediated experience of nature. Such observations resonate with Daniel Boorstin's (1963) concept of the pseudo-event i.e., a sanitised, circumscribed touristic encounter that is mediated by commercial exchange-value, and sanitized from any form of embodied engagement. Beck's (1995:37) pseudo-events account for the natural world in late modernity and the appearance of "centrally administered museums of real nature". In the staging of nature for touristic appeal, Beck describes a subconscious desire to create for humanity "arks of civilization for dying natural species". In the aftermath of environmental exploitation and neglect emerges a "mixed forest of administered nature …dying species of songbirds and plants are offered an appropriate breeding ground for their civilisation" (ibid). Touristic domestication of nature dissipates the philosophical notion that nature is counter to the civilized world of human activity and ingenuity; for "today more than ever…'nature' is being rediscovered and pampered" (ibid:38). Environmental hazard impels reflexive engagements with our existence as "natural entities" (ibid:50). The threats presented by heightened environmental risk propel into consciousness the insecurity of our existence as human beings. Environmental hazard introduces into everyday life a profound sense of human fragility; and this ontological crisis is amplified by the imperceptibility of the extent and magnitude of environmental risk. Moreover, imperceptibility individuates the consciousness of risk. Beck (ibid:52) describes the consciousness of environmental risk as overwhelming rational cognition, and in turn precipitating "a kind of self-expropriation of the senses". In the context of conflicting scientific discourse "One need only trust one's own senses, throw aside all extraneous data, and the spectre appears to vanish into the unreality whence it came" (ibid). Herein reside additional contradictions, which confront the subject of reflexive modernity, for modernization displaces sensory perception as a basis for knowledge, and in so doing, dislodges the anchoring of human beings' sensory experience. The imperceptibility of environmental risk establishes irrational, erudite knowledge and re-introduces an appeal to the mystical realm of the symbolic. It is easy to see how, in these circumstances, symbolic mediations are attractive innovations of assurance in the age of imperceptible risk.

> Natural despoliation also links the general to the particular in that it permits the foundation of an 'alternative lifestyle' unsupported by any foundations. 'Nature' is the testable form of a concrete utopia, beyond the province of utopias riddled with doubt and turned into their contrary. It is a way of normalizing life while pretending to live in freedom from norms – a way, that is, not posing but of answering the question of how we intend to live. (Beck 1995:55-56)

The concept of nature as "pampered" infers that the "social integration of nature" through environmental conservation involves cosseting nature, indulging nature, and mothering nature. Indeed, Beck (ibid:52) states that part of the integration of nature as a realm of calculative social action produces dangers which "even where they remain impossible to calculate, involve greatly increased labour for housewives and mothers". Beck (ibid:39) argues that the integration of nature into the social sphere transforms it into "a kind of anchor by whose means the ship of civilization, sailing over the open sea, conjures up, cultivates, its contrary: dry land, the harbour, the approaching reef". In other words, the symbolically mediated nature of ecological activism is a transformed, malleable resource; a mystifying inversion of a naturalistic terrain. Beck (ibid:46) is keen not to render ecological activism as mere apparition, enfolded into the symbolic mediation of nature, for "it is not some mysterious sort of alchemy that gives the protest over certain poisons social resonance". While it is evident that ecological activism and the social integration of nature share a symbolic mediation of nature, these acts "are isolated from one another by the cultural readiness to accept despoliation" (ibid). Beck observes that the sign sharing of "destruction and protest" fissure where individuals' lives are confronted with the local realisation of the hazards of despoliation (ibid). But this does not immediately explain why ecological crisis in and of itself seems removed from the selection of activist engagements. Extracting non-biodegradable plastic materials from the foreshore is a perpetual task which multiplies along the strands of the millions of polymer threads that constitute plastic. Nevertheless, hundreds of people invest their time into participating in river clean-up events, in order to make a difference. According to Beck ecological activism is reflexively mobilised:

> To put it systematically, cultural indignation chooses between matters of the highest 'objective' urgency, and this choice is not guided by the issues themselves, but by cultural symbols and experiences that govern the way people think and act, having their origin in their history and in societal living conditions. (ibid:47)

Beck's (ibid) observation here provides credence to a focus on the circumstances in which subjects reclaim the constellation of gendered meaning attributed to river conservation, and sustain "social resonances and the political protest of the environment movement". River conservation along the foreshore clearly involves symbolically mediated connotations that are associated with the domestic sphere. Nevertheless, these constellations of significations "touch a cultural nerve and cause alarm, shattering and making comprehensible the unreality and hyperreality of hazards in everyday life" (ibid). Environmental degeneration enters into academic discourse at levels of calculation that are imperceptible to the non-scientific world, and this is why visually perceptible local ecological crisis can easily catalyse public interest. As Beck expresses it:

> As reality becomes increasingly unreal and inhospitable, one needs to find symbols of the change – pictures in the mass media of dying forests and seals (the latter are also a favourite mascot of tourist resorts and islands), or of the tree withering in our back garden – in order to manufacture culturally the comprehensibility of the incomprehensible. Cultural condensation, magnifying lenses as it were, are required to allow one to extrapolate from the small and everyday to the large, thus bestowing control in everyday life over that which takes its course beyond the horizon of the perceivable and imaginable. (ibid:48)

River conservation community outreach programmes aimed at extracting invasive materials are problematicized and discerned as protests against the cultural contamination that non-biodegradable materials present to the metaphoric symbolic ideal of the river foreshore's natural ecology. In this sense, mobilizing volunteers into river restoration events can be conceived as a rebellion with feminist undertones; it reclaims the gendered connotations ascribed to the domestic division of labour, and redirects symbolic mediations of nature and domestic labour, as part of a counter offensive against the denaturalisation of the river foreshore. Central to this offensive is a feminine ideal discursively constructed through the ideals and representations of embodied care. Just as ideal femininity denotes a selfless care for others, so it is that the temporality of the feminine ideal is associational i.e., existing as part of an embodied reception and association with the care needs of others. In ecological activism, the linguistic repertoires of care, embodiment and the volunteering of our time constitute a nurturing feminine ideal. As part of my ethnographic observations and interviews with the three river restoration charities, I became increasingly intrigued by the manifold ways in which the feminine ideal was mobilized by agents as part of complex translations between governing agencies, river restoration charities, and volunteers keen to cultivate their investiture of social capital. In this respect I agree with Beck's (ibid) observation that ecological protest needs to be discerned through a bracketing of nature; for nature is constructed through ecological activism, and in relation to its symbolic mediations. Furthermore, in the milieu in which statistical modelling of climate change has become an indecipherable algorithmic congestion, symbolic mediation of gendered care ethics are transforming the phenomenological imperceptibility of pan-scientific environmental risk analysis into locally discerning realities. Statistical forecasts spanning hundreds of years are imperceptible experiential realities when compared with the astonishing materiality of non-biodegradable plastic bags accumulating along the river foreshore. The primary concern of this chapter is to explore gender and the mediation of the symbolic. In its appeal to an ethic of care, river restoration programmes reclaim the sign and signification of domestic reproductive labour. River conservation activism is not a matter of natural, but of cultural linguistic symbolic mediations; "a phenomenon of cultural sensibility and of the attentiveness of institutions" (ibid). Framed within this context, the contamination of river ecology "is experienced as cultural and political alarm and then becomes the subject of social actions and protests" (ibid.). It is my contention

Figure 6.2., "I sacrificed to duty, and was turn'd, pale and unthank'd away" ((Ann Yearsley *Familiar Poem to Milo*, 1796). Photographic Image, South London, River Thames, September 2013

Figure 6.3., "Truth, angel of the soul when undismay'd, Illumes thy face, th' attentive thought to charm: To hail thee, liberty forsakes her shade", (Ann Yearsley *Elegy,* 1796). Photographic Image, South London, River Thames, September 2013

that the cultural magnifying lenses through which degradation of river ecology is symbolically mediated, refracts cultural and political hazards through distress signals that signify ecological crisis as a crisis in the ethics of care. Indeed, a prominent finding of my empirical research highlights processes through which constructs of the feminine ideal are enrolled into a domestication of the symbolic mediation of nature as part of their translation into dialogic programmes of environmental education.

Reclaiming Domestication as a Gendered "Ethics of Care"

> ***Interviewer: It might be suggested that the focus on individual activity and collecting litter, for example, or looking after the waterways; it might be argued that those activities are very much about a domestication of the environment or degradation. And what we are seeing here is a response in terms of the environment in relation to making domestic the solution. Making it something we look after through domestic forms of labour?***
>
> **Interviewee:** *I think we have so over domesticated what is a natural system. We have a system, in the Thames in London, for example, quite honestly it wasn't... well, who ever thought it would be dealing with however [many] million people there are living in London...When he [Joseph Bazalgette] built it in the 1860s it was built for two million people, now we have something like eight million people using that system. So in terms of a natural waterway, which I see it as a natural drainage system, yes; but it's within [the] balance of the environment. If we come and we build our city beside it, that river can't sustain what it was meant to do, it can't do it. So, I believe yes we might be domesticating, we might be kind of like trying to put back what we put in already. But I think it's critical that we do that, because we are over using what nature has been put there for. And that poor river, yes it's there to clear things away and it has a natural place where it might deposit silts somewhere and then pick [up] sand somewhere else [in] another part of the river. It will do that itself naturally. And the environment department might come in and alter its water course and all those sorts of things, for a human need, in which case we have domesticated the river, but you know that's where volunteering can come back and actually put it back. Or actually just modify it a little bit, so you know in the 60s we put these concrete culverts in a lot of waterways and we thought we were being really clever by making it straight and it would flow faster, things would drain more quickly. And now a lot of projects we are doing these days in conservation volunteering and so forth is actually taking concrete culverts out and 're-meandering' the river so that the river can go back to its natural format. And that in a sense is undomesticating what we have previously domesticated...*

Interviewer: So, for you, domestication is part of the activities that are conducted in terms of environmental sustainability? Is environmental sustainability undoing the domestication or adding to the domestication?

Interviewee: *I think that it's undoing the domestication, bringing it back to a natural, a more natural former...There's a project I worked with called the Fobney Wetland Nature Reserve which, again, has its own Facebook page. And that was an area of disused land which we've made it into wetland nature reserve if you like. We've re-meandered the river and we've created a wild flower meadow; we've reseeded it. So in a sense you can argue that we've tried to put back wild nature. You can argue we've domesticated it. (Volunteer Projects Manager, London River Thames Conservation Charity II, September 2013).*

The interviewee provides insights into the nature of care as an ethic of the domestication of river ecology. In feminist analyses of the voluntary sector, the nature of care is prefigured by gender identity. While it is evident that the previous interview extract does not directly refer to male and female experience, an "ethics of care" consistently informs the interviewee's conception of domestication; rivers are domesticated to care for and serve populations; rivers are cared for through restoration programmes; rivers require domestication as part of their restoration into a natural condition. The centrality of care to the social construction of feminine identity has relevance in understanding the practices through which the symbolic mediation of domestication provides avenues for enrolling volunteers into critical dialogue about river ecology. In feminist accounts of the voluntary sector, the "ethics of care" refers frequently to Carol Gilligan's (1993) psychological framing of care as motivated by the psycho-dynamics of maternal (de)attachment. Notwithstanding essentialist connotations, it is difficult to sustain a feminist analysis of care-based archaic models of the normative family and familial roles. Nevertheless, the psychological framework of Carol Gilligan has been, and continues to be, influential to second wave feminist accounts of care, and thus warrants further analysis.

Figure 6.4., "Time alone, Must stamp the value of our spirit's wealth" (Ann Yearsley *The Consul..,* 1796). Photographic Image, South London, River Thames, September 2013

Psychological Mediations of Gender and Care

> Attachment and separation anchor the cycle of human life, describing the biology of human reproduction and the psychology of human development. (Gilligan 2000:151)

According to Gilligan, the figuration of emotional attachment and separation are sequenced in one's early childhood, formulated in adolescence "as identity and intimacy", and during adulthood anchored "as love and work" (ibid). Writing as a feminist psychologist, Gilligan directly engaged with an absence of adult female experience in attachment theory, whereby the developmental ordering of emotional development removes the adult female as part of its pursuit of a linear archetype of emotional development through detachment. Gilligan observed that the invisibility of adult female experience in psychological attachment theory had the effect of instrumentalizing childhood maternal relations, and assuming these bonds to be mere stages in the teleological development of adult emotions. As Gilligan expresses it: "The significant relationships of early adulthood are thus construed as the means to an end of individual achievement, and these 'transitional figures' must be cast off or reconstructed following the realization of success" (ibid:152). Gilligan critiques mainstream psychology's dependence on a modeling of psychological development framed with a process of ongoing individuation. According to Gilligan's previous studies, they "convey a view of adulthood where relationships are subordinated to the ongoing process of individuation and achievement, whose progress, however, is predicated on prior attachments and thought to enhance the capacity for intimacy" (ibid:154). The experience of women rarely appears in this modeling. Thus, psychological theory presents the processes of ongoing detachment and individuation as the normative modelling of emotional development, and in so doing, denigrates attachment as pathological. Gilligan notes that psychology has an entire trajectory missing from its accounts of emotional development: "a failure to describe the progression of relationships toward a maturity of interdependence" (ibid: 155). In absenting female experiences, defined by interdependence, mainstream psychology rendered female emotional development as in a sense incomplete because of its inability to achieve separation and for being overwhelmed by the quagmire of relational dependency. Based on her studies of young adults, Gilligan observed female emotional development as functionally distinct rather than pathologically different:

> From the different dynamics of separation and attachment in their gender identity formation through the divergence of identity and intimacy that marks their experience in the adolescent years, male and female voices typically speak of the importance of different truths, the former of the role of separation as it defines and empowers the self, the latter of the ongoing process of attachment that creates and sustains the human community. (ibid:156)

Responding to these different articulations of attachment, Gilligan was motivated to "restore in part the missing text of women's development, as they describe their conception of self and morality in the early adult years" (ibid). A feature of her interviews with young adults suggested to Gilligan that interdependence is an integral aspect of the moral development of females. When asked to describe themselves, Gilligan's female respondents were apparently inclined towards descriptions of their self in connection with maternal and caring roles. It is from this observation that Gilligan (ibid: 159) claimed that "the standard of moral judgment that informs their assessment of self is a standard of relationship, an ethic of nurturance, responsibility and care". A curious feature of her findings was that the successful academic professional women interviewed tended to measure their success in relation to their ability to achieve nurturing relationships. Professional careers were assumed to be an obstruction to or even jeopardize a care ethic. Consequently Gilligan claims:

> Thus in all of the women's descriptions, identity is defined in a context of relation and judged by a standard of responsibility and care. Similarly, morality is seen by these women as arising from the experience of connection and conceived as a problem of inclusion rather than one of balancing claims. (ibid:160)

Conversely, for males, the articulation of moral identity is, according to Gilligan, a different form of psychology. While it was evident that the male respondents in her study did refer to other people and relations of attachment "no particular person or relationship is mentioned, nor is the activity of relationship portrayed in the context of self-description" (ibid: 160-161). At the level of the analysis of linguistic repertoires, it was evident that the language of male psychological development preferences adjectives associated with separation, and was less likely to describe the self in terms of attachment and interdependence. In this sense, a key aspect of male self description "is tied to quantification of identity" (ibid:163). Rather than express identity in relation to an attachment to significant others, Gilligan observed that the males in her study described themselves in terms of quantifiable achievement of distinctive activity and measurable standards of success. These observations led Gilligan to argue that "the sequential ordering of identity and intimacy in the transition from adolescence to adulthood better fits the development of men than it does the development of women" (ibid). Objectifiable power, status and success define male identity and mediate their relationship to others. Knowledge and capital derived from a man's individuation from attachment are believed to form the basis of a care ethic predicated on "the adult ethic of taking care" (ibid:164). This form of intimacy involves judicious evaluations of situations in terms of information available and balancing actions in relation to the effect these might have on the individual's own security. Gilligan (ibid:163) describes "taking care" as a form of intimacy which:

> becomes the critical experience that brings the self back into connection with others, making it possible to see both sides – to discover the effects of actions on others as well as their cost to the self. The experience of relationships brings an end to isolation, which otherwise hardens into indifference, an absence of active concern for others, though perhaps a willingness to respect their rights.

According to Gilligan, given that women derive their psychological development through interdependent relations, they encounter different forms of moral dilemmas. This is because women define their psychology through "intimacy and care". A moral dilemma emerges from situations in which "relationships are secured by masking desire", and conflict is deflected through equivocation then brings to light the problem of "the locus of responsibility and truth" (ibid: 164). It is Gilligan's assertion that the transition from adolescence involves similar dilemmas for males and females, but that they are resolved differently. In males, the sequencing of development into adulthood requires reconciling detachment with intimacy and for females having to reconcile attachment with choice. With regards to the latter, Gilligan (ibid) argues that: "The critical experience then becomes not intimacy but choice, creating an encounter with self that clarifies the understanding of responsibility and truth". For female subjects, this involves adopting "an ethic of care" as opposed to the male's "ethic of rights" (ibid). As Gilligan further says:

> The morality of rights is predicated on equality and centered on the understanding of fairness, while the ethic of responsibility relies on the concept of equity, the recognition of differences in need. While the ethic of rights is a manifestation of equal respect, balancing the claims of other and self, the ethic of responsibility rests on an understanding that gives rise to compassion and care. Thus the counterpoint of identity and intimacy that

> marks the time between childhood and adulthood is articulated through two different moralities whose complementarity is the discovery of maturity. (ibid:165)

Gilligan's formulation of the female "ethic of care" and male "ethic of justice" is based on an earlier critique of the phallocentric language of Freud's account of male and female psycho-sexual development. The Oedipus complex is central to Freud's theory of development. The phallocentricity of Freud's analysis relates to his depiction of male morality in terms of the threat of castration and castration anxiety. As Gilligan, referring to Freud states:

> From this observation of difference, that 'for women the level of what is ethically normal is different from that it is in men', Freud concluded that women 'show less sense of justice than men, that they are more often influenced in their judgments by feelings of affection or hostility' ([Freud] 1925, pp., 257-258). (Gilligan 2000:7)

In contrast to Freud's phallocentric conception of male and female morality, Gilligan develops a feminist psychology of morality informed by the psycho-sexual theory of Nancy Chodorow (among others). Thus, Gilligan argues that because women are predominantly responsible for childcare "early social environment differs for and is experienced differently by male and female children" leading to "basic sex differences" that "recur in personality development" (ibid: 7). The initial social environments to which Gilligan refers are the first three years of childhood, and despite the tendency for the primary carer to be female, it is Gilligan's observation that "the interpersonal dynamics of gender identity formation are different for boys and girls" (ibid). For girls, the dynamic is one of continuity, whereby the girl child's sequencing of identity development is congruous with the gender identity of the primary carer (the mother). Thus, "girls, in identifying themselves as female, experience themselves as like their mothers, thus fusing the experience of attachment with the process of identity formation" (ibid:7-8). Conversely the sequencing of psychological development for the male child requires detachment from identification with the primary carer (mother). In summary:

> Consequently, relationships, and particularly issues of dependency, are experienced differently by women and men. For boys and men, separation and individuation are critically tied to gender identity since separation and individuation are critically tied to gender identity since separation from the mother is essential for the development of masculinity. For girls and women, issues of femininity or feminine identity do not depend on the

> achievement of separation from the mother or on the progress of individuation. Since masculinity is defined through separation while femininity is defined through attachment, male gender identity is threatened by intimacy while female gender identity is threatened by separation. Thus males tend to have difficulty with relationships, while females tend to have problems with individuation. (ibid: 8)

While it is evident that Gilligan's theoretical framework has coherence and rhetorical logic, irrepressible inconsistencies disrupt the plausibility of attachment theory as a basis of analysis. Central to these theoretical problematics is the normative heterosexual model of the family as utilized by Gilligan; for the predominance of the nuclear family to Gilligan's attachment theory renders inoperable other family forms as a basis for the child's moral development. Furthermore, the archetypical representation of the mother essentializes maternal qualities of care and unconditional love, and in so doing, marginalizes difference. Nevertheless this framework has resilient endurance and has, elsewhere, been adopted as a basis for feminist analysis of environment activism. For example, Zelezny *et al.,* (2000), in their empirical analysis of gender and environmental attitudes, state that when "compared to males, females had higher levels of socialization to be other oriented and socially responsible" (ibid: 443). A further aspect of their findings relates to the concept of the "conceptualized other". Zelezny *et al.,* claim that female respondents reported consistently higher levels of ecocentrism when compared to their male counterparts. Alongside these traits, Zelezny *et al.,* (ibid) suggest that age combines with gender in the constitution of the "other orientation" that inclines females towards perceiving environmentalism as personal responsibility. Zelezny *et al.,* (ibid: 455), state that an aim of their research is to challenge the stereotypical view that females are acquiescent, compliant and apathetic about environment issues. In contrast Zelezny *et al.,* state that gender is an indicator of environmental responsibility "females, regardless of age (i.e., youth or adult) reported more concern for the environment and proenvironmental behaviours than men" (ibid: 450). Zelezny *et al.,* (ibid) refer to Gilligan's feminist psychology and her account of the ethics of care. Thus gender differentials in environmental attitudes are described as thus: "because females, compared to males, are socialized to value the needs of others, women exhibit more helping behaviour and altruism" (ibid: 445). When accounting for methodological issues that might have skewed the statistical findings, Zelezny *et al.,* state that the self-reporting method utilized can produce "socially desirable reporting". They also recognize the impact of "generational effects", however they conclude that "differences in environmentalism are likely due to socialization" (ibid: 455). Despite attempting to control for social factors Zelezny *et al.,* (ibid) reinforce the a-socio-historical subject of Gilligan's feminist psychology and ethic of care. Gilligan's psychology of care ethics insufficiently attends to the issues of multiplicity and agency; political economy, and governance. Issues of multiplicity and change can be derived from an alternative conception of care provided by a feminist postmodern social work tradition.

Postmodern Mediations of Gender and Care

Postmodern frameworks in social work and clinical care challenge presuppositions based on essential gender identity, fixed cultural differences, and the transcendental patterns that govern social life. Theorists engender care through the postmodern paradigms dedicated to process, on-going negotiations of gender and caring identities. Instead of essential ideas of gender and caring competency, postmodern writers encourage conceptions of individual rights to care. For example McEldowney and Connor (2011) in their work on nursing and the ethic of care, present a "praxiological" model which focuses on decentered cultural competencies. They introduce a postmodern emphasis of difference and distinction as part of their focus on the concept of "cultural safety". The latter nursing concept contrasts with cultural competence, for it refers to a postmodern "praxiological process" which "takes place in and is shaped by…the general context of the encounter" (ibid:343). The postmodern features of care in this framework relate to the "praxiology" of practice, which refers to the process whereby "new knowledge and understanding develops from critical and structured reflection in and on action in the practice context" (ibid: 344).

Elsewhere, feminists have explored care ethics within the context of postmodern time and space compression. Informed by Harvey's (1990), concept of "flexible accumulation", Joan Tronto (2003) describes the commodification of time and space in advanced capitalist society. Volatility characterizes the temporality of advanced capitalist accumulation and the accelerated time of capitalism presents challenges to the practice of caring. Tronto explores this in terms of the practicalities of caring for the self and others. With regards to the latter, capitalism's extension of commodity capitalism into the realm of care provides Tronto with some basis for critical debate. Separating female from male experiences of caring in the context of time intensification, Tronto identifies time-saving technologies as part of a "steering time" that is apposite for a capitalist regime of accumulation intent on extending the commodity form into care

Figure 6.5., "The path I chose unknowingly, by chance" (Ann Yearsley *The Consul..,* 1796). Photographic Image, South London, River Thames, September 2013

relationships (ibid:123). An additional realm of care affected by the intensification of time in capitalist accumulation is "the construction of the care for the self" (ibid). Tronto supports feminists, raising concerns regarding the perpetual state of readiness that pervades social relations in advanced capitalist society. An integral feature of the culture of readiness is control, whereby the presentism of instantaneous time demands a subject that is available to respond just-in-time. As Tronto (ibid) expresses it: "An interesting thing happens when members of the capitalist class come to see themselves as perpetually 'on', ready at a moment's notice to spring into action and take advantage of capitalist volatility. There is a change in the relationship of agency and control". The presentism of work in advanced capitalism erodes accumulated expertise: "No amount of past preparation can be 'stored up' to make the world conform to one's current actions. Instead, there is a different form of presentism that individuals must practice best to take care of themselves" (ibid). Taking control of one's schedule relaxes time and enables subjects to take care of their social needs and relationships. For, "those who are not in total control" success is achievable "if they are constantly at the ready" (ibid). Several consequences follow from this postmodern condition: first "those who are able to become capitalist successes require and can demand much more constant care and service for themselves. There is a new competitive advantage to be gained, for example, by having a fit body that can ignore pressures of time and space"; second "there is much greater difficulty [in] providing care for others who do not conform to this model of preparation for volatile and compressed time and space" (ibid: 124). The suggestion being that, in the competitive culture of perpetual availability, care needs that exceed designated allocations of time become onerous obstacles to the capitalist subject's availability. It is evident that Tronto, presupposes a subject that can succeed in this culture of readiness; and that their survival is predicated on a zero-sum conception of power. Tronto states that there exist subjects "not in total control" (ibid: 123). Implied in this assumption is a feminine subject that lacks control/power. Tronto's framing of power universally transfers passivity to a (female) cohort that is assumed to lack power. An alternative conception of power is provided in post-structuralist theories of power. In the *Subject and Power*, Michel Foucault (1982) argues that: "Power exists only when it is put into action, even if, of course, it is integrated into a disparate field of possibilities brought to bear upon permanent structures" (ibid:219). As I will discuss later in this chapter, Foucault develops a conception of power as relational; rather than zero-sum. Recognising power as a relationship encourages reconsideration of Tronto's (2003) claim that in the postmodern condition, an intensification of time imputes "total control" to some at the expense of others. Further areas of critical engagement relate to Tronto's supposition that postmodernist feminists have largely abandoned social remembering as a form of praxis; and this is evident in the tendency for postmodernist feminists to emphasise futurology in their ethics of care. Tronto describes how feminists in Western Europe have attempted to challenge the future time of advanced capitalism; as Tronto states:

> In Italy feminists have agitated to change the 'city's time' (tempi della citta). That official government offices were open during the hours when women have had other responsibilities meant that women had been effectively unable to take advantage of public offices or to conduct public business. The result has been a change in the opening and closing hours of government businesses so that women's lives could be more easily arranged. (ibid:124)

Tronto advocates a rolling-out of feminist activism geared towards aligning female time schedules with those of the public sphere. A theme that increasingly unites these strategies is the intention to utilize time-space compression as a resource in feminist activism against the excesses of time-space compression. Tronto at times applauds this strategy, for "When it appears that the contemporary time-space compression may diminish the controlling power of past oppressive structures, using this account of time becomes attractive for those who would end oppression" (ibid: 125). Tronto is also exceptionally keen to reclaim the past as a basis for postmodern feminist praxis, and thus challenges aspects of the futurology of postmodernist feminism. In the accelerated future present of advanced capitalism, past and future converge in an instantaneous present. A primary outcome of this "is that the future looms larger than the past as a guide for our action and thought" (ibid). Postmodernism can invite emphasis on a future present, and its formulation of feminism has often embraced a notion of time untethered from the past; indeed "The very ideal of feminism as a progressive political movement is to claim a basis to throw off sexist assumptions about women, gender and sexuality that were generally accepted in the past" (ibid). In their quest for equality and advancement, progressive feminists have been disinclined to accept "tradition" as a basis for advancing their politics. Tronto cautions against "privileging the future", instead advocating an "understanding of time's place" in political developments (ibid: 132). Tronto's advocacy for the "power of remembrance" constitutes part of an ethics of care predicated on assisting Western feminists who "feel that their legitimacy in making future and utopian claims … [was] undercut by their positions of privilege" (ibid: 132). Adopting a postmodern condition of time in which the present is a mere "prelude to the future" have encouraged feminists to desire detachment from "seemingly old-fashioned ideas" (ibid). Tronto's attempt to reclaim remembrance as feminist praxis has relevance for ecology politics. Nevertheless, without embedding the act of remembering within the structural relations of power, Tronto's advocacy for memory slips easily into, what elsewhere has been described as "imperialist nostalgia". Rosaldo (1993) defines this process as a curious feature of colonialism in which the agents of colonialism "often display nostalgia for the colonized culture as it was "traditionally" (that is, when they first encountered it)" (ibid:69). Rosaldo describes the paradoxical features of this process as thus: "my concern resides with a particular kind of nostalgia, often found under imperialism, where people mourn the passing of what they themselves have transformed" (ibid). When applied to an analysis of river restoration, imperialist nostalgia is possibly evident when "someone deliberately alters a form of life, and then regrets that things have not remained as they were prior to the intervention" (ibid:70). If one accepts the predominant narrative of river pollution as a consequence of urbanization and irresponsible domestic waste management, then river restoration involves a situation in which "people destroy their environment, and then worship nature" (ibid). In the absence of a systematic embedding of remembering into structured relations of power and inequality, feminist praxis intent on uniting history with present futures engages with imperialist nostalgia. And in so doing transforms the subject of nostalgia into "stable worlds" that provide landmarks from which to valorize the apparent civilization of all that is present. River restoration can be depicted in this framework as a response to people feeling "vulnerable to their future" (Michelle Rosaldo quoted in Rosaldo 1993:86). Thus to frame river restoration from a postmodern feminist framework dedicated to causation and the remembering in the present of the historical past, might have the effect of "Mourning the passing of traditional society" and this partakes of an imperialist nostalgia (ibid). Both the mourning of a transition from natural ecology and the yearning for a return to a natural state "use a mask of innocence to cover their involvement with processes of domination" (ibid). Nostalgia engaged in endless forms of postmodern reflexivity invariably disinclines engagement with the structures of systematic inequality. As Rosaldo (ibid: 87) expresses it: "Nostalgia at play with domination" expresses compelling benevolence in its endeavour to make specific appeals to the natural (ibid). But these symbolic mediations provide idealized impressions of nature, and in so doing add to the processes by which "ideological discourses work more through selective attention than outright suppression" (ibid). In this respect, Tronto is plausible in her claim that "feminist ethics is ultimately tied to whatever feminist vision of the future we can describe" (2003:132).

Much of the problem with postmodernist feminist ethics of care is their reluctance to systematically engage with structured relations of power and inequality; and this is particularly evident in postmodernist feminist discourses of time and care. Decontextualised from class, race and other social structures the treatment of time in postmodernist feminist ethics of care appears at first glance to present an attractive universal position.

Baudrillard (1968/2005) describes how promotional culture "constitutes less a language, whose living syntax it lacks, than a set of significations" (ibid:212). Although impecunious in form, postmodern language is remarkably "efficient" in its ability to manifest semiosis; for "it is basically a code" (ibid). Parallels can be drawn between the postmodern feminist language of care and the "objects-cum-advertising system" of promotional culture, for they share a "living syntax" which "does not structure the personality, but designates and classifies it"; and furthermore: "It does not structure social relationships, but breaks them down into a hierarchical repertoire" (ibid). For example, Tronto (2003) argues that postmodernist feminists should not celebrate the future at the expense of history. However, Tronto's apolitical account of time constitutes history as an object-position in chronological time. In so doing, Tronto's discourse at first glance appears universally accessible; indeed as Baudrillard might observe: "In its formal expression it constitutes a universal system for the identification of social rank: the code of 'status'" (ibid). Tronto's discourse appears to be suggesting that all history can be approached from a feminist perspective i.e. conceived as (her)Story. In the context of promotional culture "the code of status" derived from advertising language, operates "to bring everyone under its sway" (Baudrillard 1968/2005:212).

The "code of status" also constitutes "a moral code, for it is sanctioned by the group, and any infraction of it entails the apportionment of some measure of guilt" (ibid). Similarly the notion that feminists should adopt an approach to history which conceives the past as an object-position, introduces into feminist discourse the imposition to remember and as such "is a totalitarian code, for no one escapes it [and] escaping it in a private sense cannot prevent us from participating every day in its collective development…Even actions intended as resistance to it must be defined in terms of a society that conforms to it" (ibid:212-213). Thus, on the one hand Tronto's focus on history offers a universally accessible discourse; but on the other hand the treatment of history in terms of an a-political object-position in chronological time constitutes a hierarchical repertoire against which all other conceptions of history are evaluated and defined. Baudrillard (1968/2005:214) might observe this as a detachment of the "code of status" such that it has come to embody a postmodernist language of time in which "the signs of value are entirely socialized and objectivized" but this "by no means implies any true 'democratization'". Rather, as Baudrillard observes "it would appear that the insistence on a univocal reference merely exacerbates the desire to discriminate: within the very framework of this homogenous system, a perpetually renewed obsession with hierarchies and distinction is to be observed". It is my contention that the singularity of sign value manifest in the discourse of a history, conceived as an object-position in chronological time, constitutes a "code of status" which as Baudrillard observes "produces an illusion of transparency, an illusion of readable social relations, behind which the real structure of production and real social relationships remain illegible" (ibid:214).

Suffice to say, it is not sufficient for us to care to remember, if we also care to forget our privileged and differential positioning in structured relations of power and inequality. Reflecting on civil rights and inequality, Martin Luther King Jr., once said: "The arc of the moral universe, although long, is bending toward justice" (2014:96).

Taja Poso (2010) is similarly inclined to conceive of care, in postmodernity, as a process of plural temporalities. Paso introduces the concept of "revisiting" as part of a methodological approach to researching residential care. The concept of "revisiting" relates "to a variety of ways in which the researchers visits the research site again and in which the relations between the first and later visits are given some consideration substantially and methodologically" (ibid:27). While providing engaging and highly innovative research technologies, postmodern frameworks have also been subject to criticism. Orme (2003) contrasts this standpoint, along with the postmodern models of social work care; the former is said "to contribute to the perpetuation of …gendered polarities by highlighting differences between women and men" and the latter "espouses an individualism, that disregards collective continuities" (ibid:137). In defense of the application of postmodern theory to feminist analysis of care, Orme suggests postmodern approaches have been highly successful in their deconstruction of the hegemony of scientific rationality as an approach to theory building (ibid). Similar questions are addressed in Lena Dominelli's (2006) reluctant advocacy of postmodern feminist community action. Dominelli praises the ability of the postmodernist feminists to empower identity politics, but she challenges the propensity for a heightened individualism to undermine collective feminist consciousness and dissipate the formulation of political strategies based on shared experience. Dominelli nevertheless does present a cautionary tale worthy of consideration; for in the postmodern milieu of flux, multiples and changes to signifier feminism constitutes a powerful constellation of meaning. Dominelli (ibid:86) describes how "Women's position as signifiers of the nation is being reaffirmed as a site of women's oppression". The postmodern feminist framework clearly has resonance with my study of the feminine ideal and its role as a symbolic mediation of river restoration. Nevertheless, postmodernism's negation of the subject (in terms of a discerningly coherent self) and displacement of inter-subjective time for instantaneous nostalgia and future presents, and its insufferable refusal to situate meaning within a discernible political economy present difficulties when considering the application of a postmodern framework. An alternative position is presented by post-structuralism.

Figure 6.6., "When contemplation puts my woes to flight. Nay, but Thou art familiar to my thought" (Ann Yearsley *Prayer and Resignation,* 1796). Photographic Image, South London, River Thames, September 2013

Figure 6.7., "The ills of life fall off me as I stand. A dauntless spirit gazing up to thee; And when this weary frame may press the sand, Or my light ashes strew the roaring sea" (Ann Yearsley *Prayer and Resignation,* 1796). Photographic Image, South London, River Thames, September 2013

Post-structural Mediations of Gender and Care

In *The Subject and Power*, Michel Foucault (1982) explores "the question of the subject". Foucault committed decades of scholarship to composing a history of the contrasting techniques "by which, in our culture, human beings are made subjects" (ibid: 208). Central to Foucault's theoretical framework is his attempt to elicit processes of "objectification" and "subjectification". With regards to the former, Foucault (ibid) asserts that his entire oeuvre has been concerned with the exploration of three modes of rendering human subjects into calculable objects amenable to surveillance and programmes of administration. Forms of investigation that attempt to ascribe scientific status to their operation constitute the first of Foucault's three modes of objectification. Foucault presents "the objectivising of the speaking subject" in structural linguistics: "*grammaire générale*, philology and linguistics" (ibid), as examples of processes through which human beings are transformed into objective phenomena. Also significant are the objective rationalities of finance, economics, and "the analysis of wealth". Foucault highlights the scientific prerogative of the natural sciences as powerful enclaves in which the existence of the human, as a living being, is subject to processes of nomenclature, aggregation and allocation. These processes through which scientific discourse is mobilized as part of a classificatory analysis of the subject were apparent in Chapter Four's analysis of the European Union's *Water Framework Directive*. The latter dictate necessitates that local governance agencies work with river conservation charities and environmental specialists as part of a determining of river conservation and community prerequisites. Such programmes resonate with "the objectivizing of the speaking subject" for they attempt to render knowable the opinions of environmental citizens and subject these ideals to governmental techniques of scientific enquiry. Classification is a feature of the second of Foucault's investigations into objectivizing technologies. Foucault identifies modern governance as pervaded by "dividing practices" through which subjects are distinguished from themselves and each other. "Dividing practices" are discernible in the technologies of the advertising sign, used to inscribe meaning into space, place and social practices. Consider the following transcription extract, for it provides an illustration of how the sign value of visibility can act as both a unifying and a "dividing practice" in its advertising of good citizenship.

Interviewer: So you were a nurse and you decided to, what made you go into conservation then?
Interviewee: *Because I always wanted to do it since I was twelve.*

Interviewer: How did you know from the age of twelve?
Interviewee: *I used to go to Grizedale Forest in the lakes and stay at my parents' cottage...Anyway next door was Grizedale Forest at the back and I wrote to the Forestry Commission and asked: what do I have to do to work in the Forestry Commission? And I was twelve I remember they said: get secretarial skills. That was how it was then.*

Interviewer: Not any more?
Interviewee: ...
Interviewer: And so what sort of conservation work did you do?
Interviewee: *I did a course a conservation skills course. We learnt about fencing...felling trees; IDing. All sorts of stuff. But bit too late now.*

Interviewer: But now you're doing it?
Interviewee: *There's no way I'd get a job in it. I can volunteer, can't I?*

Interviewer: Do you think the voluntary work; it is making a difference, isn't it?
Interviewee: *I know, this is what I say with people here. People do hear: "Oh it makes a difference". But it does it's not just a saying. It really does ...Can you imagine if each person here, what the amount they must pick up; and then telling other people about it. And being seen, you know. Education, that's the thing.*

Interviewer: Being seen that's an interesting one. Is there something about being seen?
Interviewee: Yes being seen, doing things. Just being seen.

Interviewer: Yes
Interviewee: Being visible

Interviewer: I never thought of that [in relation to river restoration]
Interviewee: You know you don't have to add; something might lodge in somebody's mind: "Oh what are they doing?" You never know, on the other hand maybe not, people just, you know, walk past
Interviewer: Yea
Interviewee: *If you just grab one person that's good. One person's better than none* ***(Volunteer, London River Thames Conservation Charity III, September 2013)***

The interview extract provides insights into the visual culture of river restoration. The issue of visibility is illustrative of the operation of sign technologies. Foucault (1982:208) describes how "dividing practices" have the effect of rendering the subject "either divided inside himself or divided from others". Visibility is an inscriptive sign technology that objectivizes subjects. At the same time, visibility provides an axis through which "a human being turns him – or herself into a subject" (ibid). The interviewee refers to visibility as indexed to processes of self-learning and consciousness raising. Visibility is a technology of the sign; a mechanism through which individuals come to learn to recognize themselves as both subject and object of the sign. This suggests that the focus of post-structural analysis is not power *per se*, but rather, the dynamic interplay between the subject and power. In my interviews with the river restoration volunteers, it was evident that individuals are integrally involved in the sign mediation of ecological activism. Foucault (ibid:209) states that "the human subject is placed in relations of production and of signification". Consider the following interview extract, for it provides some illustration of how the subject of river restoration "is equally placed in power relations" (ibid):

> ***Interviewee:*** *...I have been reading this book and it's called The Oceans of Life and it's really depressing, despite the quite optimistic title. It's all about how plastic is poisoning the sea and stuff and then I was reading it and it was like, there are these clean-up acts all over England. And I was like: I quite like to do that!*
>
> ***Interviewer: Do you think a community is starting to grow around these issues?***
> ***Interviewee:*** *I don't know, I think my generation definitely, like everyone recycles everything and if you don't it's a bit weird. So I think that is definitely becoming more prevalent.*
>
> ***Interviewer: And do you feel part of the community here?***
> ***Interviewee:*** *I mean, I live quite a way, away so not like the social community. But everyone's been really nice, cos I came by myself. So everyone's been really nice and stuff, so yea*
>
> ***Interviewer: What will you take away from today's event?***
> ***Interviewee:*** *I don't know. I think I feel good I think. And that really nice stone I found, it's got holes in. I'm going to use it as a pen tidy, I'm gonna take that away. But I don't know I just feel like I've done my bit. Although the most of the plastic that is in the water is from exfoliators and you can't actually pick that up because it's so small. So I feel the difference we are making is quite small but it's better than no difference. (Female, South London, River Thames Foreshore, September 2013).*

The interviewee's impetus to act without certainty illustrates the complexities of power. It is clearly misconceived to assume that the interviewee is merely a passive object of power; for so many aspects of the interviewee's actions suggest self-directed agency. Indeed, one can discern the interviewee's actions as manifesting resistance against an anatomy of power, in which the mediations of environmental degeneration are so devastating that they debilitate the subject. Thus, the complexities of the subject of ecology and power are discernible when the analysis is concerned not so much to individuate a possessor of power or identify power as residing within an institution. Foucault (1982:212) defines power as a "technique":

> This form of power applies itself to immediate everyday life which categorizes the individual, marks him by his own individuality, attaches him to his own identity, imposes a law of truth on him which he must recognize and which others have to recognize in him. It is a form of power which makes individuals subjects. (ibid)

Nothing surely clarifies so well the subject of symbolic mediation than a focus on power as existing "when it is put into action, even if, of course, it is integrated into a disparate field of possibilities brought to bear upon permanent structures" (ibid:219). Focusing in this way on power can open up the symbolic mediations through which ecological activism can be communicated. Relations of definition and meaning are certainly informed by power, but an integral feature of the circulation of signs is the subject. The power of symbolic mediations is relational; and thus the mobilization of the feminine ideal in visual cultures of ecological activism, needs to be comprehended in terms of an intercourse of power. As Foucault states:

> In effect, what defines a relationship of power is that it is a mode of action which does not act directly and immediately on others. Instead it acts upon their actions: an action upon action an action, on existing actions or on those which may arise in the present or the future. (ibid:220)

When we consider the previous interviewee's description of their motivation to partake in ecological activism, it is evident that symbolic mediations of river restoration act indirectly. The interviewee describes reading about environmental devastation; by happenchance the book contained details about river restoration and a further sequence of chance encounters precipitated her journey to the river foreshore of South London. A further example of the symbolic mediation of ecological activism in terms of "an action upon an action, on existing actions" (ibid), is evident in the following interview extract:

> ***Interviewer: In terms of describing your commitment to be here: what does it mean to you to be here?***

Interviewee: *It's nice to be in the fresh air, close to the river Thames the mighty river running through London. You don't often get a chance to be up this close to it and be able to meet new people and doing something worthwhile. It's rewarding.*

Interviewer: Do you think it enables a sort of activism or sort of commitment to trying to restore the river?
Interviewee: *It does, yes of course ... And as the lady said in the beginning there doesn't seem to be any body, any official body who is responsible for clearing up the river. So get out there and do it*

Interviewer: And do you feel that a sense of community is developing?
Interviewee: *I hope so. At the moment I don't know. There's always the potential for that to happen. (Male, South London, River Thames Foreshore, September 2013).*

The symbolic mediation of ecological activism is clearly allied to the exercise of power as action on the existing on-going symbolic mediations of subjects. Signs engage subjects that are themselves engaged in the production of signification. Consequently, the relationality of signs opens up the possibility of resistance. As Foucault (1982:221) expresses it: "Power is exercised only over free subjects, and only insofar as they are free". Sign constellations mediate free actions and thus individuals are presented with a "field of possibilities in which several ways of behaving, several reactions and diverse comportments maybe realized" (ibid). Freedom is axiomatic to the power of symbolic mediations: "In this game, freedom may well appear as the condition for the exercise of power" (ibid). The decisive problem is not that of acquiescence to symbolic mediations; but rather the provocation of resistance to the "hail" (Althusser 1971) of the sign. As Foucault (ibid:221–222) expresses it: "At the very heart of the power relationship, and constantly provoking it, are the recalcitrance of the will and the intransigence of freedom". Signs are mediated by the intransigence of the spirit, and relationships of antagonism best describe the sign mediations of ecological activism. In this sense: "It would be better to speak of an 'agonism' – of a relation which is at the same time reciprocal incitation and struggle" (ibid:222). Consider the following extract derived from an interview with a male volunteer involved in river restoration:

Interviewer: What does it mean to you to be here?
Interviewee: *Well, I think it is important that we can have the opportunity to get involved in cleaning up the environment. This obviously does that. Just as at the start when they introduced the fact that nobody is actually*

responsible for cleaning up the dirt from the river. So they make these opportunities available and people that care about the environment can do something about it. I do and so it's a good opportunity to come and make a difference. You know you clear away all these cages full of rubbish. You know tonnes of it

Interviewer: Have you done this before?
Interviewee: *Yea I have yea*

Interviewer: Do you think a community is building around these issues?
Interviewee: *Yes, I think so, cos you'll find, I think that quite a lot of people that come to each particular event are fairly local to the area so they are very aware of the river, erhm, and the problems from litter and stuff. Stuff that gets blown in, stuff that people flush down the drain, when the sewers overflow; all the rubbish gets swept into the river. Then people live around it can see it happening. So when you have an opportunity to come down and clean it up, then I think we probably take that chance. But it's a good opportunity to see London from different angles as well. Because most of the time you're stuck on top of an embankment and the river is behind a wall and it's sort of along way away and it's just kind of cut off from everywhere... Because when you get into central London, you've the really high flood walls, obviously it's what they are there for and you can't see over them. And, you know, people are head down going from one place to another. They don't really see the river. Although if you go onto the Southbank there are bits that you can get down onto. Like outside the National Theatre and stuff like that and you see people going down there, mucking around...*

Interviewer: What do you think about the fact that a lot of it is plastic?
Interviewee: *Well that's modern rubbish, I mean, a lot of the stuff we see down here is broken glass as well. People have been using it as a dumping ground ever since London was first created as a place to live. But the plastic's a real problem. As they say, it doesn't biodegrade. You know, fish, you know ducks, birds, water fowl etc., eat it, you know, it can cause them serious problems. So, yea, it is good that we can clean it up. People don't really pay much attention to, you know, don't put X,Y,Z down the drain; you know, don't leave your bin bags out before the time they are collected other wise it flows around the streets and it inevitably ends up in the river by accident rather than by design. (Male, South London, River Thames Foreshore, September 2013).*

Foucault (1982:222) describes "the exercise of power as a way in which certain actions may structure the field of other possible actions". It has been a consistent theme of this chapter that river restoration is symbolically mediated by the spectre of the feminine ideal. Where it is evident, as in the previous interview, that river restoration is made meaningful through recourse to the feminine ideal, it is ever more apparent that gender identity is contiguous with the relationality of power. Gender identities are embedded in the social milieu; they are actively composed by conscious human beings, and thus "not reconstituted 'above' society as a supplementary structure whose radical effacement one could perhaps dream of" (ibid). Central to my theoretical formulation, of a post-structural ethics of care, is an appreciation of gender identities as discursive, i.e., constituted in and through language; social practice; institutions; economy; politics and regimes of governance. The integration of the symbolic into the discursivity of gender identity far exceeds a notion of discourse as communication. Indeed:

> It is necessary to distinguish power relations from relationships of communication which transmit information by means of a language, a system of signs, or any other symbolic medium. No doubt communicating is always a certain way of acting upon another person or persons. But the production and circulation of elements of meaning can have as their objective or as their consequence certain results in the realm of power; the latter are not simply an aspect of the former. (Foucault 1982:217)

To this end, symbolic mediations should not be merely conflated with power as part of the construction of gendered identity; but this does not mean that power relationships and sign relationships exist in different domains. As Foucault expresses it:

> Power relations, relationships of communication, objective capacities should not therefore be confused. This is not to say that there is a question of three separate domains. Nor that there is on one hand the field of things, of perfected technique, work, and the transformation of the real; on the other that of signs, communication, reciprocity, and the production of meaning; finally that of the domination of the means of constraint, of inequality and the action of men upon other men. It is a question of three types of relationships which in fact always overlap one another, support one another reciprocally, and use each other mutually as means to an end. (ibid:217-218)

Foucault (1988:18) identifies inter-relationships between sign technologies (signs, meanings, signification); technologies of production (transformation of material things); technologies of power (objectivising categorization); and technologies of the self (self-regulating abilities of subjects to impact discourse on the cultivation of body and soul). Formulating an account of gender and ecological activism therefore requires examining "the application of objective capacities" in terms of the "relationships of communication" that circulate in and through the discursive constructs of gender identity. Symbolic mediations are inextricably bound to the objectivizing capacity of technologies of power ("whether they consist of obligatory tasks" or processes of visibility) (Foucault 1982:218). Symbolic mediations are meaningful communications; for they "imply finalized activities (even if only the correct putting into operation of elements of meaning) and, by virtue of the modifying the field of information between partners, produce effects of power" (ibid:218). Consequently,

> [Symbolic mediations] can scarcely be dissociated from activities brought to their final term, be they those which permit the exercise of this power (such as training techniques, processes of domination, the means by which obedience is obtained) or those which in order to develop their potential call upon relations of power (the division of labor and the hierarchy of tasks). (ibid:218)

In this analysis of the subject of gender and ecological activism, it is appreciated that the "transformation of the real" (technologies of production); "signs, communication, reciprocity, and the production of meaning" (technologies of the sign); and "domination of the means of constraint" (technologies of power) are interrelated features of symbolic mediations (ibid). Nevertheless, their application is never uniform and invariant; this is because "in a given society there is no general type of equilibrium between finalized activities, systems of communication, and power relations" (ibid). Rather,

> The activity which ensures … the acquisition of aptitudes or types of behaviour is developed thereby by means of a whole ensemble of regulated communication (lessons, questions and answers, orders, exhortations, coded signs of obedience, differentiation marks of the 'value' of each person and of the levels of knowledge) and by the means of a whole series of power processes (enclosure, surveillance, reward and punishment, the pyramidal hierarchy). (ibid:218-219)

Part of Foucault's thinking here derives from his advocacy for a definition of "government" that extends beyond the formal institutions and structures of politics and the state. Rather "government" refers to a shifting assemblage of formal and informal agencies, practices and institutions that variously and differentially align the self-regulating ability of subjects with the design, objectives and scope of a regime of governance. As Foucault expresses it:

> The exercise of power consists in guiding the possibility of conduct and putting in order the possible outcome. Basically power is less a confrontation between two adversaries or the linking of one to the other than a question of government. This word must be allowed the very broad meaning which it had in the sixteenth century. 'Government' did not refer only to political structures or to the management of states; rather it designated the way in which the conduct of individuals or of groups might be directed: the government of children, of souls, of communities, of families, of the sick. (ibid: 221)

Symbolic mediations are inextricably linked to the exercise of power in government, i.e., they are involved in the designation of "the way in which the conduct of individuals or of groups might be directed" (ibid). A consistent feature of my research into the feminine ideal, as part of a symbolic mediation of river restoration, has involved focusing on governance both in terms of conventional political structures and "also modes of action, more or less considered and calculated, which were destined to act upon the possibilities of action of other people" (ibid). Thus, for example, I have empirically researched the impact of the European Union's *Water Framework Directive* on river conservation charities. My interviews with three river conservation charities produced primary data which has been triangulated with data derived from my interviews along the foreshore of the River Thames. Both sets of data have trace elements of government in the sense of "political structures or …the managements of states" (ibid); but the data also illustrates "the exercise of power as a mode of action upon the actions of others" (ibid). Emerging from this triangulation of data is the suggestion that the coordination of technologies of production, signs, and power constitutes a shifting and irregular field of practice. Symbolic mediations of ecological activism tend not to easily slot into the formal apparatuses of political institutions and environmental organisations. Rather, and as Foucault observes:

> [T]here are diverse forms, diverse places, diverse circumstances or occasions in which these interrelationships establish themselves according to a specific model. But there are also 'blocks' in which the adjustment of abilities, the resources of communication, and power relations constitute regulated and concerted systems. (ibid:218)

The symbolic mediation of ecological activism is constituted through a complex milieu of moderated communications (environmental citizenship programmes, advertising signs and scheduled events) and, less formally, "by the means of a whole series of power processes (enclosure, surveillance …)" (ibid:219). Further to the operation of assemblages of actions upon actions, it is evident that these symbolic mediations of ecological activism, in terms of the feminine ideal, were not prescriptive or directly imposed upon the subject. Rather, symbolic mediations of ecological activism exemplify the manifold ways in which the exercise of power opens up a field of possibilities in the cultivation of gender identities. In this respect, consider the following extracts of interviews with river restoration volunteers:

> ***Interviewer: Have you been here often?***
> **Interviewee:** *No, I've just done one before a couple of years ago and I just quite enjoy it, it's really interesting. This side, it was more interesting, the other side you found more sort of bits of clay pipe, this is all plastic bags but it's still nice to get down here. You don't see it, when you're just travelling through London, you don't see the Thames, it's nice.*
>
> ***Interviewer: And what does it mean to you to be here?***
> **Interviewee:** *Erhm, just trying to make, to get out of the house; trying to make myself useful you know; meet new people. So that's about it, yea*
>
> ***Interviewer: And you said useful: what is the use and value?***
> **Interviewee:** *Erhm, I do think that if you keep areas nice, people treat them better and so there is less likely to be; less people sort of throwing stuff in the river. So it's just part of that, of looking after the environment (Female, South London, River Thames Foreshore, September 2013).*

Figure 6.8a., "Thy country's wrongs, the hopeless peasant's woe, Science and knowledge banished from the crowd" (Ann Yearsley *Elegy*, 1796). Photographic Image, South London, River Thames, September 2013

Figure 6.8b., "Thy country's wrongs, the hopeless peasant's woe, Science and knowledge banished from the crowd" (Ann Yearsley *Elegy*, 1796). Photographic Image, South London, River Thames, September 2013

Figure 6.9., “Than as an atom ‘mid the vast profound, Impell’d, not swerving from the whole” (Ann Yearsley *Soliloquy*, 1796). Photographic Image, South London, River Thames, September 2013

Interviewee: *This is the main waterway in London and it's such an important [feature], [and so] necessary to keep it clean. But it's so easily polluted as you can see. Plastic, things that don't biodegrade in the water and it's harmful to the wild life, you know. And it's harmful to the water quality, which we are trying to improve the quality of the water obviously, all the time. And gradually with the help of all these wonderful volunteers that's what's happening. The Thames is getting cleaner; the wild life is proliferating in many places and you just feel as if you've done, put something back to society, really. Whereas most people seem to take, we put something back. Although it seems a little, but it all adds up over the years. So do what you can! (Male, South London, River Thames Foreshore, September 2013).*

Interviewer: What did you call them [referring to sentimental objects placed in the river]?
Interviewee: *Votive offering*

Interviewer: Votive?
Interviewee: *Votive, kind of, to the gods or whatever. Or just, people just think, oh it might travel to their relatives. It might be safe travel; erhm pregnancy; be happy. We find a lot in the Brent actually. Erhm and there's messages in bottles. I haven't, but some people found them and said they can be quite, erhm don't open them. But it's just somebody else's thoughts.*

Interviewer: And they are called votive offerings? And that's to the gods?
Interviewee: *Yea, best wishes on the water. I mean, water is something that's constant. You know it never goes away, it's always there; but it moves, it's constant but it moves. It's great ...Can I just say about the trolleys? We were at Erith a while ago. We actually had a group of sea scouts I think and we had so many shopping trolleys there we had them down this pier and we had them down the ...we had them all lined up pulling these shopping trolleys out and pulleys and all this mud. Now that would have taken some good pictures [she laughs]. (Female, Organizational Member of London River Thames Conservation Charity III).*

Figure 6.10a., "This frame dissolving, to the busy winds" (Ann Yearsley *Soliloquy*, 1796). Photographic Image, South London, River Thames, September 2013

Figure 6.10b., "This frame dissolving, to the busy winds" (Ann Yearsley *Soliloquy*, 1796). Photographic Image, South London, River Thames, September 2013

The previous interview extracts are illustrative of how relationships of communication are neither formulaic nor immutable. Rather, it is evident from the varieties of descriptions of an ethic of embodied care, that the feminine ideal is constructed through "diverse forms, diverse places, diverse circumstances or occasions" (Foucault 1982:218). But these varied descriptions are embedded within the context of a structured event (the river restoration), and thus in contrast to the postmodern conception of symbolic mediation as free-floating forms of signification, it is evident that sign systems "are also 'blocks' in which the adjustments of abilities, the resources of communication, and power relations constitute regulated and concerted systems" (ibid). In Chapter Four, I discussed the translation of EU Directives into community out-reach environmental education programmes. In Chapter Five, I discussed the appendage, to voluntary sector programmes of river restoration, and sub-political agendas. Both of these chapters suggest that the activity of river restoration, and its mediation through the feminine ideal, are "developed there by means of a whole ensemble of regulated communications" (ibid). Moreover, in their varied and distinctly textured articulations of the care ethic, the interviewees illustrate that "there is no general type of equilibrium between finalized activities, systems of communication and power relations" (ibid). Rather, what is significant here, is the manner in which the feminine ideal of care "is integrated into a disparate field of possibilities, brought to bear upon permanent structures" (ibid:219). Thus, the power of symbolic mediation is not achieved through an abjuration of freedom. Indeed, as Foucault asserts: "What is to be understood by the disciplining of societies in Europe since the eighteenth century is not, of course, that the individuals who are part of them become more and more obedient" (ibid). But rather: "In effect, between a relationship of power and a strategy of struggle there is a reciprocal appeal, a perpetual linking and a perpetual reversal" (ibid: 226).

This chapter has been concerned with the examination of the institutional and personal mechanisms, actions upon actions, and strategies that are brought into play in river restoration; "Rather than analyzing power from the point of view of its international rationality", this chapter has been guided by the analysis of "power relations through the antagonism of strategies" (ibid:211). The interpretations that have emerged from this analysis consistently illustrate that "the strategy of struggle also constitutes a frontier for the relationship of power" (ibid: 225).

Figure 6.11a., "I frown'd indignant on my chains, and tun'd, My rural lay to universal love" (Ann Yearsley *Remonstrance*, 1796). Photographic Image, South London, River Thames, September 2013

Figure 6.11b., "I frown'd indignant on my chains, and tun'd, My rural lay to universal love" (Ann Yearsley *Remonstrance*, 1796). Photographic Image, South London, River Thames, September 2013

Figure 6.12., "Love e'en the soil that nurses them, refresh" (Ann Yearsley *Remonstrance*, 1796). Photographic Image, South London, River Thames, September 2013

Timely reflections

In the sub-political geographies of river restoration, the feminine ideal provides a potent and compelling symbolic mediation. Unconditional care enfolded into pockets of leisurely time, ambling along the river foreshore, is an appealing simulacra when juxtaposed against the intensities of work in capitalist societies. As is evident from the transcription extracts, the linguistic repertoires and visual metaphors used by volunteers, they consistently describe river restoration as a retreat from the fervency of capitalist society. Furthermore, it was apparent that for river conservation organizational members, the "away days" offered respite from the impact of managerial policy initiatives. In recent decades, the voluntary sector has been subject to intense managerial reform as part of the national and international neoliberal restructuring of local government and politics. In utilizing the feminine ideal as a symbolic mediation of river restoration, organisations actively frame river clean-up as the antithesis to the prevailing neoliberal regime of market-based managerialism.

Elsewhere, feminists have explored aspects of these dynamics as part of an analysis of resistance to managerialism. Baines, Cunningham and Fraser (2010) focus on case study data of two voluntary agencies (situated in Scotland and Australia) and they explore processes through which workers actively respond and resist the encroachment and risks to social obligations presented by the imposition of managerialist policy. Chapter Four of this book extensively discussed the impact, on the river conservation voluntary sector, of European Union Directives. The *Water Framework Directive* facilitates existing community outreach educational programmes, but it also constitutes an exogenous managerialist imposition. Conversations and interviews with organizational members, while out on the river foreshore, expressed their liberation from the extent to which the exogenous impositions of market-based managerialism encroach on what Baines et al (ibid: 347) conceptualize as "professional forms of participation". Elsewhere, the pervasion of managerialism in the voluntary sector is said to have been instantiated in the 1980s as part of the New Right's market reform of local authority provision (Lyon and Glucksmann 2008). Third sector voluntary organisations were linked to the Conservative government's establishment of a separation between purchasers and providers of local authority services. Alongside these exchange relations "local authorities were obliged to purchase more from independent providers" (ibid:111). Social services providers were transformed into "enablers, contracting with the independent sector rather than direct providers of care" (ibid).

In recent years the not-for-profit sector has been positioned centrally in the neoliberal exchange matrix of social service public and private partnerships. Milbourne (2013:39) convincingly argues that the coalition's "rationale for market competition" extends and exceeds New Labour's "Third Way solutions". As a consequence, the Third Sector is beset by a contract culture that provides charities with short-term ad-hoc funding in return for increased subjectification to enhanced performance measures and monitoring. Milbourne (ibid:44) states: "What is emerging … as the public sphere contracts, is the greater reliance on ad-hoc voluntary actions, combined with progressive privatization of public service provision … in a somewhat unregulated market". Alongside these changes, recent policy initiatives have attempted to reverse the centralist top-down management of the previous administration, whereby, the Coalition's Big Society philosophy emphasizes entrepreneurially motivated micro-markets of local service providers. Thus, we see that "Recent political programmes and priorities have emphasised the potential for self-help communities, civic and civil action, and the shift from government towards more fluid governance" (ibid). The community-based micro-market approach presents opportunities for entrepreneurial endeavor, but there are signs that inequities are presenting barriers to market competition. The sector is currently besieged by market-driven monopolistic practices. For example:

> Procurement decisions are increasingly using financial criteria and business reputation to allocate giant contracts, ostensibly to judge anticipated performance and reduce risks … and local service experience and expertise are being sidelined … Effectively, market values and reputation are supplanting social values and local knowledge as indicators of organizational legitimacy. (Milbourne 2013:46)

Recall how in Chapter Four, the Executive Director of London River Thames Conservation Charity II spoke passionately about the frustrating encounters with the Environment Agency, as part of the formulation of River Basin Management programmes. In these circumstances, the symbolic mediation of river conservation as an act of care provided valuable advocacy for translating the agenda of the Environment Agency with the agenda of the London river charities. The symbolic mediation of a feminine care ethic enabled the river charities to present the encumbered knowledge of volunteers as a basis for appeals to the authentic experiences of citizens. In this respect, I agree with Baines *et al's* (2010:347) comment that "Affinity or basic humanistic care processes seem to be the most resilient and near omnipresent form of participation in this sector". In the ascription of the feminine ideal to river restoration activities, the organizational culture of these conservation charities extends into the river foreshore, an ethic of care that espouses interconnection and embodiment. In this sense of care, it is not merely the taken-for-granted foundation of the organisation, but it is also mobilized in the creation of sub-political geographies cultivated as sites for symbolic resistance to the atomization of working practices and the erosion of working communities in the context of neo-liberal market-based managerialism. As Baines *et al.,* (ibid:348) describe "While managerialism and Taylorized practice models may remove or reduce opportunities for workers' participation, care is a theme or storyline that gave workers other ways to understand their work and why they did it".

The quixotic phenomenon of nature makes even more probable the existence of resistance to the technocratic managerialist frameworks of neo-liberal third sector market reform. In environmental activism, neoliberalism persistently encounters limits to the extent to which humanity will capitulate to the ravages of capital accumulation. The symbolic mediation of the river, as a feminine ideal, provides conservation charities with, what Baines *et al.,* (ibid) describe as, an "entry point from which to nurture, legitimize and expand their resistance and opportunities for participation". But describing river restoration as a feminine ideal of care is tenuous and risky, for the feminine ideal of care provides both an axis for meaningful existence and a basis for the exploitation of the selfless devotion to care. As Baines *et al.,* (ibid) express it:

> [C]are is part of a contradictory and contested set of social relations. Reflecting the vulnerability of affection and interdependence, care provides deep meaning and interconnectivity between workers, clients and larger communities and smoothes the way to further participation; however, it provides an axis on which exploitation…of the largely female work force is tolerated.

Baines *et al.,* (ibid) suggest that a possible way to proceed is to centre stage the concept of "care and its affinity processes at the heart of their efforts to mobilize the sector". Elsewhere, Joan Orme (2003) introduces ideas relevant to a postmodern feminism of social work. Postmodernism provides an apposite vehicle for critiquing the appropriation of the feminine ideal as a form of symbolic mediation in river restoration. Postmodernism proposes decentered, multiplicities as a basis for discerning gendered identities. In so doing, postmodernism challenges notions of a fixed feminine and masculine identity. In feminist social work, postmodernism challenges an advocacy for female-centred practices based on an assumption of essential affinity garnered from physiology, female sex roles and childrearing. In this latter conception, women are formulated as knowledgeable beings motivated to engage in river restoration as part of their shared experience and collective consciousness. An example of this framework and its translation in care work is as follows:

> [W]omen's contributions to community life have sustained social capital in the UK and … their involvement in community associations has risen at a higher rate than men's. In highlighting women's caring capacities, feminists have made visible women's contribution to public social capital. This has highlighted that women's caring work, including forming tenants' groups around housing issues and creating playgroups to develop resources for children, belongs in the public arena and acknowledged activities carried out in the home or informal networking, not as private matters, but as part of the wider civil society that is embedded in communities. This approach counters the devaluation of women's contribution to social capital. (Dominelli 2006:83)

Orme (2001:137) states: "Standpoint feminism, the experience of women reflexively engaged in struggle, was fundamental to women centred practice. Hence, feminist social work focused on working with women, and was seen to be a project for women workers". Conversely, postmodern feminist social work theorists shun meta-narratives, preferring the deconstruction of experience over recourse to a shared phenomenological basis for womanhood. In so doing, postmodern feminist social work engages with the complexity of gender identity and emphasizes the ambiguity of fixed representations. Universal claims to women-centred practice are said to produce regimes of truth and subjugation; for they negate difference and essentialize similarities. This is evident in the advocacy of a separation of public from private; with gender-essential attributes linked to these polarities (ibid). Lena Dominelli (2006) refers to Olive Banks' (1981) account of three distinct waves in feminist activism. The first wave sought to secure equality of opportunity for women within existing political structures and institutions; the second wave argued that the system was entirely pervaded by patriarchal values, necessitating women-centered provision; and the third wave advocates individual freedoms based on a politics of difference. Third-wave feminists: "like Jennifer Baumgardner argue that their innovations are more person-focused than those of the second wave, but as socially significant"

(Dominelli 2006:83). In this sense, third wave feminists are defined as guided by "diffuse range of issues with women as both subjects and objects of social activism". Viewed through a third wave lens the appeal of the feminine ideal symbolic mediation of river restoration might be perceived in terms of autonomous individual distinction. The feminine ideal, from a third wave lens, is "specific to each individual woman who attaches the term 'feminist' to herself". (ibid:84). To this end, women and males who identify with the feminine ideal actively engage in reframing "the boundaries between personal and political, the private and public; and theorizing from daily routines" (ibid). From this perspective, one must reconceive the "calling" to volunteer and participate in river restoration. For the feminine ideal symbolic mediation of the river, from a postmodern feminist perspective, is no longer indexed to universally essential standpoint and "mobilised around caring responsibilities" (ibid:82). Rather, when viewed through third-wave feminist lens women and men interpolate the feminine ideal of river restoration; they act as "private citizens" (Baumgardner and Richard, quoted in Dominelli 2006:84) with the same intensity and demands for rights as "yesterday's public heroines" (ibid). In their appeal to a feminine ideal of symbolic mediation of river restoration, it is my contention that river charities implicitly "hail" (Althusser 1971) forth subjects to interpolate (Ashcroft 2001) the feminine ideal of river restoration. It is my contention that the calling to participate in river restoration affirms women's and men's increased possibility to accentuate individual identities as part of a geopolitics of care for the environment. Dominelli has observed: "Postmodernist feminist's focus on individual identities has strengthened women's ability to address identity politics and celebrate diversity" (2006:84). Dominelli also raises significant concerns regarding the possibilities of feminist praxis in the absence of and hastened by the "loss of the strategic essentialism initiated by the women's movement" (ibid).

One might question whether feminism based on essential dichotomies ever succeeded in unifying difference. Conversely, third wave feminism, as Dominelli (ibid:85) recognizes, has "sought to address ambiguities and multiple positioning at the individual level rather than suppress these in the interests of a forced or false unity". In this sense, it can be argued that the symbolic mediation of river restoration as a postmodern feminine ideal, engages at the level of individual identities. There are, however, limitations with this appeal to an individualistic identity politics, for it merely depicts volunteers as enrolling into "constant and unfinished processes of becoming rather than reaching an end point, and engage in negotiating differences to rethink meanings of identity and community" (ibid). As a postmodern feminine ideal, river restoration encourages the "holding and negotiating contradictory positions in the desire to accommodate and resist oppression" (ibid). River restoration emerges from a postmodern framework as floating signification, which although meaningful is quixotic in its collusions and collisions as and between sign systems.

Conversely post-structural feminist accounts of ecology and gender interrogate difference as part a deconstruction of the metaphysics of gender and in so doing raise issues about how gender identities come into being. Examples include; Nagington, *et al.,* (2013) in their study of patients in the context of palliative and direct nursing care, who utilized a procedure of post-structuralist discourse analysis "to examine how discourses operate on a moral level" (ibid:893). This chapter has been concerned to explore the symbolic mediation of the feminine ideal, in river restoration community-outreach programmes. Foucauldian feminist analytics were developed so as to explore gender identification as a condition and consequence of the sign(s) of river restoration. Gendered subjectivities are constituted in and through a multiplicity of institutions, social practices and actions upon actions. Symbolic mediations of the feminine ideal enter into the complex multiplicities of gender and the subject of ecological activism. Suffice to say that gender identities are multiple, and this is partly because "Power relations are rooted in the system of social networks" (Foucault 1982:224).

Consequently, multiplicity characterizes the power/knowledge relations of gendered identity constructions. Discourses of gender identity "are multiple: they are superimposed, they cross, impose their own limits, sometimes cancel one another out, sometimes reinforce one another" (ibid). Multiple and often contradictory discourses of gender are actively interpolated by subjects as part of the cultivation of gender identity. Among the myriad of feminine and masculine identities, one can discern ideal formulations. The discursively constituted constructs of ideal femininity define meaningful existence in terms of embodied connectivity, manifested in an unconditional, relational devotion to time in the care for significant others (Odih 2007). In western society, the feminine ideal exudes limitless nurturing capacities, which are unmotivated by instrumental reward. The feminine ideal espouses the cultivation of meaningful existence through a timeless devotion to care. Emotional fulfillment, through affective behaviour defines the feminine ideal.

Affirmation derives from cultivating connections with others who are in need of care and this "joins the need for approval with the wish to care for and help others – capturing women between the passivity of dependence and the activity of care" (Orme 2001:107). Conversely, hegemonic masculinity eschews the affective passivity, embodiment and caring ethic of the feminine ideal. Masculinity in its hegemonic form gains meaningful existence from the expectation that it is an active initiator, self-contained and self-directed in the competitive pursuit of goals. Affirmation and reward impact in particular upon the masculine self, differently from the feminine self's ceaseless pursuit of care as a basis of self-affirmation. Hegemonic masculinity is not devoid of the motivation to care, but for manifold reasons, ideal masculinity configures care in terms of instrumental rationalities. Consequently, the particularity of the moment of care is dissuasive as a meaningful exhibition of care. Rather, ideal masculinity affirms its ability to care through the pursuit of calculative reparation and demonstrable righteousness.

In the just-in-time casino culture of post-Fordist "flexible accumulation" (Harvey 1990), it is evident that both the masculine ideal and the feminine ideal are in crisis. This is because the just-in-time flexibility of employment in advanced capitalist society corrodes duration and undermines the cultivation of gender identity through the longue durée of narrative time. Nowhere is this more evident than in the rise of "zero hours contracts". In 2013 the UK's Trade Union Congress (TUC), among other agencies, published news briefings condemning the rise in "zero hours contracts". According to Gov.UK (2013) a "zero hours contract" refers to a form of work contract in which the employee is "on call to work when you need them"; but the employer does not "have to give them work". The TUC quoting statistics published by the then Shadow Secretary for Health, revealed that "over 300,000 workers in the care sector alone are employed on zero hours contracts" (2013). Given that these contracts are premised on extreme forms of flexibility, in which periods of time might lapse between the hours of paid employment, workers are required to financially and cognitively become used to acute conditions of uncertainty. Ideal masculinity struggles to formulate long-term goals in these conditions, and likewise, ideal femininity encounters insurmountable obstacles to its narrative endeavour of care. "Flexible accumulation" (Harvey 1990) is acutely corrosive in the just-in-time casino culture of paid employment in advanced capitalist society. Set in the context of work intensification and extenuated working hours, the symbolic mediation of the female ideal is an appealing mediation.

In their promotional appeal to a care ethic, river restoration programmes collude with gendered symbolic mediations of guardianship. But this appeal to a feminine nurturing ethic is also a consequence of the extent to which the "cultural symbols and experiences that govern the way people think and act hav[e] their origin in their history and in societal living conditions" (Beck 1995: 47). Symbolic mediations of river restoration in terms of a feminine ideal provides the promise of an ethics of care embedded in embodied materiality; this is appealing, when compared to the disembodied incomprehensibility of conflicting scientific proclamations. Indeed, the scientific community is increasingly cognizant of this, and thus hastens to illustrate the complexities of global climate change in terms of melting glaciers and their impact on the habitat of settled communities, the wilderness and wildlife. Nevertheless, our increasingly local experiences of global environmental degradation will continue to mediate our self-confrontation with the congested algorithmic signs governing scientific discourse on climate change, for "Unreality and hyperreality" characterize the lay experience of scientific abstractness (ibid: 47). As the representations of environmental hazard become "increasingly unreal and inhospitable" (ibid: 48) the feminine ideal's ethic of embodied environmental care becomes evermore appealing.

Figure 6.13a., "Friendship! Thou noblest ardour of the soul, Immortal effence, languor's best support...Long have I searched the human heart for thee...Firm cement of the world, endearing tie" (Ann Yearsley *Address to Friendship*, 1796). Photographic Image, South London, River Thames, September 2013

Figure 6.13b., "Friendship! Thou noblest ardour of the soul, Immortal effence, languor's best support…Long have I searched the human heart for thee…Firm cement of the world, endearing tie" (Ann Yearsley *Address to Friendship*, 1796). Photographic Image, South London, River Thames, September 2013

CONCLUSION

WATERSHEDS IN ECOFEMINIST ANALYSIS OF HETEROTOPIAS

Interviewee: *Well, the position I applied for, they've got funding for a new project on a river, not far from here actually. It just flows from Eltham through to Hall Place, about five and a quarter miles long. And they wanted somebody to basically run the project and get local people to volunteer, and get in the river and pull the rubbish out, and teach in local schools and that kind of thing. And so that was the position that I originally applied for and so as part of that application I had to create a presentation on what I would do in the first six months if I was to get the job. So I had to think what activities would I set up? How would I attract local people? How would I engage with them; and that kind of thing? That was part of the interview process.*

Interviewer: So what did you say during the interview? What was your plan, strategy?
Interviewee: *What, the kind of presentation?*

Interviewer: Yes
Interviewee: *Well I focused on a few different things. Like, I am.. I think all the way from the beginning my personal passion is teaching. But I've never wanted to classroom teach. I have never wanted to be like a schoolroom teacher. I wanna teach outside. So I went, did my research, and because I was a cycling instructor I had worked in quite a lot of the local schools. So I had some good contacts to start with. And I thought up a few different programmes that I could teach in the river, that wouldn't cost the school or the kids anything. And found out about local, like youth groups like Brownies, Scouts, Guides, that kind of thing; and sort of local charity groups where people with learning difficulties and disabilities. So I kind of went for the more challenging, like trying to tackle the more challenging roots, rather than mass volunteers you know. That's kind of how I did it.*

Figure C.1., previous page, "Those hours are flown! The transport's o'er!, Yet memory from her heav'nly store, My lasting grief beguiles: Amid the vigils of the night, Thus tun'd by thee to fine delight, O'er her pale vision smiles", (Ann Yearsley *Extempore*, 1796). Photographic Image, South London, River Thames, September 2013

Interviewer: How would you describe the kind of teaching that you would like to do? You said that it is out door teaching?
Interviewee: *Yea, yea, so every event that I deliver is outside. So I don't do... well, rarely; I mean I can count on one hand the amount of times I have taught inside. I try and make all the sessions outdoors. So I take the kids out of the classroom and teach them about the river in the river, so that they can see it and understand it a bit better. So one of the... I think one of the better sessions I did was on downstream change. And you know: how does the flow change the width, the depth, that kind of thing? And I identified three sites along the river. It was year fives so we are not talking like high class scientific techniques. Like theory techniques. It was like a meter ruler, an orange and a stopwatch; that kind of thing. But we got to the third site and the river was visibly faster and wider. And I just got them standing in the river and it was like: now you can see it! Cos I always, personally, I'm a visual learner. And I used to look in text books and be like; it never really sunk in very well for me, unless I could see it. But what is tricky about that is keeping their attention while they are standing in the river. Keeping them focused, cos there's so many distractions.*

Interviewer: So it's a sort of experiential, they have to experience the actual subject of focus i.e., they have to be in the river. So how would you describe that as a pedagogy; as a teaching style? It can be different?
Interviewee: *It's very different. Because quite commonly I find that the kids are a bit apprehensive; very apprehensive at first because now-a-days it's really difficult for teachers to make time to get kids out of the classroom. Because of pressures that they are under: red tape; health and safety all that kind of stuff. And so I find that also kids these days have got so many distractions like video games and like phones and stuff. They're quite; it's like an alien environment to them when they first come out: they don't wanna touch anything; oh it's dirty. And by the end of it, ah they love it. They are like, you know they are really engaged and they really, and they've got soaken feet and they're all like excitable. And they've learnt something you know. So it's a challenge in a way teaching, because there's a lot of, you have to be delivering what you are delivering information wise but, also I'm constantly like: is everyone here?; don't do that; like don't put your hands in your mouth. It's tricky but I think it's worth while, like it's worth the effort. And we do, like all of our events are risk assessed as well and I often meet with the teachers before I take the kids out. So I find out what they are doing in class. So I link the activity that I write for that group, I'll link it to the national curriculum so that it's not like an additional thing that a teacher's got to do. It's like a, it complements what they are teaching in the classroom... If you know what I mean - rather than it being a burden?* ***(Education Coordinator, London River Thames Conservation Charity III, August 2013)***

It is commonly asserted that Marx's dialectical materialism designates relations between nature and humans in Promethean terms, whereby the historical trajectory of advances in modes of productive technology, ultimately entail the subjugation of nature to human rationality. There is evidence to suggest that, alongside this mediation of nature, conceived in a dystopian Promethean view, Marx's position on the question of nature articulated echoes of romantic eco-centrism. Explanations for this oscillation are enigmatically sparse, nevertheless, several writers foreground Marx's bifurcation of nature, as consistent with his critique of Hegel's dialectic. Of significance here is Alfred Schmidt's (1971) *The Concept of Nature in Marx*. According to Schmidt, the Germany of the 1850s provides some indication of why Marx's consideration of human relations to nature is surprisingly manifold and dimensionally complex. During the 1850s, obscurantist opposition to technical progress in Germany was based around "the deification of the immediacy of elemental nature" and the wish "to maintain precapitalist forms of production" (ibid:132). Continuous with this sentiment an ideology of nature existed which deplored the destructive advancement of capitalism, and in which "nature was glorified as a refuge in face of capitalism's ever more ruthless pillaging" (ibid). According to Schmidt (1971), some insight into Marx's theoretical consciousness is gained when one considers these ecological debates to have foregrounded Marx's material investigation. Measured against this view, Marx's position on the question of nature was also indebted to prominent philosophical debates concerning the dialectic materiality of freedom from necessity. Of particular significance were the synergies and disjunctures between Marx and Hegel's nature of freedom. Hegel conceived of human freedom as being actualised through the unfettering of humans from the capricious control of necessity. Freedom "is a conception completely freed from the necessity of Nature, a conception which leaves that necessity behind and operates spontaneously above it" (Hegel 1998:156). Schmidt's (1971) suggestion that the Hegelian dialectic evident in Marx's account of transcendence from nature was inspired by Marx's observation of Germany in the 1850s, further suggests that personal biography, politics and society influenced Marx's writings. Indeed, in the young Marx, it is evident that his deliberations and circumspection about maturity and manhood influenced his earliest work regarding the dialectics of nature time and destiny. For example, Marx, in 1835, in his earliest writing as a teenager, wrote with passion and reverence about nature, humanity and vocation:

> Nature has assigned to the animal the sphere of its activity, and the animal acts calmly within it, not striving beyond, not even surmising that there is

> another. To man, too the Deity gave a general goal, to improve mankind and himself, but left it up to him to seek the means by which he can attain this goal, left it up to him to choose the position in society which is most appropriate and from which he can best elevate both himself and society... But we cannot always choose the vocation to which we believe we are called. Our social relations, to some extent, have already begun to form before we are in a position to determine them. Even our physical nature often threateningly opposes us, and no one dare mock its rights!...But the high opinion we have of the ideas on which our vocation is based bestows on us a higher standpoint in society, enlarges our own dignity, makes our actions unwavering ... If a person works only for himself he can perhaps be a famous scholar, a great wise man, a distinguished poet, but never a complete, genuinely great man. History calls those the greatest men who ennobled themselves by working for the universal. Experience praises as the most happy the one who made the most people happy...When we have chosen the vocation in which we can contribute most to humanity, burdens cannot bend us because they are not sacrifices for all. Then we experience no meagre, limited, egoistic joy, but our happiness belongs to millions, our deeds live on quietly but eternally effective, and glowing tears of noble men will fall on our ashes. (Marx 1835/1997:35-39)

This book has been concerned with the issues of care for nature and humanity. The denaturing of environmental sustainability is observably accelerating and continuously bringing into question the proficiency of capitalism, to superintend the proliferating risks of global climate change. Industrial modernization involves exploitative subjugation, and the capitalization of the natural environment is an integral feature of this capitalist dynamic. It is evident that preceding modes of industrial capitalism modernized in disregard of issues of sustainability in the expenditure of renewable resources. "Risk society" is the unintended outcome or reverse side consequence of industrial capitalist opulence. According to Ulrich Beck, the conditions of risk society precipitate changes at the level of individual apprehension. In the risk society, insecurities erupt through the disjunctures, contradictions and conflicts that capillary from industrial society into subsequent orders of modernity. The current milieu of privatized securities, erode collective solidarities, individuating identities and undermining social defense against local experiences of global risks. Nowhere is this clearer than in the paralyzing grip of environmental crisis. Overwhelmed by unprecedented magnitudes of environmental risk, political institutions no longer profess to guarantee present and future wellbeing. Faced with the local immediacy of international environmental risks, individuals are compelled into defensive action. Beck describes "sub-politics" as confrontational and immediate, a direct political engagement indexed to the particular locality of disjuncture. It is a politics of the moment, in the moment. This book has explored the sub-political geographies of river restoration so as to critically examine the symbolic mediations of ecological activism. Michel Foucault identifies inter-relationships between sign technologies (signs, meanings, signification); technologies of production (transformation of material things); technologies of power (objectivising categorization); and technologies of the self (self-regulating abilities of subjects to impact discourse on the cultivation of body and soul). Primarily informed by ethnographic observations and interviews, this book examines how symbolic mediations of gender and ecological activism are inextricably bound up with "government" i.e., shifting assemblages of formal and informal agencies, practices and institutions that variously and differentially align the self-regulating ability of subjects with the design, objectives and scope of a regime of governance. But the operation of assemblages of actions upon actions through symbolic mediations of ecological activism, in terms of the feminine ideal, is neither prescriptive nor directly imposed upon subjects. This book has extensively illustrated how the symbolic mediations of ecological activism exemplify the manifold ways in which the exercise of power opens up a field of possibilities for the cultivation of gender identities.

Neoliberal environmental governance of river conservation, coupled with the organizational modernization, imposed and sustained by the European Union's water directives, engender *Other Spaces* of feminist ecological alignment. The riparian landscapes of our urban cities are manifestations of political and ideological rationalities operating under the constraints of capitalist markets, and saturated by the contradictions of neoliberal environmental science. Neoliberal rationalities configure river waterways as "sites", the dimensions of which are analogous to Michel Foucault's account of spatial heterotopias, as polymerous relations of propinquity between junctures. Many of the modernising initiatives instituted by the European Union's Water Framework Directive can be discerned as biopolitical neoliberal regimes governing local river space, through the enfolding into the "spaces of emplacement", of the "sites" of programmatic calculation, classification and market-based neoliberal environmental science. Primarily informed by organizational ethnographies, extensive interviews and ethnographic observations of river restorations, this book has empirically examined how the relationally embodied heterochronies of ecological activism challenge the programmatic rationalities of the European Union's river "government" i.e., its shifting assemblages of formal and informal agencies, practices and institutions that variously and differentially align the self-regulating ability of subjects with the design, objectives and scope of the European Union's neoliberal regime of river governance. This book's analysis of complex inter-governmental networking eliding the local governance of rivers with voluntary sector community-outreach and European Union directives, identifies new locations of ecological activism precipitated by political affinities, which have become simultaneously public and private. The capacity of river heterotopias to intersect the public and private spheres of urban cities, emphasises the intrinsic reproductive labour time of river restoration; for as Michel Foucault (1986:26) suggests, the heterochronies of urban heterotopia are at one and the same time "outside of time", while also constituting "a sort of perpetual and indefinite accumulation of time in an immobile place". It is my belief that the intersecting biopolitical heterochronies of the urban river space confirm this *Other Space* as an intriguing gendered heterotopia.

Figure C.2a., "On air my fancy seems to fly, Spirits I long have mourn'd are nigh, And worlds beneath me roll" (Ann Yearsley *Extempore*, 1796). Photographic Image, South London, River Thames, September 2013

Figure C.2b., "On air my fancy seems to fly, Spirits I long have mourn'd are nigh, And worlds beneath me roll" (Ann Yearsley *Extempore*, 1796). Photographic Image, South London, River Thames, September 2013

Figure C.3a., "Permit me, pensive friend, who long have known, A mother's duty, pleasing cares to own…To every motion happy Nature tries, Whilst life seems fluid, and from pressure flies" (Ann Yearsley *To Mira*, 1796). Photographic Image, South London, River Thames, September 2013

Figure C.3b., "Permit me, pensive friend, who long have known, A mother's duty, pleasing cares to own…To every motion happy Nature tries, Whilst life seems fluid, and from pressure flies" (Ann Yearsley *To Mira*, 1796). Photographic Image, South London, River Thames, September 2013

Figure C.3c., “Permit me, pensive friend, who long have known, A mother’s duty, pleasing cares to own…To every motion happy Nature tries, Whilst life seems fluid, and from pressure flies” (Ann Yearsley *To Mira*, 1796). Photographic Image, South London, River Thames, September 2013

Figure C.3d., "Permit me, pensive friend, who long have known, A mother's duty, pleasing cares to own…To every motion happy Nature tries, Whilst life seems fluid, and from pressure flies" (Ann Yearsley *To Mira*, 1796). Photographic Image, South London, River Thames, September 2013

Figure C.4., "So a lady…dressed … from head to foot like a jointed baby by her woman, and the assistance of another maid or two, is a more helpless creature than doll the diary-maid, who, all the winter long, dresses herself in the dark in less time than the other bestows in placing of her patches" (Bernard Mandeville, *The Fable of the Bees*, 1795). Photographic Image, South London, River Thames, September 2013

Figure C.5a., "Perfection, be it trifling as the mote, Which reveals in the Sunbeam, cannot own. Its essence self-originating … Vain are thy toils in Science! Vain the web. Hoary Philosophy shall ever spin, If in thy future views, thou ne'er canst form, Some good to hope for!" (Ann Yearsley *The Materialist*, 1787). Photographic Image, South London, River Thames, September 2013

Figure C.5b., "Perfection, be it trifling as the mote, Which reveals in the Sun-beam, cannot own. Its essence self-originating … Vain are thy toils in Science! Vain the web. Hoary Philosophy shall ever spin, If in thy future views, thou ne'er canst form, Some good to hope for!" (Ann Yearsley *The Materialist*, 1787). Photographic Image, South London, River Thames, September 2013

Figure C.5c., "Perfection, be it trifling as the mote, Which reve'ls in the Sun-beam, cannot own. Its essence self-originating … Vain are thy toils in Science! Vain the web. Hoary Philosophy shall ever spin, If in thy future views, thou ne'er canst form, Some good to hope for!" (Ann Yearsley *The Materialist*, 1787). Photographic Image, South London, River Thames, September 2013

References

Abbott, P., and Meerabeau, L., (1998); *The Sociology of the Caring Professions*, London: UCL.

Ackroyd, P., (2007); *Thames: Sacred River*, London: Chatto & Windus.

Althusser, L. (1971); Ideology and Ideological State Apparatuses, in *Lenin and Philosophy and Other Essays by Louis Althusser*, trans Brewster, B., New York: Monthly Review Press.

Anderson, B., (1991); Imagined *communities: reflections on the origin and spread of nationalism,* London: Verso.

Andrew, B. Keller, J. and Schwartzman, L. (2005); *Feminist Interventions in Ethics and Politics*, Oxford: Rowman and Littlefield Publishers.

Andrews, K., (2013); No More than as an Atom 'Mid the Vast Profound': Conceptions of Time in the Poetry of William Cowper, William Wordsworth, and Ann Yearsley, in (eds) Blair, K., and Gorji, M., *Class and the Canon. Constructing Labouring-Class Poetry and Poetics 1750-1900*, Basingstoke:Palgrave Macmillan

Apple, M., (2005); Education, markets and an Audit Culture, *Critical Quarterly*, Volume 47, Issue 1-2, pp., 11-29.

Ashcroft, B. (2001); *Post-colonial Transformation*, London: Routledge.

Bailey, N., (1731); *An Universal Etymological English Dictionary: Comprehending the Derivations of the Generality of Words in the English Tongue, Either Ancient Or Modern from the Ancient British, Saxon, Danish, Norman and Modern French*, London: J. Knapton.

Baines, D., Cunningham, I. and Fraser, H. (2010) Constrained by Managerialism: Caring as Participation in the Voluntary Social Services, *Economic and Industrial Democracy*, 32: 329–52.

Banks, L., and Stanley, C., (1990); *The Thames, A History from the Air*, Oxford: Oxford University Press.

Banks, O., (1981); *Faces of Feminism: A Study of Feminism as a Social Movement,* Oxford: Martin Robertson.

Baudrillard, J., (1968/2005); *The System of Objects*, trans., Benedict, J., London: Verso.

Baumgardner, J., and Richards, A., (2000); *Manifesto: Young Women, Feminism and the Future*, New York, NY: Farrar, Straus and Giroux.

Beck, U., (1995); *Ecological Politics in an Age of Risk*, Cambridge: Polity Press.

Beck, U., (2007); The Reinvention of Politics: Towards a Theory of Reflexive Modernization, in Beck, U., Giddens, A. and Lash, S. (eds) *Reflexive Modernization: Politics, Tradition and Aesthetics in the Modern Social Order*, Cambridge: Polity Press.

Bernauer, T., and Kuhn, P., (2010): Is There an Environmental Version of the Kantian Peace? Insights From Water Pollution in Europe. *European Journal of International Relations* 16/1 pp., 77-102.

Bjorkland, R., and Pringle, C., (2001); Educating Our Communities and Ourselves about Conservation of Aquatic Resources Through Environmental Outreach, *BioScience*, Vol., 51, No., 4, pp., 279-282.

Blake, W., (2008); *The Complete Poetry and Prose of William Blake*, (ed) Erdman, D., California: University of California Press.

Boorstin, D. (1963); *The Image, or, What Happened to the American Dream?* Harmondsworth: Penguin Books.

Bourdieu, P., (1984/2010); Distinction. *A Social Critique of the Judgement of Taste*, London: Routledge.

—. (1995); *The Field of Cultural Production*, Cambridge: Polity Press.

Buckingham-Hatfield, S., (2000); *Gender and Environment*, London: Routledge.

Camden, W., (1722); *Britannia: or a Chorographical Description of Great Britain and Ireland, Together with the Adjacent Lands*, London: Mathews and Churchill.

Campaña Costa Rica te Quiero Limpia (2010); *Costa Rica te Quiero Limpia Rescatemos el rio Virilla*. Available at: http://investiga.uned.ac.cr/protored/index.php?option=com_content&view=article&id=170:campana-costa-rica-te-quiero-limpia&catid=74:informacion-general-cu-orotina&Itemid=74

Cavanagh, K., and Cree, V., (1995); *Working with Men: Feminism and Social Work*, New York: Routledge.

Cecil, R., (1822); *Memoirs of John Bacon*, Esq, R.A., London; L.B.Seeley, Fleet-Street.

Chodorow, N., (1978); *The Reproduction Of Mothering: Psychoanalysis and The Sociology Of Gender*, London: University of California Press.

Church, A., (1880); *Isis and Thamesis, Hours on the River from Oxford to Henley*, London: Seeley & Co.

Clarke, J., (1998); Doing the Right Thing/ Managerialism and Social Welfare., in (eds) Abbott, P., and Meerabeau, L., (1998); *The Sociology of the Caring Professions*, London: UCL.

ClimateUK (2012); *A Summary of Climate Change Risks for London. To coincide with the publication of the UK Climate Change Risk Assessment (CCRA) 2012*. Available at:

http://climatelondon.org.uk/wp-content/uploads/2012/11/ccra.pdf.

Cohen, W., and Cohen, J., (2009); *Race and Reconciliation in America*, Lanham, MD: Lexington Books.

Collins, S., Reitsma-Street, M., Porter, E., and Neysmith, S., (2011); Women's Community Work Challenges Market Citizenship, *Community Development*, Vol., 42, Issue 3, pp., 297-313.

Combe, W., (1794); *An History of the River Thames, Vol., 1*, London: Bulmer,W., and Co for Boydell, J., and Boydell, J., Reproduced by British Library.

Cree, V., (1995); Why do Men Care? in (eds) Cavanagh, K., and Cree, V., *Working with Men: Feminism and Social Work*, New York: Routledge.

Cusack, T., (2007); Introduction: Riverscapes and the Formation of National Identity, *National Identities*, Vol., 9, No., 2, June, pp., 101-104.

—. (2010); *Riverscapes and National Identities*, Syracuse: Syracuse University Press.

Davies, K,. (1990); *Women, Time and the Weaving of the Strands of Everyday Life*. Aldershot: Avebury.

DEFRA (2013); *Securing the Future of Flood Insurance. An Introductory Guide, 27 June 2013, Department for Environment, Food and Rural Affairs*. Available at: https://consult.defra.gov.uk/flooding/floodinsurance/consult_view.

—. (2013a); *Catchment Restoration Fund: Environment Agency Summary Report 2012-2013, Presented to Parliament Pursuant to Section 70 (80 and section 70 (9) of the Charities Act 2006*. Available at: https://www.gov.uk/government/uploads/system/uploads/attachment_data/file/248857/pb14032-catchment-restoration-fund-report.pdf

—. (2014); *Policy, Improving Water Quality*, Available at: https://www.gov.uk/government/policies/improving-water-quality

Dominelli, L., (2006); *Women and Community Action*, Bristol: Polity Press.

Dordoy, A., and Mellor, M., (2000); Ecosocialism and Feminism: Deep materialism and the Contradictions of Capitalism, *Capitalism Nature Socialism*, 11:3, pp., 41-61.

Eaton, H. and Lorentzen, L. (2003); *Ecofeminism and Globalization. Exploring Culture, Context and Religion*, Oxford: Rowman and Littlefield Publishers.

Eliot, T.S., (2001); *Four Quartets*, London: Faber and Faber

Environment Agency (2013a); *Thames Water Ordered to Pay £41,000 for Polluting River Ash in Ashford*. Available at: http://www.environment-agency.gov.uk/news/148314.aspx?page=3&month=6&year=2013.

—. (2013b); *Thames Water Fines for Sewage Sludge Pollution of Local River in Weybridge*. Available at: http://www.environment-agency.gov.uk/news/148022.aspx?page=2&month=5& year=2013.

—. (2013c); *Thames Water Ordered to Pay Over £14,000 for Sewage Pollution in Oxfordshire*. Available at: http://www.environment-agency.gov.uk/news/146043.aspx

—. (2013d); *Richmond-upon-Thames, London Borough Environmental Fact Sheet*, November 2011. Available at: http://www.environment-agency.gov.uk/static/documents/Research/Richmond_2011.pdf.

—. (2013e); *Hounslow, London Borough Environmental Fact Sheet*, November 2011. Available at: http://www.environment-agency.gov.uk/static/documents/Research/Hounslow_2011.pdf.

—. (2013f); *Brent Catchment Summary*. Available at: http://www.environment-agency.gov.uk/static/documents/Research/Brent_summary.pdf.

—. (2013g); *Little Marlow Sewage Treatment Works Update*. Available at: http://www.environment-agency.gov.uk/static/documents/Leisure/Little_Marlow_update_v_3.pdf.

—. (2013h); *Lower Lee Catchment Summary*. Available at: http://www.environment-agency.gov.uk/static/documents/Research/Lower_Lee_summary.pdf.

—. (2013i); *Introduction to the Water Framework Directive*. Available at: http://www.environment-agency.gov.uk/research/planning/33362.aspx?style=print.

—. (2013j); *River Basin Management Planning*. Available at: http://www.environment-agency.gov.uk/research/planning/33240.aspx.

—. (2014a*); 2013 and 2014 Winter Flooding*. Avalable at: http://www.environment-agency.gov.uk/homeandleisure/floods/152039.aspx

—. (2014b) *The Thames Barrier*. Available at: http://www.environment-agency.gov.uk/homeandleisure/floods/38353.aspx

—. (2014c); *Dredging*. Available at: http://www.environment-agency.gov.uk/homeandleisure/floods/152033.aspx

—. (2014d); *Dredging Begins on the River Parrett in Somerset*. Available at: http://www.environment-agency.gov.uk/news/152411.aspx

European Commission (2008); *Water Note 9. Integrating Water Policy: Linking all EU Water Legislation within a Single Framework.*, Water Notes on the Implementation of the Water Framework Directive, Water Information System for Europe (WISE). Available at:

http://ec.europa.eu/environment/water/participation/pdf/waternotes/water_note9_other_water_legislation.pdf

—. (2008a); *Water Note 12, A Common Task: Public Participation in River Basin Management Planning*. Available at: http://ec.europa.eu/environment/water/participation/pdf/waternotes/water_note12_public_participation_plans.pdf.

—. (2008b); *Joining Forces for Europe's Shared Waters: Coordination in International River Districts, Water Information System for Europe, Water Note 1*. Available at: http://ec.europa.eu/environment/water/participation/pdf/waternotes/water_note1_joining_forces.pdf.

—. (2008i); *Cleaning up Europe's Waters: Identifying and Assessing Surface Water Bodies at Risk*. Available at: http://ec.europa.eu/environment/water/participation/pdf/waternotes/water_note2_cleaning_up.pdf

—. (2008ii); *Pollution: Reducing Dangerous Chemicals in Europe's Waters*. Available at: http://ec.europa.eu/environment/water/participation/pdf/waternotes/water_note8_chemical_pollution.pdf

—. (2013); *Introduction to the New EU Water Framework Directive*. Available at: http://ec.europa.eu/environment/water/water-framework/info/intro_en.htm.

—. (2014); *Infringement Cases*. Available at: http://ec.europa.eu/environment/legal/law/press_en.htm

—. (2014a); *Environment: Commission Takes Denmark To Court Over Water Management Problems*. Available at: http://europa.eu/rapid/press-release_IP-14-157_en.htm

—. (2014b); *Environment: Commission takes Austria to Court over Failure to Protect Water Quality on Schwarze Sulm River* . Available at: http://europa.eu/rapid/press-release_IP-14-448_en.htm

European Parliament (2000); *Directive 2000/60/EC of the European Parliament and of the Council of 23 October 2000, Establishing a Framework for Community Action in the Field of Water Policy*. Available at: http://ec.europa.eu/environment/water/water-framework.

Ewald, F. (1991); 'Insurance and risk', in J. Burchell, C. Gordon and P. Miller, *The Foucault Effect: Studies in Governmentality* Hemel Hempstead: Harvester.

FAO United Nations (2012); *Passport to Mainstreaming Gender in Water Programmes. Key Questions for Interventions in the Agricultural Sector. Food and Agriculture Organization of the United Nations*. Available at: http://www.fao.org/docrep/017/i3173e/i3173e.pdf.

Fairclough, N., (2001); The Discourse of New labour: critical Discourse Analysis, in (ed) Wetherell, M, Taylor, S., and Yates, S., *Discourse as Data*, London: Sage.

Fearnside, W., (1834); *Tombleson's Thames*, Oxford: Oxford University Press.

Feinstein, B., (1968); The Faerie Queene and Cosmogonies of the Near East, *Journal of the History of Ideas*, Vol., 29, No., 4., pp., 531-550.

Felski, R., (2002); Telling Time in Feminist Theory., *Tulsa Studies in Women's Literature*, Vol., 21, No., 1 (Spring), pp., 21-28.

Fenves, P., (1986); Marx's Doctoral Thesis on Two Greek Atomists and the Post-Kantian Interpretations, *Journal of History of Ideas*, Vol., 47, No.,3, (Jul. – Sep.,) pp., 433-452

Ferguson, M., (1995); *Eighteenth-Century Women Poets: Nation, Class and Gender*, Albany: State University of New York Press

Foster, B. J., (2000); *Marx's Ecology: Materialism and Nature*, New York: Monthly Review Press.

Foster, J., and Burkett, P., (2001); Marx and the Dialectic of Organic/Inorganic Relations: A Rejoinder to Salleh and Clark, *Organization Environment*, Vol., 14, pp., 451- 462.

Foucault, M., (1982); The Subject and Power, in Dreyfus, H. and Rabinow, P. (eds) *Michel Foucault. Beyond Structuralism and Hermeneutics*, London: Harvester Wheatsheaf.

—. (1983/2008); *This is Not a Pipe*, Trans., Harkness, J., Berkeley: University of California Press.

—. (1986); Of Other Spaces, *Diacritics*, Vol. 16, No. 1, pp., 22-27.

—. (1988); Technologies of the Self, in Martin, L., Gutman, H. and Hutton, P., (eds), *Technologies of the Self; A Seminar with Michel Foucault*, London: Tavistock Publication.

Fraser, I., (2000); Hegel and Marx on Needs; The Making of a Monster, in (ed) Burns, T., and Fraser, I., *The Hegel-Marx Connection*, London: MacMillan Press.

Freire, P., (1985); *The Politics of Education: Culture, Power and Liberation*, trans. Macedo, D., Westport, CT: Greenwood Publishing Group.

—. (1996); *Pedagogy of the Oppressed*, London: Penguin.

Freud, S., (1930/2010); *Civilisation and its Discontents*, CT: Martino Publishing.

Gaylin, W., (1986); *Rediscovering Love*, New York: Viking

Gentleman of Cambridge (1755); *The Praises of Isis; a Poem*, London: R., and J., Dodsley.

Gilligan, C., (1993); *In A Different Voice. Psychological Theory and Women's Development*, London: Harvard University Press.
—. (2000); *In A Different Voice. Psychological Theory and Women's Development*, London: Harvard University Press.
Goetz, A., (2008); *Who Answers to Women? Gender and Accountability*, New York: United Nations Development Fund for Women.
Goldman, R., (1995); *Reading Ads Socially*. London: Routledge.
Gov.UK (2013); *Contract Types and Employer Responsibilities*. Available at: https://www.gov.uk/contract-types-and-employer-responsibilities/ zero-hour-contracts.
Greater London Authority (2013); *The Greatest City on Earth: Ambitions for London by Boris Johnson*, *http://www.london.gov.uk/sites/default/files/2020_vision_web.pdf*
Hall, S., (1859); *The Book of the Thames: from its Rise to its Fall*, London: Arthur Hall, Virtue, and Co.
Hall, S., (1978); The Politics of 'Mugging'. In Hall S., Critcher, C., Jefferson, T., Clarke, J., and Roberts, B., (eds.) *Policing the Crisis: Mugging, the State, and Law and Order*, London: Macmillan Press.
—. (1987); Minimal Selves, in (eds), Bhabha, H., and Appignanesi, L., *Identity: The Real Me: Postmodernism and the Question of Identity*, London: ICA (Institute of Contemporary Arts).
Halliday, S., (2013); *The Great Stink of London, Sir Joseph Bazalgette and the Cleansing of the Victorian Metropolis.*, Gloucestershire: The History Press.
Harvey, D., (1990); *The Condition of Postmodernity*, Oxford: Blackwell.
Hegel, G.W. F., (1998); *Phenomenology of Spirit*, Delhi: Motilal Banarsidass.
Heraclitus (2003); *Heraclitus, Fragments*, London : Penguin.
Herbert, A., (1966); *The Thames*, London: Weidenfeld and Nicolson.
Heron Maps (2011); *River Thames and the Thames Path, Waterways Series*, London: Phoenix Maps.
Hobson, C., (1987); *Exploring the World of the Pharaohs*, London: Thames and Hudson.
Hoekstra, A., (2013); *The Water Footprint of Modern Consumer Society*, London: Routledge.
House of Commons (2013); *House of Commons Science and Technology Committee. Water Quality: Priority Substances*. First Report of Session 2013-2014. Available at: http://www.publications.parliament.uk/pa/ cm201314/cmselect/cmsctech/272/272i.pdf
House of Commons Environment, Food and Rural Affairs Committee (2014); *Winter Floods 2013-2014. First Report of Session 2014-2015.*

Report, Together with Formal Minutes Relating to the Report. Ordered by the House of Commons to be Printed 11 June 2014. Available at: http://www.publications.parliament.uk/pa/cm201415/cmselect/cmenvfru/240/240.pdf

IPCC (2013a); *Working Group 1 Contribution to the IPCC Fifth Assessment Report Climate Change 2013: The Physical Science Basis. Summary for Policymakers*. Intergovernmental Panel on Climate Change. Available at: http://www.ipcc.ch/.

—. (2013b); *Headline Statements from the Summary for Policy Makers. Climate Change 2013 The Physical Science Basis*, The Intergovernmental Panel on Climate Change. Available at: http://www.ipcc.ch/news_and_events/docs/ar5/ar5_wg1_headlines.pdf.

—. (2014); *Climate Change 2014: Impacts, Adaption and Vulnerability*. Available at: http://ipcc-wg2.gov/AR5/images/uploads/IPCC_WG2AR5_SPM_Approved.pdf

Isla, A. (2005); Conservation as Enclosure: An Ecofeminist Perspective on Sustainable Development and Biopiracy in Costa Rica, *Capitalism, Nature, Socialism*, 16(3): 49–61.

—. (2005b); *The Tragedy of the Enclosures: an Eco-feminist Perspective on Selling Oxygen and Prostitution in Costa Rica*, Available at: www.wrm.org.uy

—. (2006b); Women, Enclosure, and Accumulation: A Rejoinder to Robert Chapman, *Capitalism, Nature, Socialism*, 17(4): 58–65.

—. (2009); The Eco-Class-Race Struggles in the Peruvian Amazon Basin: An Ecofeminist Perspective. *Capitalism, Nature, Socialism*, 20(3): 21–48.

JNCC (2014); *Council Directive 2000/60/EC Establishing a Framework for Community Action in the Field of Water Policy (Water Framework Directive)*. Joint Nature Conservation Committee. Available at: http://jncc.defra.gov.uk/page-1375

Johnson, B., (2011); *Securing London's Water Future. The Mayor's Water Strategy*, October 2011. Available at: http://www.london.gov.uk/sites/default/files/water-strategy-oct11.pdf

Kendall, J., (2003); The *Voluntary Sector: Competitive Perspectives in the UK*, with a foreword by Lord Dahrendorf, London: Routledge.

Kenyon, W. (2005); A Critical Review of Citizens' Juries: How Useful are they in Facilitating Public Participation in the EU Water Framework Directive? *Journal of Environmental Planning and Management*, 48(3), pp., 431–43.

King, M, L., (2014); *The Autobiography of Martin Luther King, JR.,* (ed) Carson, C., London: Abacus.

Kvale, S., (2009); *Interviews: Learning The Craft Of Qualitative Research Interviewing*, Los Angeles: Sage.

Lane, E., (1955); The Importance of Fluvial Morphology in Hydraulic Engineering, *American Society of Civil Engineers*, Vol., 81, Paper No., 745, pp., 1-17.

Lave, R., (2012); *Fields and Streams. Stream Restoration, Neoliberalism, and the Future of Environmental Science*, London: The University of Georgia Press.

Lawson, W., (1859); The Geography of River Systems, London: J.R.Dickons, Publishers, Market Place

Lee, N., (2000); The Feminist Standpoint Revisited and Other Essays, *Capital and Class*, 24: 215–17.

Levy, D., (1987); Female Reigns: The Faerie Queene and the Journey to the West, *Comparative Literature*, Vol., 39, No.,3, pp.,218-236.

London Assembly (2011); *For a Rainy Day the Mayor's Role in Managing London's Flood Risk in Case of Severe Rainfall*, London Assembly Environment Committee, Available at: http://www.london.gov.uk/sites/default/files/For%20a%20Rainy%20Day%20-Draft%207%20Final.pdf

—. (2013); *Submissions to our Work on Biodiversity*. Available at: https://www.london.gov.uk/sites/default/files/Combined%20Responses%20final.pdf

—. (2013a); *London's Environment Revealed. State of the Environment Report for London*, June 2011, Data Updates, February 2013. Data London, http://data.london.gov.uk/documents/SOE-2013-report.pdf

—. (2014a); *24,000 London Homes at Risk from Flooding, London Assembly*. Available at: https://www.london.gov.uk/media/assembly-press-releases/2014/01/24000-london-homes-at-risk-from-flooding

—. (2014b); *Flood Risk, London Assembly*. Available at: http://www.london.gov.uk/mayor-assembly/london-assembly/meetings-whole-assembly/motions/flood-risk

—. (2014c); Where *Would London be Without the Thames Barrier – Under Water?* London Assembly. Available at: https://www.london.gov.uk/media/assembly-press-releases/2014/01/where-would-london-be-without-the-thames-barrier-under-water

—. (2014d); *Flood Risks in London, Summary of Findings*, Available at: http://www.london.gov.uk/sites/default/files/14-04-07-Flood%20risk%20slide%20pack%20-%20FINAL.pdf

Lundmark, C and Jonsson, G., (2014); Prospects for Learning in River Management: Exploring the Initial Implementation of the Water Framework Directive in a Swedish River BASIN, *Environment Education Research*, Vol., 20, No., 2, pp., 161-176.

Luxemburg, R., (1971); *The Accumulation of Capital*, London: Routledge and Kegan Paul.

—. (1913/2003); *The Accumulation of Capital,* London: Routledge and Kegan Paul Ltd.

Lyon, D. and Glucksmann, M., (2008); Comparative Configuration of Care Work Across Europe, *Sociology*, 42: 101–18.

MacGregor, S., (2006); No Sustainability without Justice: A Feminist Critique of Environmental Citizenship, In Dobson, A., and Bell, D., *Environmental Citizenship*, Boston: Massachusetts Institute of Technology (MIT).

Mandeville, B., (1795); The Fable of the Bees; or Private Vices, Public Benefits. With an Essay on Charity and Charity Schools, And A Search into the Nature of Society. London: Bathurst, C., et al.

Martin, B., (1759); *The Natural History of England: or, A Description of Each Particular County, in Regard to the Curious Productions of Nature and Art, Volume 1*, London: Owen, W.

Marx, K., (1835/1997); Reflections of a Youth on Choosing an Occupation, in (ed) Easton, L., and Guddat, K., *Writings of the Young Marx on Philosophy and Society*, Cambridge: Hackett Publishing Company.

—. (1841/2006); *Difference Between the Democritean and Epicurean Philosophy of Nature.*, in (ed) Schafer, P., The First Writings of Karl Marx., Brooklyn:Ig Publishing

—. (1844/ 2007); *Economic and Philosophic Manuscripts of 1844*, New York: Dover.

—. (1857/1973); *Grundrisse; Foundations of the Critique of Political Economy* (Rough Draft), Translated with a Forward by Nicolaus, M., Middlesex: Penguin Books.

—. (1887/2003); *Capital, Vol. 1: A Critical Analysis of Capitalist Production*, London: Lawrence & Wishart.

—. (1885/1978); *Capital: A Critique Of Political Economy*, Vol., II; introduced by Ernest Mandel; translated by David Fernbach. Harmondsworth: Penguin Books in association with New Left Review.

—. (1894/1981); *Capital, A Critique of Political Economy Volume Three*, Intro., Mandel, E., and Trans., Fernbach, D., Middlesex:Penguin.

Marx, K, and Engels, F., (1846/2007); *The German Ideology*, London: Lawrence & Wishart.

Marx, K., and Engels, F., (1848/1970); *The Communist Manifesto*, trans Moore, S., Harmondsworth: Penguin Books.

Matamoros, R., Meza, R., Morales, M., (2012) ; *Heroes Del Riciclaje, Universidad De Costa Rica*, Facultad De Ciencias Sociales, Escuela De Psicologia, Escuela De Ciencias de la Communicacion Colectiva. Available at : http://s3images.coroflot.com/user_files/individual_files/478824_4nl0if3OXx2x7e4oaQTdPiupK.pdf

Mawhinney, M., (2009); Marx, Nature, and the Ethics of Nonidentity, *Rethinking Marxism: A Journal of Economics, Culture and Society*, Vol., 12, Issue, 1, pp., 47-64.

Mayor of London, London Assembly (2011); *Securing London's Water Future*. Available at: http://www.london.gov.uk/sites/default/files/water-strategy-oct11.pdf

—. (2013b); *Maximising Water Efficiency in London*. Available at: http://www.london.gov.uk/priorities/environment/looking-after-londons-water.

—. (2013c); *Securing London's Water Future*. Available at: https://www.london.gov.uk/priorities/environment/vision-strategy/water

—. (2013d); *Managing Rainwater and Drainage*. Available at: http://www.london.gov.uk/priorities/environment/looking-after-londons-water/managing-rainwater-and-drainage.

—. (2014a); *London's Response to Climate Change*, London Assembly. Available at: https://www.london.gov.uk/sites/default/files/FALP%20chapter%205%20-%20London%27s%20response%20to%20climate%20change.pdf

McEldowney, R., and Connor, M., (2011); Cultural Safety as an Ethic of Care: A Praxiological Process, *Journal of Transcultural Nursing*, 22(4): 342–9.

McCarthy, G., (2003); *Classical Horizons. The Origins of Sociology in Ancient Greece.*, Albany:State University of New York

McLaughlin, E., (1974); The Heresy of the Free Spirit in the Later Middle Ages by Robert E Lerner., *Speculum*, Vol., 49, No., 4, Oct., pp., 747-751.

Mellor, M., (1992); Eco-feminism and Eco-Socialism: Dilemmas of Essentialism and Materialism, *Capitalism, Nature, Socialism*, 3(2): 43–62.

—. (1994); Varieties of Ecofeminism, *Capital, Nature, Socialism*, 5(4): 117–25.

—. (1996); Ecofeminism and Ecosocialism. Dilemmas of Essentialism and Materialism, in (ed) Benton, T., *The Greening of Marxism*, The Guildford Press: London.

Mellor, M., (1997); *Feminism and Ecology*, Oxford: Polity Press.

—. (2003); Gender and the Environment, in Eaton, H. and Lorentzen, L. (eds) *Ecofeminism and Globalization: Exploring Culture, Context and Religion*, Oxford: Rowman and Littlefield Publishers.

—. (2009); Ecofeminist Political Economy and the Politics of Money, in (ed) Salleh, A., *Eco-Sufficiency and Global Justice, Women Write Political Ecology.*, London:Pluto Press

Middleton, N., (2012); *Rivers, A Very Short Introduction*, Oxford: Oxford University Press

Mies, M., (1998); *Patriarchy and Accumulation on a World Scale: Women in the International Division of Labour*. 2nd rev. edn, London: Zed Books.

—. (2007); Patriarchy and Accumulation on a World Scale – Revisited. Keynote lecture at the Green Economics Institute, Reading, 29 October 2005, *International Journal Green Economics*, 1(3/4).

Mies, M., and Shiva, V., (1993); Introduction: Why We Wrote this Book Together, in (ed) Mies, M., and Shiva, V., *Ecofeminism*, London: Zed Books.

Milbourne, L., (2013); *Voluntary Sector in Transition: Hard Times or New Opportunities*? Bristol: Policy Press.

Milne, A., (1982); *London's Drowning*, London: Thames Methuen.

Ministra de Salud Costa Rica (2010); *Ministra De Salud Participó En Campaña De Limpieza Del Rio Virilla , Comunicado De Prensa*. Available at: http://www.ministeriodesalud.go.cr/index.php/centro-de-prensa/inicio-prensa-comunicados-de-prensa-ms/doc_view/165-ministra-de-salud-participo-en-campana-de-limpieza-del-rio-virilla

Ministerio De Salud (2010a); *Campana De Limpieza De Ríos Costa Rica Te Quiero Limpia*. Available at : http://www.ministeriodesalud.go.cr/index.php/centro-de-prensa/inicio-prensa-comunicados-de-prensa-ms/doc_view/184-campana-de-limpieza-de-rios-costa-rica-te-quiero-limpia

Mitchell, M., and Pentzer, S., (2008); *Costa Rica: A Global Studies Handbook*, ABC-CLIO Ltd

Monica, J., F., (2010); *Montaña de Esperanza Rescate del Rio Virilla.* Published by Wild Water Tours Costa Rica., http://www.wildwatertours.com

Morrow, R., (1994); *Critical Theory*, London: Sage.

Mountain of Hope (2013); *Montana de Esperanza*, www.facebook.com

Nagington, M., Luker, K., and Walshe, C., (2013); 'Busyness' and the Preclusion of Quality Palliative District Nursing Care., *Nursing Care.*, Vol., 20, Issue 8, pp., 893-903.

Newman, J., (2013); Spaces of Power: Feminism, Neoliberalism and Gendered Labor, *Social Politics: International Studies in Gender, State and Society*, Vol., 20, Number 2, Summer, pp., 200-221.

Nieuwenhove, R., (2000); Ruusbroec: Apophatic Theologian or Phenomenologist of the Mystical Experience? *The Journal of Religion*, Vol., 80, No., 1, Jan., pp., 83-105.

Odih, P., (2007); *Gender and Work in Capitalist Economies*, Maidenhead: Open University Press.

—. (2010); *Advertising and Cultural Politics in Global Times*, Farnham: Ashgate.

—. (2013); *Visual Media and Culture of Occupy*, Newcastle Upon Tyne: Cambridge Scholars Publishing.

Olsson, J., Jonsson, A. and Andersson, B., (2011); A Model-supported Participatory Process for Nutrient Management: A Socio-legal Analysis of a Bottom-up Implementation, *International Journal of Agricultural Sustainability*, 2(2): 379–88.

Orme, J., (2001); *Gender and Community Care. Social Work and Social Care Perspectives*, Basingstoke: Palgrave.

—. (2003); 'It's Feminist because I Say So!' Feminism, Social Work and Critical Practice, *Qualitative Social Work*, 2(2): 131–53.

Parliamentary Papers (1884); *Pollution of the River Thames, No., 56 Metropolitan Police. Thames Division Special Report, 15th August 1884*. Available at: http://www.forgottenbooks.com/readbook_text/Parliamentary_Papers_v1_1000751101/499

—. (1857); *Acts of Conservancy of the River Thames CLA/037/TC/02 /011 1857 & 1894*. Available at: http://apps.nationalarchives.gov.uk/a2a/records.aspx?cat=075-cla037&cid=2-2-32#2-2-32

—. (2014a); *Winter Floods 2013/14, Hartwell-Naguib, S., and Roberts, N., Standard Note: SN/SC/06809, Section: Science and Environment*, Available at: http://www.parliament.uk/business/publications/research/briefing-papers/SN06809/winter-floods-201314

Petts, G., and Foster, I., (1985); *Rivers and Landscape*, London: Edward Arnold.

Phillips, J., (1871); *Geology of Oxford and the Valley of the Thames*, Oxford: MacMillan and Co, Publishers to the University of Oxford.

Plumwood, V., (1996); Nature, Self, and Gender: Feminism, Environmental Philosophy, and the Critique of Rationalism, in Warren, K. (ed.) *Ecological Feminist Philosophies*, Bloomington and Indianapolis: Indiana University Press.

Pope, A., (1826); *The Poetical Works of Alexander Pope*, (ed) Johnson, S., London: Jones and Company.

Poso, T., (2010); Revisiting Residential Care. Methodological Considerations, *Qualitative Social Work*, 9(1): 27–42.

Restore (2013); *Rivers by Design. Rethinking Development and River Restoration*. Available at: http://www.restorerivers.eu/Portals/27/Publications/Rivers%20by%20design.pdf

Reciclando En Espiral Costa Rica (2010); *Limpieza Río Virilla ~ Montaña de Esperanza*. Available at: http://reciclandoenespiral.com/blogreciclaje/category/campanas-de-reciclaje/

Richardson, R., (1898); A Trace of Egypt at Eleusis, *American Journal of Archaeology*, Vol., 2, No., 3/4 (May - Aug 1898), pp., 223-232.

Rist, M., (1972); *Epicurus. An Introduction*, London: Cambridge at the University Press

Roberts, C., (2003); Drowning in a Sea of Estrogens: Sex Hormones, Sexual Reproduction and Sex, *Sexualities*, Vol., 6, No., 2, pp., 195-213.

Robertson, A., (1784); *The Diary Maid or Vocal Miscellany. Being a Collection of Choice Songs, Scots and English. With a Variety of Toasts and Sentiments*., Edinburgh: A Robertson.

Rosaldo, R., (1993); *Culture and Truth: The Remaking of Social Analysis*, London: Routledge.

Rudy, A., (2006); On the Ecofeminist Editorial: 'Moving to an Embodied Materialism', *Capitalism, Nature, Socialism*, 17(4): 105–14.

Salleh, A., (1991); Eco-Socialism/Eco-Feminism, *Capitalism, Nature, Socialism*, 2(1):129–37.

—. (1995); Nature, Woman, Labour, Capital: Living the Deepest Contradiction, *Capitalism, Nature, Socialism*, Vol., 6, No., 1, pp., 21–39.

—. (2001); Sustaining Marx or Sustaining Nature?. An Ecofeminist Response to Foster and Burkett, *Organization and Environment*, 14(4): 443–50.

—. (2003); Ecofeminism as Sociology, Feminism and Ecology, *Capitalism, Nature, Socialism*, 14(1): 61–74.

—. (2005); Moving to an Embodied Materialism, *Capitalism, Nature, Socialism*, 16(2): 9–14.

—. (2006); Embodying the Deepest Contradiction: A Rejoinder to Alan Rudy, *Capitalism Nature Socialism*, Vol., 17, Number 14, pp., 115-124.

—. (2010); From Metabolic Rift to 'Metabolic Value': Reflections on Environmental Sociology and the Alternative Globalization Movement, *Organization and Environment*, 23(2): 205–19.

—. (2014); Forward, in (eds) Mies, M., and Vandana S., *Ecofeminism*, London: Zed Books.

Salleh, A., and O'Connor, M., (1991); Eco-socialism/Eco-feminism, *Capitalism Nature Socialism*, Vol., 2, Issue 1, pp., 129-137

Sandler, B., (1994); Grow or Die: Marxist Theories of Capitalism and the Environment, *Rethinking Marxism*, Vol., 7, No., 2, pp., 38-57.

San Martin (2002); Water Resources in Latin America and the Caribbean: Issues and Options. Inter-American Development Bank, Sustainable Development Department, Environment Division. Available at: http://info.worldbank.org/etools/docs/library/80628/IWRM4_SanMartinWater.pdf

Sarton, G., (1952); *Ancient Science Through the Golden Age of Greece*, Courier Dover Publications.

Schmidt, A., (1971); *The Concept of Nature*, translated from the German by Fowkes, B., London: New Left Book.

Schmidt, J., and Mitchell, K., (2013); Property and the Right to Water: Toward a Non-Liberal Commons, *Review of Radical Political Economics*, Vol., 46, Issue, 1, pp., 54-69.

Seccombe, T., and Arber, E., (1593); *An English Garner: Elizabethan Sonnets*, NewYork: E.P.Dutton and Co.

Sheasby, W., (1997); Inverted World: Karl Marx on Estrangement of Nature and Society, *Capitalism, Nature, Socialism, 8 (4): 31-46.*

Shaw, G., (2014); *The Egyptian Myths. A Guide to the Ancient Gods and Legends*, London: Thames & Hudson

Shiva, V., (2002); *Water Wars, Privatization, Pollution and Profit*, London: Pluto Press.

Singh, N., (2006); Women's Participation in Local Water Governance: Understanding Institutional Contradictions, *Gender, Technology and Development*, 10(1): 61–76.

Smith, B., Clifford, N., and Mant, J., (2014); The Changing Nature of River Restoration, *WIREs Water* Vol., 1, Issue 3., May/June pp., 249-261. Available at: http://onlinelibrary.wiley.com/doi/10.1002/wat2.1021/pdf

Spenser, E., (1590/1758); *The Fairy Queen, Volume 2*, London: J& R Tonson in the Strand.

—. (1590/1825); *The Poetical Works of Edmund Spenser in Five Volumes, Volume 2*, London: W.Pickering.

Stanley, J., (2002); *Mainlining Marx*, London:Transaction Publishers

Steady, F., (2009); Environmental Justice Cross-Culturally. Theory and Praxis in the African Diaspora and in Africa, in Steady, F. (ed.) *Environmental Justice in the New Millennium: Global Perspectives on Race, Ethnicity and Human Rights*, New York: Palgrave Macmillan.

Steedman, C., (2005); Poetical Maids and Cooks Who Wrote, *Eighteenth-Century Studies*, Vol., 39, No., 1 (Fall), pp., 1-27

Suddoth, E.B., Hassett, B.A., Cada, P., and Bernhardt, E.S., (2011); Testing the Field of Dreams Hypothesis: Functional Responses to Urbanization and Restoration in Stream Ecosystems, *Ecological Applications*, Sept., 21 96), pp., 1972-88.

Tapsell, S., (1997); Rivers and River Restoration: a Child's Eye View, *Landscape Research*, Vol., 22, No., 1, pp., 45-65.

Thames Water (2012); *Why Does London Need the Thames Tideway Tunnel?* Available at: http://www.thameswater.co.uk/tw/common/downloads/aboutus-projects-tidal-thames/london-tideway-improvements-the-case-for-the-thames-tunnel.pdf.

THERRC (2009); *The London Rivers Action Plan*. The River Restoration Centre, Bedford: Cranfield University Available at: http://www.therrc.co.uk/lrap/lplan.pdf

Thomas, A., (1986); *Egyptian Gods and Myths.*, Aylesbury, Shire Publications.

Tickell, T., (1808); Oxford, Inscribed to Lord Lonsdale, in *The Cabinet of Poetry, Containing The Best Entire Pieces to be Found in The Works of the British Poets., Vol., III*, London: Richard Philips

Tompkins, S.M.J., (1938); *The Polite Marriage*, London:Cambridge University Press

Tronto, J., (2003); Time's Place, *Feminist Theory*, Vol., 4, No., 2, pp., 119–139.

TUC (2013); *Concern at 'Shocking' Rise in Zero Hours Contracts*. Available at: http://www.tuc.org.uk/workplace-issues/working-time/concern-shocking-rise-zero-hours-contracts.

United Nations (2010); *Resolution Adopted by the General Assembly. 64/292. The Human Right to Water and Sanitation*, United Nations General Assembly. Available at: http://apps.who.int/gb/ebwha/pdf_files/WHA64/A64_R24-en.pdf

United Nations Department of Economic and Social Affairs (2013); *International Decade for Action 'WATER FOR LIFE' 2005-2015. Gender and Water*. UNDESA Available at: http://www.un.org/waterforlifedecade/gender.shtml.

UNEP (1999/2009); Global *Environment Outlook 2000*, United Nations Environment Programme (UNEP), USA: Earthscan Library Collection

—. (2000); *GEO Latin America and the Caribbean. Environment Outlook 2000, United Nations Environment Programme*. Available at: http://www.unep.org/geo/pdfs/geolac-2000.pdf

—. (2006); *Global International Waters Assessment. Eastern Equatorial Pacific, GIWA Regional Assessment 65, United Nations Environment Programme*, Kalmar (Sweden): University of Kalmar. Available at; http://apps.unep.org/publications/pmtdocuments/giwa_regional_assessment_65.pdf

UN-Habitat (2013); *State of Women in Cities 2012–2013. Gender and the Prosperity of Cities*. Available at: http://www.zaragoza.es/contenidos/medioambiente/onu/994-eng.pdf.

United Nations Human Settlements Programme UNHSP (2011); *Final Report World Water Day 2011: Water and Urbanization. Water for Cities: Responding to the Urban Challenge*. Available at: http://www.ips.org/africa/library/WWD-TV-2011-FINAL-lowres.pdf

UNICEF (2012); *Progress on Drinking Water and Sanitation 2012 Update*. Available at: http://www.wssinfo.org/fileadmin/user_upload/resources/JMP-report-2012-en.pdf.

UN Org (2013); *International Decade for Action 'Water For life' 2005–2015*. Available at: https://www.un.org/waterforlifedecade/gender.shtml.

UN Water (2006); *Gender, Water and Sanitation: A Policy Brief. UN-Water Task Force on Gender and Water*. Available at: http://www.unwater.org/downloads/unwpolbrief230606.pdf.

—. (2007); 2007 *World Water Day. Coping with Water Scarcity Challenge of the Twenty-first Century*. Available at: http://www.unwater.org/water-cooperation-2013/get-involved/campaign-materials/water-factsheets/en/.

—. (2011a); *Water Quality Challenges and Impacts. Available at:* http://www.unwater.org/downloads/waterquality_policybrief.pdf.

—. (2011b); *Final Report World Water Day 2011. Water and Urbanization. Water for Cities: Responding to the Urban Challenge*. Available at:

http://www.unwater.org/fileadmin/user_upload/watercooperation2013/doc/Factsheets/water_and_urbanization.pdf.

—. (2012); *Managing Water Under Uncertainty and Risk. The United Nations World Water Development Report 4, Volume 4., UN Water.* Available at:
http://unesdoc.unesco.org/images/0021/002156/215644e.pdf

UN World Water Development Report 4 (2012a); *Facing the Challenges*, The United Nations World Water Development Report 4, Volume 3. Available at:
http://unesdoc.unesco.org/images/0021/002156/215644e.pdf#page=812

UN Water (2013); *Water Quality*. Available at;
http://www.unwater.org/downloads/water_quality.pdf

—. (2013a); *Water for Food. Available at:*
http://www.unwater.org/fileadmin/user_upload/watercooperation2013/doc/Factsheets/water_for_food.pdf.

—. (2013b); *Transboundary Waters.* Available at:
http://www.unwater.org/fileadmin/user_upload/watercooperation2013/doc/Factsheets/transboundary_waters.pdf.

—. (2013c); *Sanitation.* Available at:
http://www.unwater.org/fileadmin/user_upload/watercooperation2013/doc/Factsheets/sanitation.pdf.

—. (2013d); *Climate Change*. Available at:
http://www.unwater.org/fileadmin/user_upload/watercooperation2013/doc/Factsheets/climate_change.pdf,

—. (2013e); *Water Biodiversity.* Available at:
http://www.unwater.org/fileadmin/user_upload/watercooperation2013/doc/Factsheets/water_and_biodiversity.pdf.

—. (2013f); *Water Scarcity.* Available at:
http://www.unwater.org/fileadmin/user_u.load/watercooperation2013/doc/Factsheets/water_scarcity.pdf,

—. (2013g); *Water and Urbanization*. Available at:
http://www.unwater.org/fileadmin/user_upload/watercooperation2013/doc/Factsheets/water_and_urbanization.pdf.

—. (2013h); *Water and Disaster.* Available at:
http://www.unwater.org/fileadmin/user_upload/watercooperation2013/doc/Factsheets/water_disasters.pdf.

—. (2013i); *An Increasing Demand.* Available at:
http://www.unwater.org/water-cooperation-2013/water-cooperation/facts-and-figures/en/.

—. (2013j); *The Post 2015 Water Thematic Consultation.* Available at:

http://www.unwater.org/downloads/Final9Aug2013_WATER_THEMATIC_CONSULTATION_REPORT.pdf.
—. (2013k); *UN Water Activities.* Available at: http://www.unwater.org/TFgender.html.
—. (2013l); *Water Quality. World Water Day 2013.* Available at: http://www.unwater.org/water-cooperation-2013/get-involved/campaign-materials/water-factsheets/en/.
—. (2014); *Water and Energy; The United NATIONS World Water Development Report 2014.* Available at: http://www.unwater.org/publications/publications-detail/en/c/218614/
Walford, E., (1987); *London Recollected. Its History, lore and Legend, Volume VI*, London: The Alderman Press .
Walicki, A., (1988); Karl Marx as Philosopher of Freedom, *Critical Review* Vol., 2, No., 4, pp., 10-58.
Warren, K., (1996a); Ecological Feminist Philosophies An Overview of the Issues, in Warren, K. (ed.) *Ecological Feminist Philosophies*, Bloomington and Indianapolis: Indiana University Press.
Warren, K., (ed.) (1996b); *Ecological Feminist Philosophies*, Bloomington and Indianapolis: Indiana University Press.
Warren, K., (1996c); The Power and the Promise of Ecological Feminism, in Warren, K. (ed.) Ecological *Feminist Philosophies*, Bloomington and Indianapolis: Indiana University Press.
WSP (2010a); About *Water and Sanitation Program*. Available at: https://www.wsp.org/about
—. (2010); *Mainstreaming Gender in Water and Sanitation. Gender in Water and Sanitation, November 2010*, Water and Sanitation Program (WSP). Available at: http://www.wsp.org/sites/wsp.org/files/publications/WSP-gender-water-sanitation.pdf
Waldron, M., (1996); *Lactilla, Milkwoman of Clifton. The Life and Writings of Ann Yearsley 1753-1806*, Georgia:The University of Georgia Press
Water Wise (2013); *2013 International Year of Water Cooperation.* Available at: http://www.waterwise.co.za/site/news/environmental/2013_International_year_of_water_cooperation.html
Weinreb, B., and Hibbert, C., (1988); *The London Encyclopaedia*, London: Macmillan.
West, P., and Carrier, J., (2004); Ecotourism and Authenticity: Getting Away from It All? *Current Anthropology*, Vol., 45, No.,4, pp., 483-498.

Wheeler, A., (1979); *The Tidal Thames, The History of a River and Its Fishes*, London: Routledge & Kegan Paul.

Wild Water Tours (2013); *Wild Water Tours Costa Rica*, Available at: http://www.wildwatertours.com

Witt, R.E., (1997); *Isis in the Ancient World*, London: The Johns Hopkins University Press.

Wiseman, J., (1990); "To be God with God": The Autotheistic Sayings of the Mystics, *Theological Studies*, Vol., 51, pp., 230-251.

WWDR4 (2012); *Managing Water under Uncertainty and Risk. The United Nations World Water Development Report 4, Volume 1,* World Water Assessment Programme (WWAP). Available at: http://www.unesco.org/new/en/natural-sciences/environment/water/wwap/wwdr/wwdr4-2012.

Yearsley, A., (1796/2011); *The Rural Lyre; a Volume of Poems*, British Library, Historical Print Editions.

—. (1787); *Poems on Various Subjects*, British Library, Historical Print Editions.

Zelezny, L., Chua, P. and Aldrich, C. (2000); Elaborating on Gender Differences in Environmentalism, *Journal of Social Issues*, 56(3): 443–57.

INDEX

"Order and beauty with their passion play, Visions arise from interchange of thought, With dear refinement and instruction fraught. To love alone society must owe…" (Ann Yearsley *The Rural Lyre,* 1796). Photographic Image, South London, River Thames, September 2013

"The cream of the *Dairy*" (William Makepeace Thackeray, 1856). Photographic Image, South London, River Thames, September 2013

"The arc of the moral universe, although long, is bending toward justice" (Martin Luther King, Jr.,). Photographic Image, South London, River Thames, September 2013